Urban Nature

This book showcases the diversity of ways in which urban residents from varying cultural contexts view, interact and engage with and give meaning to urban nature, aiming to counterbalance the dominance of Western depictions and values of urban nature and design.

Urban nature has up until now largely been defined, planned and managed in a way that is heavily dominated by Western understandings, values and appreciations, which has spread through colonialism and globalisation. As cities increasingly represent a diversity of cultures and urban nature is being increasingly recognised as contributing to residents' wellbeing, belonging and overall quality of life, it is important to consider the numerous ways in which urban nature is understood and appreciated. This collection of case studies includes examples from Africa, Asia, Europe, and North and South America and reflects on the multi-dimensional aspects of engagements with urban nature through a biocultural diversity lens. The chapters cover several themes such as how engagements with nature contribute to a sense of wellbeing and belonging; the implications that diversity has on the provision, design and management of urban environments; and the threats inhibiting residents' abilities to engage meaningfully with nature. The book challenges the dominant discourse, Western ideological understandings and meta-narratives of modernisation and unilineal urban transitions.

A timely addition to the literature, *Urban Nature: Enriching Belonging, Wellbeing and Bioculture* offers an alternative to Western ideological understandings of nature and values and will be of great interest to those working in human and environmental urban ecology. It will also be key reading for students in the relevant fields of anthropology, development studies, geography, social ecology and urban studies.

Michelle L. Cocks is Associate Professor in the Department of Anthropology, Rhodes University, South Africa.

Charlie M. Shackleton is Professor and DST/NRF Research Chair of Interdisciplinary Science in Land & Natural Resource Use for Sustainable Livelihoods in the Department of Environmental Science at Rhodes University, South Africa.

Routledge Studies in Urban Ecology

Urban Ecosystems
Understanding the Human Environment
By Robert A. Francis and Michael A. Chadwick

Urban Forests, Trees, and Greenspace
A Political Ecology Perspective
Edited by L. Anders Sandberg, Adrina Bardekjian and Sadia Butt

Green Infrastructure
Incorporating Plants and Enhancing Biodiversity in Buildings and Urban Environments
By John W. Dover

Urban Landscape Ecology
Science, Policy and Practice
Edited by Robert A. Francis, James D.A. Millington and Michael A. Chadwick

Urban Expansion, Land Cover and Soil Ecosystem Services
Edited by Ciro Gardi

Urban Biodiversity
From Research to Practice
Edited by Alessandro Ossola and Jari Niemelä

Urban Nature
Enriching Belonging, Wellbeing and Bioculture
Edited by Michelle L. Cocks and Charlie M. Shackleton

For more information about this series, please visit: https://www.routledge.com/Routledge-Studies-in-Urban-Ecology/book-series/RSUE

Urban Nature

Enriching Belonging, Wellbeing and Bioculture

Edited by

Michelle L. Cocks and Charlie M. Shackleton

Routledge
Taylor & Francis Group
LONDON AND NEW YORK

earthscan
from Routledge

First published 2021
by Routledge
2 Park Square, Milton Park, Abingdon, Oxon OX14 4RN

and by Routledge
52 Vanderbilt Avenue, New York, NY 10017

Routledge is an imprint of the Taylor & Francis Group, an informa business

British Library Cataloguing-in-Publication Data
A catalogue record for this book is available from the British Library

Library of Congress Cataloging-in-Publication Data
Names: Cocks, Michelle, editor. | Shackleton, C. M.
(Charles Michael), editor.
Title: Urban nature: enriching belonging, wellbeing and bioculture/
edited by Michelle L. Cocks and Charlie M. Shackleton.
Identifiers: LCCN 2020025647 (print) | LCCN 2020025648 (ebook) |
ISBN 9780367427573 (hbk) | ISBN 9780367854898 (ebk)
Subjects: LCSH: Urban ecology (Sociology)–Case studies. |
Cities and towns–Growth–Case studies. | Biodiversity–Case studies. |
Sustainable development–Case studies.
Classification: LCC HT241 .U7259 2021 (print) | LCC HT241 (ebook) |
DDC 307.76–dc23
LC record available at https://lccn.loc.gov/2020025647
LC ebook record available at https://lccn.loc.gov/2020025648

ISBN: 978-0-367-42757-3 (hbk)
ISBN: 978-0-367-85489-8 (ebk)

Typeset in Bembo
by Deanta Global Publishing Services, Chennai, India

This book is dedicated to all the participants who have shared their stories about their relationship with nature with each of us. In return we hope we have captured their essence to reflect the significance that they hold for their owners.

Contents

Figures

Tables

Contributors

Adedeji, Joseph Adeniran: Chapter 2

Joseph Adeniran Adedeji holds B.Tech., M.Tech. and PhD degrees in Architecture. He is currently a senior lecturer and registered architect with the Architects Registration Council of Nigeria, and a member of the Nigerian Institute of Architects in the Department of Architecture, The Federal University of Technology, Akure, Nigeria. He is a former postdoctoral Carson Fellow at Rachel Carson Centre for Environment and Society, Ludwig-Maximilians-University, Munich, Germany. His teaching, research works and publications are situated at the intersection of spatial considerations for the comfortable use of urban open spaces, cultural morphology of cityscapes, and more intensely, landscape hermeneutics of the urban grain in African context.

Ambrose-Oji, Bianca: Chapter 11

Bianca Ambrose-Oji is an environmental sociologist working at Forest Research, UK for the Social and Economic Research Group (SERG). She has experience of operational practice, action research and applied research in four key areas: adaptive and innovative forest and greenspace governance; public and community engagement, equality and environmental justice; nature connection and wellbeing; and the behaviour of land owners and woodland managers. Bianca takes an actor-oriented and political ecology approach to her research, as well as models of co-production and co-management of space, place and policy.

Campbell, Lindsey: Chapter 8 and 12

Lindsay K. Campbell is a research social scientist with the USDA Forest Service Northern Research Station, based at the New York City Urban Field Station (www.nrs.fs.fed.us/nyc). Her current research explores the dynamics of urban environmental governance, natural resource stewardship and sustainability policymaking. In addition to her research, Lindsay helps direct the Science of the Living City programme for the Urban Field Station, including fellows, seminars, salons and artists in residence. Dr. Campbell holds a BA in Public Policy from Princeton University, a Masters in City Planning from MIT and a PhD in Geography from Rutgers University.

Cocks, Michelle: Chapters 1, 6 and 13)

Michelle Cocks is an associate professor in the Anthropology Department, Rhodes University. Her research interests lie within the realm of human and nature studies. She has used the concept of biocultural diversity to frame her studies and hopes her work contributes to creating a more bioculturally diverse world.

de Mendonça, Guilherme: Chapter 9

Guilherme Cruz de Mendonça is a professor of Environmental Law at Rio Federal Institute of Education, Science and Technology and has a PhD in Environment from the State University of Rio de Janeiro. He has a Master in Law from the State University of Rio de Janeiro – UERJ and specialised in cultural heritage with the National Artistic Heritage Institute – IPHAN in partnership with UNESCO. He has experience in the area of Law, with emphasis on Environmental Law, Cultural Heritage and Critical Theory. He is the leader of the Institutional Research Group Biocultural DiverCity.

Elands, Birgit: Chapter 11

Birgit Elands is an assistant professor at the Forest and Nature Conservation Policy Group (FNP) of Wageningen University, the Netherlands. Her research deals with different aspects of human-nature interaction, such as recreational behaviour, biocultural diversity, public participation and engagement with nature, and nature experiences and connection to nature, both in an urban and rural context. Her work contributes to research of FNP that focuses on forest and nature conservation policies and governance, including the intended and unintended socio-economic, ecological and political consequences they produce. Birgit Elands has participated in several European projects, including Green Surge (FP7 WP-leadership) and Multifor.RD project (FP3 coordinator).

Franciscoc, Kainana, S.: Chapter 12

Ke aloha nō kākou. My name is **Kainana Francisco**. I come from the Puna and Hilo districts on eastern Hawai'i island. Kīlauea, Maunaloa, and Maunakea are my foundational mountains, Kapoho and Peiwe my special ocean places, Moanianilehua and Kanilehua my life-giving waters, and, Waikahekahe and Waiākea the forests that help me grow. I am a student of hula (traditional dance) with the hālau (traditional school), Unulau. It is this training and lifeway that has given me the skills and knowledge to mature in my potential of serving my community as a natural resource specialist with the USDA Forest Service. Mahalo!

Garcia, Zach: Chapter 8

Zach Garcia is a graduate of the Yale School of the Environment (formerly Yale School of Forestry and Environmental Studies) where he focussed on religious ecologies and environmental stewardship. Zach is currently the Director of Environmental Education, Natural Resources, and Research at Wesselman Woods Nature Preserve in southern Indiana.

Giardinac, Christian P.: Chapter 12

Christian Giardina is a research ecologist with the USDA Forest Service's Institute of Pacific Islands Forestry in Hilo, Hawai'i, where he studies global change impacts on forests, biocultural approaches to conservation and restoration, and the ocean from any form of outrigger canoe. He serves as a handling editor at the Journals *Ecosystems* and *Sustainability* and represents the Forest Service at the Hawaii Conservation Alliance and the USDA's Southwest Climate Hub. He is an avid learner in Hālau 'Ōhi'a and an inspired agency partner in the Pu'uwa'awa'a community-based subsistence forest area in the Moku of Kona on Hawai'i Island.

Haase, Annegret: Chapter 11

Annegret Haase works at Helmholtz Centre for Environmental Research – UFZ, Leipzig, Germany, at the Department of Urban and Environmental Sociology. Her main research foci are sustainable urban development and urban transformation, in particular socio-spatial and socio-environmental processes in cities, urban shrinkage and regrowth, inequalities, justice issues and segregation, urban land use change/conflicts, urban green spaces, urban diversity and migration, social conflicts and goal conflicts of sustainable development, urban governance, participation and actor perspective.

Haynes, Duncan: Chapter 6

Duncan Haynes is currently completing his master's degree in Anthropology at Rhodes University, South Africa.

Jaatsi, Mai: Chapter 5

Mia Jaatsi is a PhD student in Human Geography at the University of Helsinki. She also works with multidisciplinary education in urban studies and planning and fosters societal interactions through the Urban Academy network. Mia's research interest lies in examining the use of public space and community-driven activities in the city. Her Master's thesis looked at urban parks as a place for social and cultural interactions. Mia's latest intervention was an EIT Climate-KIC funded project Urban Oasis that examined the reuse of indoor spaces for communal urban gardens.

Kealiikanakaoleohaililanid, Kekuhi: Chapter 12

Aloha. I am **Kekuhi Kealiikanakaoleohaililani**, teacher of hula (Hawai'i's ritual dance form), Hawai'i ritual, and Hawai'i Life ways. I live in the forest of Pana'ewa in the moku or district of Hilo with my youngest daughter, Kekuhi, and husband, Tangaro. My mountain is Kīlauea, my water is Waiākea and my clan is the volcano clan.

I created Hālau 'Ōhi'a, a transformational Hawai'i Stewardship programme, for professionals who care for Hawai'i's first people. My curiosities, in this decade, are human nature, epigenetics, neuroscience, quantum physics and how to make coconut flour. Mahalo.

Konijnendijk, Cecil: Chapter 10

Cecil Konijnendijk, is a professor of urban forestry at the University of British Columbia in Vancouver. Cecil's research and teaching focus on urban green space governance and planning, as well as the ecosystem services and benefits that urban nature provides. He co-founded the scientific journal *Urban Forestry & Urban Greening* in 2002 and is Editor of the Future City book series. He holds a number of visiting professorships in China. In 2020, he founded the Nature-Based Solutions Institute with Johan Östberg.

Lagunes Gasca, Ana Lucía: Chapter 7

Ana Lucía Lagunes Gasca was born in Mexico City. She is a migrant and feminist scholar, both by cultural inheritance and decision. She has been trained in social psychology and in the field of education for interculturality and sustainability. She has dedicated herself to the development of creative methodologies for participatory research and community collaboration for a good collective life or "buen vivir".

Lauster, Nathanael: Chapter 10

Nathanael Lauster is a sociologist and sometime demographer at the University of British Columbia who primarily studies housing, home, households, and urban environments. In addition to numerous articles, Professor Lauster is also the award-winning author of *The Death and Life of the Single Family House: Lessons from Vancouver on Building a Livable City* (Temple University Press). He blogs about housing, immigration, policy, and city life at homefreesociology.com.

Manyani, Amanda: Chapter 6

Amanda Manyani completed her master's degree in the Environmental Science Department at Rhodes University. She is currently enrolled for her PhD at Stellenbosch University, South Africa.

McMillen, Heather: Chapter 12

Heather McMillen is an urban & community forester with the Hawai'i Department of Land & Natural Resources and an affiliate faculty member of the Department of Natural Resources and Environmental Management at the University of Hawai'i at Mānoa. Heather holds a PhD in Anthropology, and a certificate in Ecology, Evolution, and Conservation Biology from the University of Hawai'i at Mānoa, and she is a certified arborist. Her work focuses on biocultural stewardship models in indigenous and local communities and in urban settings. Heather lives on O'ahu in the foothills of Lē'ahi with her son.

Mellegård, Viveca: Chapter 4

Viveca Mellegård is a researcher and filmmaker. She integrates film and photography as methods to engage with diverse groups of people for sustainability science and international development projects. Special interests include using storytelling to explore ways that humans and nature co-create hope-inspiring futures. A recent publication investigates the knowledge and skills embedded in craftsmanship – in particular, the qualities of reciprocity and responsibility (Mellegård & Boonstra 2020). Viveca has worked at the Stockholm Resilience Centre and on development projects in East Africa. She was previously at the BBC in London where she directed long-form science, history and arts programmes.

Merçon, Juliana: Chapter 7

Juliana Merçon is a researcher at the University of Veracruz, in Mexico. She has long studied the relations between humans and nature, both through philosophical perspectives and on the ground collaborative practices. She coordinates the Social-Ecological Action Research Group at her university and participates in various Latin American networks and social movements struggling for land and indigenous rights, gender equity and collective autonomy. Juliana loves gardening and agroecology. She hopes that her work contributes to creating a more bioculturally diverse world; a world in which justice and sustainability are made reality.

Nesbitt, Lorien: Chapter 10

Lorien Nesbitt is an assistant professor of Urban Forestry in the Faculty of Forestry, University of British Columbia. Her research focuses on urban forestry and socio-ecological interactions in urban environments, with an emphasis on equity, human health and climate change. She is currently examining (1) urban green equity, intersectionality and green gentrification; (2) experiences of women in urban forestry and arboriculture; (3) the relationship between greenness exposure and public health outcomes in urban environments, with a focus on spatio-temporal metrics and climate change; and (4) urban forest governance and resilience to social and ecological stresses.

Park, Hyeone: Chapter 10

Hyeone Park is a PhD candidate at the Faculty of Forestry at the University of British Columbia. Hyeone's current research focuses on multifunctionality and management of urban green space, especially urban food forests and community gardens. She is also interested in practices linking ecological restoration and food forestry (or forest gardening) in urban landscapes.

Pastor-Ivars, Juan: Chapter 2

Juan Pastor-Ivars is a research associate at Operating Unit Ishikawa/Kanazawa, United Nations University Institute for the Advance Study of Sustainability (UNU-IAS OUIK).

Born in Denia, Spain, he moved to the ancient capital of Japan, Kyoto, in 2009 where he deepened his understanding of traditional Japanese culture. He obtained a master's in architecture from the Polytechnic University of Valencia in 2004 and earned a master's in architecture from the Kyoto Institute of Technology in 2009. He received a PhD in architecture from the Polytechnic University of Valencia in 2015 with a focus on Japanese gardens and a thesis entitled "Ma and Oku, Ogawa Jihei and the Japanese Garden of the 19th Century". From 2016, he has been a researcher with the UNU-IAS Operating Unit Ishikawa/Kanazawa (OUIK). His areas of academic expertise are the fields of architecture, landscape architecture and landscape ecology. His current research focuses on the sustainable conservation and development of urban nature pursuing the creation of a holistic blue-green infrastructure. He is author of the book *Restoring Kinship with Nature through Japanese Gardens: The Challenge to Achieve a Sustainable Commons in Kanazawa*, published by a UNU-IAS OUIK in 2019.

Peter, Karin: Chapter 11

Karin Peters is an assistant professor with the Cultural Geography Group, Wageningen University, the Netherlands. She holds a PhD on the meaning of public spaces for issues of social integration. Her research focuses on leisure, public space, mobilities, diversity and issues of inclusion and exclusion in the Netherlands and at the European level. Recently, she has focussed on sustainable mobility with the ambition to create a global bicycle network. With this network of initiatives and partners, she aims to connect relevant stakeholders to enhance sustainable mobility in urban environments.

Radebe, Dennis: Chapter 6

Dennis Radebe completed his MSc degree in the Environmental Science Department at Rhodes University, South Africa.

Sachdeva, Sonya: Chapter 8

Sonya Sachdeva is a computational social scientist with the US Forest Service in the greater Chicago area. She is part of a Forest Service research unit that studies the human component of natural resource management within urban ecosystems. She uses wide-ranging methods like field studies, behavioural experiments and automated content analyses to understand the sociocultural factors that shape environmental values and people's relationship to nature.

Shackleton, Charlie: Chapters 1, 6 and 13

Charlie Shackleton is an interdisciplinary scientist in the position of Professor and Research Chair in Interdisciplinary Science in Land and Natural Resource Use for Sustainable Livelihoods in the Department of Environmental Science, Rhodes University, Makhanda, South Africa. His work in urban contexts revolves around significance of trees in urban spaces, ecosystem services and the use of wild resources in urban species for local livelihoods.

Svendsen, Erika: Chapter 8 and 12

Erika Svendsen is a social scientist with the USDA Forest Service and an expert in the field of environmental governance. She is the team leader for the NYC Urban Field Station, a special partnership between the US Forest Service, NYC Parks and the Natural Areas Conservancy to improve human wellbeing and the environment through knowledge co-production. She is the co-founder of STEW-MAP, a sustainability tool to visualise, amplify and support the work of thousands of civic groups as they care for their local environment. Erika is a writer for *The Nature of Cities*.

Vierikko, Kati: Chapter 5

Kati Vierikko works as senior researcher at the Finnish Environment Institute (SYKE). She is biologist and environmental scientist that aims to build bridges between theories and disciplines of social and natural sciences. She has studied cultural ecosystem services and biocultural diversity by adopting a place-based sustainability research approach. Alongside her research career she has been a local political decision-maker and active volunteer on urban biodiversity protection, urban green area planning and management in the city of Helsinki.

Walsh, Lindsey: Chapter 6

Lindsey Walsh is currently a master's student within the Anthropology Department at Rhodes University.

Preface

One of the key motivations for compiling this book is to share the richness that is encapsulated in cultural and natural diversity that surrounds us. I have had the very fortunate opportunity to spend many hours listening to people share with me their stories and memories of their experiences, interactions, ceremonies and cultural practices in nature. Many of these stories and memories hold deep emotions for the narrators including ones that bring great joy and happiness as well as loss and pain when opportunities no longer exist to partake in such activities. Encapsulated in these accounts are the ingredients that contribute to the rich cultural diversity of the region.

In 2018, I had the opportunity to attend the XVI Congress of the International Society of Ethnobiology held in Belém, Brazil, some three decades since the concept of biocultural diversity was first coined. The congress was attended by over 1,600 participants from 45 countries, and it is estimated that over 500 participants were indigenous Brazilians. Conference participants enjoyed local food and beverages prepared by vendors, live music and dance performances, and explored treasures sold by local crafters. For the duration of the congress we were mesmerised by the sights, sounds and smells that emulate from cultural diversity. This experience brought home the importance of diversity in our lives and inspired me to compile this edited book with Charlie Shackleton.

With generous funding received from the National Research Foundation of South Africa, we had the opportunity to host our 13 lead authors for a ten-day writing workshop in Stutterheim, South Africa. The ideas and thoughts generated in our time together now enrich our synthesis presented here.

The editors have made a deliberate attempt to reflect different geographic regions, approaches and writing styles to enhance the portrayal of diversity in the selection of chapters. The diversity of human and nature relationships showcased will be of interest to researchers, teachers and practitioners in the field of biocultural diversity, human ecology and sustainability. The book will also appeal to the general public who have an interest in nature and urban environments. The writing style of the authors has been kept to allow for their content to be appealing and easily accessible to both academic and non-academic readers.

1 Urban nature and biocultural realities

Charlie Shackleton and Michelle Cocks

Human wellbeing and nature

There is no longer any doubt that an important component of and contributor to human wellbeing is the natural environment in which people live, work and relax (Summers et al. 2012). Whilst initial ideas of human wellbeing, early in the second half of the 20th century, focussed on objective measures that could be quantified and contribute to humans' basic needs, they have evolved a great deal since, despite the lack of consensus on a precise definition of human wellbeing (Summers et al. 2012). Over the last five decades the conceptions of human wellbeing have become more complex and inclusive of the more subjective and less tangible components of human existence, including the natural environment (King et al. 2014).

The natural environment contributes to human wellbeing in multiple ways, most recently encapsulated as ecosystem services (Millennium Ecosystem Assessment 2005) or "nature's benefits (or contributions) to people" (Diaz et al. 2015). The latter framework also illustrates the direct benefits in the form of ecosystem goods and services, of which typically three categories are identified, namely, provisioning, cultural and regulating ecosystem services. The first are the tangible goods that are harvested and consumed, such as food, medicinal plants, timber, fish, water and genetic resources (Saarikoski et al. 2015). These may be homegrown, harvested from the wild or bought in a local store or supermarket. Most are used regularly, but some may be called upon only in times of economic or spiritual hardship. Some people also rely on them for a cash income, such as farmers and fisher folk. In contrast, cultural ecosystem services are a suite of less tangible services from nature, such as inspiration, recreation, relaxation, spirituality and education (Chan et al. 2012, Dickson & Hobbs 2017). Regulating services contribute to human wellbeing through regulating and buffering of various biophysical processes, such as flood control, pollution abatement or pest control. The original Millennium Ecosystem Assessment (2005) also recognised supporting services, but they are not generally included in the classification any longer (e.g., TEEB 2010) as they contribute to and underpin the previous three in the form of primary production, soil formation and water and nutrient cycling.

Various frameworks exist to help conceive and illustrate the links between nature and human wellbeing, including those of the Intergovernmental Science-Policy Platform on Biodiversity and Ecosystem Services (IPBES) (Diaz et al. 2015), the Millennium Ecosystem Assessment (MA 2005) and Biocultural Diversity (Buizer et al. 2016). At a generic level, the first two depict nature, or natural/green space, as generating a suite of goods and services (or benefits), some of which are used by humans for their material, financial, physical, emotional and psychological benefit, consciously and subconsciously. Modification of natural areas, purposively or inadvertently, therefore changes the mix, quantity, quality and rates of supply of the goods and services provided, which in turn affects human wellbeing positively or negatively (and one group of people may be affected differently to another group of people). Thus, there is a cycle or feedback loop where changes in one part are manifest as changes throughout the system. Whilst these feedback loops are generally portrayed and interpreted to be positive, this need not necessarily be so (Cocks & Wiersum 2014). This intimate link is also subject to (1) external drivers of change which may be of human origin (e.g., land transformation, changing market prices for goods, changing governance systems) or natural (e.g., floods, pest outbreaks) and (2) the local and wider-scale governance systems. These dynamics play out at various spatial and temporal scales. Taken all together, this is referred to as a social-ecological system. Biocultural diversity, on the other hand, focuses on the more intimate and intangible relationships between humans and nature (Maffi & Woodley 2010).

Conceptions of nature

The concept of social-ecological systems is rapidly gaining credence in both theory and practice. However, some argue that it is based on a particular view of nature and of humans' relationship with it, which are both largely anthropocentric and reductionist (Skandrani 2016) and potentially exploitative in using terms such as ecosystems services that suggest or promote a market-orientated approach for quantifying, valuing and accounting for the supply of nature's goods to humans (Peterson et al. 2010, Schröter et al. 2014, Kull et al. 2015). In this view, although linked within a social-ecological system, there is still a separation between humans and nature, such that nature is there to supply benefits to humans, who in turn, need to manage and shape it, to provide the benefits in which they are most interested (Flint et al. 2013, Skandrani 2016). This is a rather simplistic critique, as it is well recognised that even within largely Western societies, a range of attitudes towards and relationships with nature exist, ranging from strongly anthropocentric through biophilic to strongly ecocentric (Milton 1998, Flint et al. 2013).

Pre-dating and paralleling this growing body of literature framed within the social-ecological systems paradigm are the deep and rich debates on the human-nature relationship contributed from a variety of disciplines working at a more individual or local scale, such as cultural anthropology, cultural ecology and environmental psychology, and more recently, environmental ethics (Pretty et al. 2009, Russell et al. 2013). These consider the deep and innate "feelings" and "connections" that humans have about and with nature and how they contribute to personal and collective identity, mental wellbeing, spirituality, sense of place and how they shape interactions with the natural environment and components and species within it; indeed, to some the inclusion of humans as part of nature. Flint et al. (2013) characterise human-nature relationships within four overlapping domains, namely (1) world views, (2) values and attitudes, (3) behaviour and lifestyles and (4) traditions, philosophy and ethics, which we find a useful advance for fostering more insightful analysis and understanding of humans' (individual, collective and societal) conceptions of, and hence engagement with, nature.

Ways of conceptualising nature have until relatively recently been dominated by Western understandings and interpretations. For example, during the 18th and 19th centuries, nature was largely conceptualised as a "wilderness", which was defined as any area (small or large) which had not been the subject of human influence or interaction (Bell et al. 2001). This wilderness was believed to contain unpredictable natural forces and wild animals that could threaten human lives (Bell et al. 2001). Thus, nature was perceived as wild, disordered, cruel, feared and uncivilised and a place to be avoided (Bell et al. 2001). This distorted perception of viewing oneself as separate from nature can be demonstrated further by the negative and even hostile attitudes and values adopted towards the wilderness. For example, in medieval Europe words such as "terrible", "horrible" or even "disgusting" were used to express attitudes towards nature (Bell et al. 2001). In fact, Europeans so "abhorred the wilderness that travellers sometimes insisted on being blindfolded so that they would not be confronted with the terror of untamed mountains and forests" (Bell et al. 2001, p. 26). Nature, during this time, was something that needed to be subdued, dominated and tamed to ensure that civilisation flourished (Bell et al. 2001, White 1967, Young 1974). During the period of Enlightenment, European attitudes towards nature began to change. This shift was influenced by scientific discoveries as people began to view natural phenomena as complex and marvellous manifestations of God's creation (Bell et al. 2001). European intellectuals became increasingly fascinated rather than repulsed by nature. The Romantic era saw nature being represented as pleasing and beautiful within the arts, literature and music (Bell et al. 2001).

As scientific discourses began to dominate, representations of nature began to hold reductionist flavour (Cobern et al. 1999, Skandrani 2016). Nature became a resource to be exploited to meet the growing demands of industrialisation and an increasing world population (Bell et al. 2001). Market-orientated approaches began to dominate the ways in which nature was conceptualised.

Urban nature

Nature is not restricted to "natural", rural or protected areas, but is found everywhere, even in hugely modified ecosystems, landscapes and spaces. However, the abundance, diversity, quality and functioning of nature varies from place to place and may be adversely affected by the effects of human activities, infrastructure and industries; no more so than in urban and peri-urban areas. Thus, some towns and cities, or areas within them, may be relatively devoid of any greenery or biodiversity save a few hardy anthropophilic species; whereas others may have an abundance of primary or human-designed green spaces supporting a variety of plant and animal species, linked with green walkways and tree-lined streets. For example, in a review of 100 cities globally, Dobbs et al. (2014) found that the percentage of green space ranged from as little as 0.4 per cent (Kolkata, India) to over 63 per cent (Winnepeg, Canada).

With a growing realisation of the importance of nature, including urban nature, for human wellbeing and the local supply of ecosystem services, the extent, diversity and quality of nature in urban settings are increasingly quantified and indicators and targets debated. However, the findings are sensitive to what measures are used and precisely where the measures are applied. Thus, there have been comparative studies of the level of green cover in cities, or the area of public green space (in total and per capita). But much depends upon if the estimates are based on only formal green spaces, or also include informal green spaces; also whether it is just public green space, or also greenery found in private or non-public space such as residential gardens, corporate grounds or institutions with restricted access such as schools or military grounds. For example, Shackleton et al. (2018) reported that less than 20 per cent of the urban area of nine towns in South Africa was public green space, but if private green space was included then the total was above 60 per cent.

Despite this diversity of measures and inclusivity, the increasing appreciation of the importance of urban green spaces for human wellbeing underlies the growing promulgation of quantitative targets and ratios regarding the minimum or desired amount of green space required at various scales within a town, city or country. For example, the World Health Organisation (WHO) recommends at least 9 m^2 per capita. Within Europe, the UK recommends that every urban citizen be no further than 300 m from a public green space (Handley et al. 2003), whilst in Germany, Berlin strives for 6 m^2 per capita, but in Leipzig it is 10 m^2 per capita (Kabisch et al. 2015). In South Africa, it is 20–40 m^2 in the large, developed cities (McConnachie et al. 2008). Whilst a start, these sorts of guides do not stipulate the quality of the spaces (and whose criteria would be used?) or their distribution and accessibility or the need for multi-functionality.

Meeting the desired or regulated levels of urban green space is not necessarily an easy task. In historical towns and cities, the streets and blocks were planned long before the importance of urban green space and trees was widely recognised. Thus, there are very real constraints on where urban green spaces can be created, and retrofitting is both expensive and would require demolition of historical and cultural buildings or built infrastructure. In more recent cities, there are ongoing debates about the relative benefits of compact and dense developments and those that spread out over a greater area but at lower infrastructural densities (Ewing & Hamidi 2015). In many Global South countries, the sheer pace of urban growth in both human populations and geographic space makes it extremely challenging to identify, designate and protect urban green spaces for human use, conservation and the benefits they provide. In many regions, they are at constant risk of being built upon by new migrants seeking somewhere to build a shelter or a home, or informal or small-scale businesses seeking a site close to potential customers, or urban elites simply grabbing land (Haarland & van den Bosch 2015). Consequently, informal green spaces may become the most common sites for maintenance of some biodiversity in such urban settings, and the most accessible setting for many urban residents to experience urban nature.

In appreciating the challenges for some urban authorities to designate, protect and maintain urban green spaces and nature prompts the realisation that urban nature is not just about formally designated and mapped public green spaces. "Nature" exists at various scales throughout public and private spaces in towns and cities. Some bird species readily roost or nest on buildings or other human built structures, as do many invertebrates. Reptiles and amphibians can be found under rocks and in drainpipes, and many plant species thrive in cracks in pavements, in gutters and in abandoned sites. These small and isolated sites or even individuals may be a poor substitute for the more viable populations supported by large urban green spaces and corridors, but they still afford some urban residents an opportunity to view, experience and perhaps relate to some biodiversity. And small changes in infrastructure design or maintenance can encourage colonisation and support of a wide range of species. For example, Gopal and Nagendra (2014) reported how slum dwellers in Bengaluru, India, maintained numerous plant species in pots outside their doors and windowsills.

Diversifying urban nature for a diversity of needs of nature

As towns and cities become more culturally diverse (e.g., Kahn 2014, Zwiers et al. 2018), the demands for different types and experiences of urban nature will grow. Large cities are increasingly home to people from multiple national and transnational cultures, as people move to find work, reliable services, education or safety from hardship or persecution. These sojourners, short-term or permanent migrants bring with them different conceptions and expectations of nature. Some even bring familiar nature with them in the form of revered or useful plant species in small pots or transplants to private gardens, and icons and symbols of nature "back home". Others seek species that are familiar to those from home. For example, 42 per cent of the medicinal species used by Bolivian migrants in London (UK) were in common with those from Bolivia (Ceuterick et al. 2011). If known species are absent or unavailable in the new setting, in fresh or dried form (perhaps imported from the migrant's country or origin), then substitutes are found that offer the sought-after traits represented through colour, form, aroma, taste or phytochemical efficacy. This is not unique to the modern era, as humans have transported species of value to them for millennia (Voeks 2018).

The key aspect of this multiplying cultural diversity in towns and cities is that it will raise the need for a greater diversity of types and opportunities to experience urban nature. For example, observers of Ifá spirituality deem themselves as being in nature and of nature, and consequently they require sacred green areas and trees wherever they live. This includes in many towns of the Caribbean and the USA, as well as in West Africa where it originated (Mowatt 2018). Many indigenous cultures assign a sacred status to features of nature or specific sites or species, which some individuals or households maintain even after translocating to urban settings; for example, recognition of tree spirits in urban India (Chandrakanth et al. 1990). But such needs may be compromised if there is insufficient urban green space, or of inadequate quality to meet these specific needs or where open-air spiritual practices might be frowned upon (e.g., Ngulani & Shackleton 2019). Conserving such sacred places or trees within the urban matrix contributes to urban sustainability through providing ecosystem services such as habitat and refugia for other biodiversity, nutrient cycling, pollination, water infiltration, buffering air temperatures and noise and improving air quality and aesthetics, amongst others (De Lacy & Shackleton 2017, Jaganmohan et al. 2018).

Most research considering the cultural benefits of urban nature has emanated from the Global North, especially Europe and North America (Roy et al. 2012; Kabisch et al. 2015). Additionally, the majority have focussed on the use of public green spaces for recreation and relaxation (Roy et al. 2012; La Rosa et al. 2016). This may be because the most common uses of urban green spaces in Europe and the USA appear to be for exercise, relaxation and socialisation. The predominance of these sorts of activities may be because these are the ones most people pursue, or it may be that other uses of public urban green spaces might be prohibited through formal or informal regulations, norms and culturally prescribed codes of behaviour.

Indeed, there are many cultural and spiritual needs of nature that receive little mention in this body of literature and consequently are typically overlooked when green space designs and policies are debated or implemented. In many regions, formal or informal green spaces are also used for spiritual purposes and outdoor worship (Jackson & Ormsby 2017, Ngulani & Shackleton 2019), urban agriculture (Orsini et al. 2013, Poulsen et al. 2015), collection of 'wild' foods and other useful biological resources (McLain et al. 2014, Shackleton et al. 2017) and education (Wolsink 2016, Knackmuhs et al. 2017), but conscious inclusion of such benefits and activities when planning urban green spaces is currently relatively rare. Consequently, in many urban centres they remain under-represented or they are concentrated in informal and interstitial or private green spaces. Yet, it is these 'other' uses and relationships that give a fuller and richer expression to the varied and intimate oneness of humans and nature, even in urban places, which is increasingly viewed through the lens of biocultural diversity.

Biocultural diversity

Biocultural diversity (BCD) is conceptualised as the intimate and inextricable links between cultural and biological diversity, such that they are indivisible (Buizer et al. 2016). Maintenance of cultural diversity supports the maintenance of biodiversity from genes to landscape levels, and maintenance of biodiversity provides the tapestry for cultural diversity and the benefits humans receive from being part of nature. In many respects, this parallels the concept of social-ecological systems but BCD emphasises more the innateness and bi-directionality of the relationship, its dynamic and constantly changing form and meanings, as well as the inclusion of different knowledge forms (and holders) and linguistic diversity as an expression of this relationship (Buizer et al. 2016). These deep connections between biodiversity, cultural diversity and linguistic diversity are attributed to the co-evolution of humans and nature over millennia (Maffi 2008) and consequently multiple, varied and varying feedbacks between the three (Pretty et al. 2009). This relationship is most apparent and most studied in indigenous communities around the world; however, it is not restricted to such (Cocks 2006, Cocks & Wiersum 2014, Elands et al. 2015). As the essence of the relationship is revealed, it is readily apparent that biocultural relationships can be found and thrive in all settings where humans exist, from the remotest forests to megacities. However, it is also increasingly apparent that maintenance and expressions of these relationships can be forgotten, suppressed or severed through domination of various worldviews and erosion or loss of bio- cultural and linguistic diversity at all scales from local to global (Pretty et al. 2009).

As globalisation extends, the domination of particular worldviews comes at the expense of declining cultural and linguistic diversity, paralleling the more widely known loss of biodiversity. Pretty et al. (2009) conceive the relationships between bio- and cultural diversity to be created and expressed in three domains, namely (i) humans' place in and of nature, (ii) humans' management and shaping of nature and (iii) humans' knowledge and description of nature. The first considers the worldviews, beliefs and meanings that humans create about nature and how it shapes their identity and, in some worldviews, even their humanity (Pretty et al. 2009). The second holds that humans shape nature by the conscious and inadvertent selections of and modifications to genes, species and landscapes to provide a multitude of tangible and intangible benefits, and that reciprocally nature shapes human identity, culture and livelihoods through what is available in local landscapes and how nature responds to human actions (Pretty et al. 2009, Cocks & Wiersum 2014, Elands et al. 2015). The last explores how knowledge of nature, species, landscapes and processes shapes attitudes towards it and how it is managed by humans. Knowledge of nature is gleaned through interaction with and experience of nature and accumulated through time by individuals, communities and societies (Pretty et al. 2009, Barthel et al. 2010). An expression of the richness and depth of knowledge of nature is the terms and words used to describe different natures and all their elements and processes. Thus, a richness of nature and knowledge of it leads to a richness of language, which is an essential part of human culture. Yet, local knowledge about nature and natural processes is eroding rapidly as urban migration, market globalisation and access to communication technology advance (e.g., Reyes-García et al.

2005, Aswani et al. 2018). This is perhaps most marked in Western societies, but not uniformly, nor is it restricted to them.

Recognition of these three domains and their interconnectedness encapsulates that the relationships between humans and nature are not just about access to nature but also interactions and real connections with it. Thus, prescriptions about how much urban green space should be provided and where, whilst a good start, are insufficient. Deep and meaningful biocultural relationships rest on a diversity of worldviews of, knowledge about and experiences in nature, which for many will transcend just passive, and often regulated, recreation or relaxation in urban green spaces. It also requires that forms and expressions of green spaces and nature are locally constructed or co-constructed rather than a one-size-fits-all approach, where particular types of public green spaces are rolled out irrespective of the local context, knowledge, views and needs. Hence, the value of the BCD framing and approach.

Elands et al. (2015) argue that adopting a BCD framing adds to and even supersedes the prevailing ecosystem services model in three ways. The first is that it elevates culture to being a significant force in shaping social-ecological systems and hence biophysical environments. Second, it places culture at the core of biodiversity conservation, both in catalysing negative effects and in presenting new and varied opportunities for biodiversity conservation embedded in local knowledge and cultural practices. The third is that because culture is not static, neither are biocultural relationships; in other words, the BCD concept emphasises the dynamic bi-directional relationships between humans and nature.

Rationale for this book

Urbanisation is a worldwide phenomenon driving social, economic and environmental change. It is estimated that two-thirds of the global population will be living in urban centres by 2030. For the first time in human history, more people in the world now live in cities and towns than in rural areas. Most future urbanisation is predicted to take place in the developing world (UN-Habitat 2006). Preserving, providing and maintaining green spaces and infrastructure within urban environments is increasingly recognised as a means of improving the liveability and quality of life in urban areas (Handley et al. 2003). It is well established that such spaces serve important social, psychological, health, aesthetic, cultural and ecological functions (Roy et al. 2012, Lovell & Taylor 2013). Within such a context, experiences of urban green spaces are overwhelmingly portrayed in academic literature and discourses as encompassing outdoor sports, recreational activities, nature-based tourism and eco-adventure. To meet these needs urban green spaces are constructed as spaces to retreat from the busyness of urban living and have taken on the form of parks, botanical gardens, cemeteries, church gardens and tree-lined streets within public spaces. Such constructs of green space generally reflect only one particular interpretation of nature and ways of being within or relating to nature. An interpretation which is embedded within a Western ideological perspective and defined by dualist epistemology is the nature versus culture divide. Within such a construct nature is perceived as being separate from oneself.

In contrast, many works carried out by anthropologists, mostly amongst rural communities, have argued that there are different ways to conceptualise and understand nature and that these conceptualisations are the dynamic products of historical and cultural contexts. For example, many indigenous peoples do not perceive themselves as being distinct from nature. The environment is perceived as being "that which surrounds [the individual], [making] all relations [and therefore knowledge] ecological" (Milton 1996, pp. 115–116). Such a perspective acknowledges that individuals and the environment "embrace an irreducible system; that is, the person is part of the environment and in the same way, the environment is part of the person" (Descola & Palsson 1996, p. 18). These studies have highlighted the need to explore more nuanced understandings of the ways that humans live in, understand and shape their environment and the non-human species with which they share their world and how these experiences contribute to cultural ways of being and meaning-making. Since the mid-1990s social scientists have challenged modern intellectual traditions working from the dualistic premise.

In relation to conceptualising and contextualising processes associated with urbanisation these remain rooted in a meta-narrative of modernisation and unilineal urban transition, which has been criticised as inadequate in describing how people come to live and be in urban environments (Ferguson 1990, 1999). Residents of urban environments include people who have lived their whole lives in the city and are descendants of city dwellers. Many may feel a deep connection and attachment to the cities they occupy. For many, such attachments often include natural places which are frequented and are given meaning by engagements carried out within such areas. There are also national and transnational migrants who seek employment opportunities within urban environments and others who feel forced to occupy such environments due to armed conflict, natural disasters and economic need. In the latter two instances some migrants may feel forced to stay as they have no means to return "home", whilst seeing it is a temporary move and will return "home" whenever the opportunity arises. Such experiences dispel notions of unilinear transitions to modernity. They also speak to how urban environments are experienced by some as places of alienation and detachment (Kloek 2015, O'Brien et al. 2017). Scholars have attributed such experiences to the Eurocentric approach adopted around urbanisation (Bouteligier 2011, Lawhon et al. 2014) which has infiltrated governance and policy-making processes within cities and towns around the world. This has resulted in urban nature still being defined and conceptualised in a limited and often Eurocentric way even within cities and towns in the Global South, additionally entrenched through processes of colonisation. The colonial legacy has been retained in many parks in terms of species planted, statues and artworks, design and layout, as well as the regulations and norms regarding what activities and practices are permitted (or not).

Restricting communities from cultural expression through natural resources can have negative effects on the wellbeing of urban residents (Botzat et al. 2016, Kabisch & Haase 2014, Rutt & Gulsrud 2016). Work in Australia examining the effects of colonial assimilation on indigenous people shows that cultural assimilation has not diminished the connection of indigenous peoples to their environment (Cunningham & Stanley 2003, Dockery 2011). Similar examples have been documented amongst Black urban residents in South Africa. During the racially defined apartheid era, Black South Africans were denied access to public parks, gardens and commonages in areas reserved for "White" South Africans, even though there was no provision of such public green spaces in the "Black" areas. After the fall of the race-based apartheid regime in the early 1990s, access to urban green spaces became easier, although the distribution still favours the former "White" areas (McConnachie & Shackleton 2010, Kuruneri-Chitepo & Shackleton 2011). However, the nature of activities and engagements permitted or viewed as acceptable in urban green spaces and parks is still generally restricted, and no explicit mandate exists to cater for the cultural needs and meanings public green areas could hold for residents of Black African descent.

An approach which presumes that nature and culture are not in opposition to each other but are interlinked is biocultural diversity (Maffi 2008) and could offer a solution to the ways in which nature is defined and conceptualised within urban environments. The concept of biocultural diversity was originally proposed to denote the intimate link between nature and culture in relation to indigenous communities, especially those living in areas of high biodiversity (Posey 1999, Maffi 2005, Maffi & Woodley 2010). Later, the concept was expanded to recognise the diversity of life in all its manifestations, i.e., biological, cultural and linguistic, which are inter-related (and likely, co-evolved) within a complex social-ecological system. More recently, it has been used to acknowledge the diversity of manifestations that are reflected in cultural heritage sites and cultural landscapes (Cocks & Wiersum 2014). Similarly, new approaches for sustainable development have emerged (Davidson-Hunt et al. 2012, Garnett et al. 2018, Sterling et al. 2017), calling for development for rural and indigenous communities to be rooted in their distinct cultural identities and so advocating that such communities have claims for greater control over land, development and identity (Merçon et al. 2019). Within such an approach the concept of biocultural diversity heritage has been included. There is a growing awareness that even in economically advanced regions of the world a great diversity of manifestations of biocultural diversity exists in both rural and urban settings. Recognition of this resulted in the Florence Declaration, in 2014, explicitly indicating the need to acknowledge "the vital importance of cultural and biological diversity for present and future generations and the wellbeing of contemporary societies in urban and rural settings".

Within this framing of world urbanisation trends, insufficient provision of public urban green space and the inadequacy of the types of green spaces provided, the purpose of this book is to:

- display and interrogate the diversity of ways in which urban residents view, interact, engage and give meaning to urban nature. Cases were carefully selected from around the world to illustrate this and especially those which offer alternative views, visions, uses of and relationships with urban nature beyond that displayed in much of the academic literature focusing mostly on passive recreation and exercise.
- reveal what residents may bring into urban environments to emulate notions of the familiar by the plants they bring to grow in such environments.
- consider if current Global North conceptualisations of urban environments may hinder certain residents or communities from forming culturally meaningful relationships within such environments.

Core terms and structure of the book

This book brings together a number of case examples from different continents and different disciplinary backgrounds. Consequently, it is necessary that core and recurring concepts and terms are placed into context from the outset.

Core terms placed into context

Biocultural diversity

The concept of biocultural diversity first emerged from case material collected amongst minority indigenous communities and was coined to give recognition to the inextricable link between biodiversity and cultural diversity amongst traditional and indigenous people (Posey 1999, Persica & Martin 2008). The concept went a long way in raising the alarm over the rapid decline in biological and cultural diversity (Rapport & Maffi 2010) and the loss of intergenerational transmission of traditional knowledge, practices and languages (Maffi 2005, Maffi & Woodley 2010, Posey 1999). Since its inception, the concept of biocultural diversity has gradually been modified and broadened in scope (Cocks 2006, Cocks & Wiersum 2014) to include other social and ecological contexts. It is now understood to include cultural dynamics as a vital component in the way in which communities and societies live and associate with biodiversity (Pretty et al. 2009, Merçon et al. 2019). It gives recognition to "the interweave of biological and cultural diversity, people and places, and the continuing adaptation and co-evolution between landscapes and ways of life" (Laird et al. 2011, p. 276). This has led to a rich body of research from the Global South revealing that rural and urban communities maintain cultural practices reliant on natural environments and high biodiversity (Cocks 2006, Chivaura 2007, WinklerPrins & de Oliveira 2010) and retain cultural and spiritual values associated with natural landscapes and vegetation in rural and urban contexts (Cocks et al. 2012, 2018). It has also been proposed by several authors that the concept can be used as a reflexive and sensitising concept to assess the different values and knowledge of different human groups living with biodiversity within different contexts (Cocks et al. 2018), including economically advanced regions (Elands et al. 2015, Buizer et al. 2016). A great diversity of manifestations of biocultural diversity have been found to exist within both rural and urban settings (Agnoletti & Rotherham 2015, Elands et al. 2015, Agnoletti & Emanueli 2016). Within European cities the use and value of nature are increasingly recognised as a vital part of a city's cultural heritage and identity (Elands et al. 2015).

Within the context of this book the concept of biocultural diversity is used to give recognition to the multiple ways in which people and communities from different cultural contexts view, live, interact and experience nature in its diversity of forms within urban areas.

Culture and ways of being (ontological understandings)

Culture as a concept has perplexed scholars for centuries. Initial attempts to define culture were couched in efforts to rank human societies on a scale from primitive to civilised and from pre-logical to logical. Authors later described such attempts as ethnocentric and arrogant (Hiebert 2008). Ways of interpreting culture became understood as a set of beliefs and practices that are made use of to order human activities. Ethnographic studies amongst indigenous communities contributed to the understanding that people do not simply live in the same world with different labels attached to it, but rather they live in radically different conceptual worlds (see de Castro (1998), Descola (1994, 2006)). This growing awareness led to investigations attempting to understand these domains, and words such as ethos, cosmology, cultural unconscious, world making and worldview have arisen (Hiebert 2008, p. 16). Worldview, in general terms, can be "understood as what individuals or communities recognise as relevant for their functioning in the world they live in, as well as the way that content is organised, through a network of relationships, into a coherent representation" (Glaz 2017, p. 35). Thus, meaning that a specific worldview need not make sense to everyone; its role is to aid the individual or community that entertains it in the process of making sense of the world (Glaz 2017). Worldviews have come to mean what people in a community take as a given reality, and the maps they use to navigate that reality (Hiebert 2008, p. 15). It is within this framing that individual's and communities' relationship with nature will be presented and reflected upon in the different chapters of this book.

Human wellbeing

Since the 1970s the concept of wellbeing has been used as a development indicator and has categorised wellbeing in terms of material wealth or lack thereof (Agarwala et al. 2014). Measures initially used for global comparison indicators included standard of living, income range, employment and service delivery (Camfield & Mcgregor 2005). It has, however, become increasingly recognised that an increase in GDP does not necessarily result in increased levels of happiness (Diener et al. 1998, Tov & Diener 2009). Recommendations are progressively being made to include subjective appraisals of how individuals assess the positive, holistic condition of their own life (Wang & Wang 2016) and how they define their wellbeing (Camfield & Mcgregor 2005). Intangible domains are increasingly acknowledged as necessary for a happy and well-balanced life. Subjective wellbeing (SWB), which is broadly defined as a person's evaluation of their life quality, is being incorporated into wellbeing measures (Diener et al. 1998, Diener & Suh 2003, Church et al. 2013).

Knowing and experiencing nature is well recognised as making humans happier and healthier (Bell et al. 2001, Chiesura 2004, Thwaites & Simkins 2007, Sanesi & Chiarello 2006). The intangible connections to nature and human wellbeing have been identified as being made up of the following constituents: physical health, mental health, spirituality, certainty and sense of control and security, learning and capability, inspiration and fulfilment of imagination, sense of place, identity and autonomy, connectedness and belonging and subjective wellbeing (Russell et al. 2013). Nature contributions to mental and physical health have has been well documented, whereas other benefits of living in nature to learning and inspiration, or the links between identity, self-sufficiency and belonging have received considerably less attention (Russell et al. 2013, Gould et al. 2014). Reductionist and empirical Western approaches have made it possible to produce an expansive knowledge on the former, but other vital components of human wellbeing have received significantly less attention such as identity and spirituality. The latter components are inherently more difficult to assess with existing approaches and methodologies (Russell et al. 2013). Many of the chapters in this book go some way in addressing these gaps by showcasing the multitude of relationships that urban residents from different cultural contexts have with nature and the impact these relationships have on their wellbeing.

Urban and urbanisation

Virtually all countries around the world have distinguished between urban and rural populations. However, the definition of what constitutes an urban area varies between countries and in some cases even varies over time within a single country. This has meant there is no global standard for the classification of urban environments. Attempts to define urban areas or communities living in urban areas have included population size, population density, administrative or political boundaries, or economic function (Cohen 2006). The use of such attempts has become increasingly problematic and contested in response to the increased distribution and movement of people, goods and services across different types of spaces (Berdegué et al. 2014). Such developments are increasingly making the distinction between urban and rural areas redundant (Cohen 2006). It is not within the scope of this book to find workable criteria to define urban areas but only to highlight the diversity of engagements and relationships that are formed within nature by urban residents. Increasingly, connections to nature within urban areas have been shown to contribute to feelings of cohesion, belonging and wellbeing (Chiesura 2004, Sanesi & Chiarello 2006, Peter et al. 2010) which speak to the importance of finding ways to improve living conditions for all, particularly the poor and more marginalised communities.

Processes associated with urbanisation began as early as the 12th century, but the impact at global levels was only felt after the industrial revolution (Tacoli et al. 2015). For example, up until the 1900s urbanisation levels were low as there were only approximately 40 cities in the world exceeding 500,000 people, of which only 16 exceeded 1,000,000. By the 21st century, well over 400 cities exceeded population levels of 1,000,000, with some exceeding ten million (Berry 2008). It is estimated that by the year 2030 more than two-thirds of the world's population will be living in urban centres, with the highest rates of urban growth currently being experienced in the Global South, particularly in sub-Saharan Africa (UN-Habitat 2006). Urbanisation of the world population has not only taken place in large cities. Almost two billion people, 27 per cent of the world's total population, or half of the world's urban population, reside in towns and small- and medium-sized cities (Berdegué et al. 2014).

Urbanisation is considered a global multidimensional process (Ernstson et al. 2010). It has been intricately interlinked with ideological processes of forward thinking and innovation, modernisation, industrialisation and rationalisation (Jayne 2005, Gries & Grundmann 2018). Conversely, rural areas have been represented as backward, archaic and a place to be exploited by others (see Braudel 1982, Wallerstein 1974). Such framings were rooted in a meta-narrative of modernisation and unilinear urban transition whereby cities have been described as "the spatial carrier of all those characteristics which are associated with development" and rural settlements as the "spatial carrier of traditionality", which had to be overcome (Lohnert & Steinbrink 2005). Such characterisations have also been used to describe cities in many developing countries which resulted in them being described as fitting the lower rungs of the development scale or "pathological deviants" from the norm (Barnett & Parnell 2016). Research in predominantly non-Western contexts has shown the analytical inappropriateness of the rural-urban divide. In contrast, a porous and diffuse relationship between rural and urban areas exists. In response, calls have been made to recognise how many household's daily life practices cut across geographical demarcations and are spread over various locations. Such calls have come from authors tracing livelihood strategies of households who have illustrated an interconnectedness between rural and urban areas (Berdegué et al. 2014).

In the past two decades, rates of urbanisation have accelerated at a phenomenal rate. This has resulted in cities becoming a central object of a range of global development and environmental policy debates. Understanding urban processes is seen as being central to securing sustainable futures for a "majority-urban future" (Barnett & Parnell 2016). Within this framing, multiple forms of knowledge are now recognised as shaping how urban processes need to be understood. Cities are no longer understood as representing bounded entities but are understood as a composition of relations, movements and networks resulting in attempts to understand how cities works (Barnett & Parnell 2016).

In tandem with these developments, calls have been made that concepts and methods used in urban research need to be reconfigured to ensure that Western experiences are not taken as a developmental norm. Such calls are informed by neo-Marxism, postcolonial cultural theory and attempts to promote "theory from the South" and "epistemologies of the South" (Barnett & Parnell 2016). It is within this framing that urbanisation is interpreted and understood as attempts are made to give recognition to ontological framing of a multitude of practices carried out in nature by residents living in cities and small to medium towns.

Biodiversity

The term "biodiversity" has become very much a part of everyday discourse and conversation, particularly in Western contexts. Reference to biodiversity first emerged within the biological sciences to give recognition to diversity found in species, between species and of ecosystems. In 1992, the UN Convention on Biological Diversity defined biological diversity' as the variability among living organisms from all sources including, inter alia, terrestrial, marine and other aquatic ecosystems and the ecological complexes of which they are part". The term has been used as an object of measurement, a basis for decisions in conservation planning and as a primary conservation objective (Santana 2019). Billions of dollars have been spent in the name of biodiversity, and nearly all national governments have signed the international Convention on Biological Diversity. Nonetheless, numerous criticisms have been levelled at the definition, notably for its vagueness. Increasingly, attempts to define it in precise terms are being labelled as futile. Authors, particularly within the fields of humanities, have highlighted that the term no longer simply refers to the variety and variability of the natural world but refers to our conceptualisation and valuation of it (Toepfer 2019). It is not within the scope of this book to discuss the merits and failures of these but to draw attention to the value that experiences with nature's diversity have brought to people living in urban settings. To highlight the biological diversity encapsulated within these experiences, scientific names have been included where relevant in individual chapters as in many instances, associations, meanings and attachments were attributed to particular species.

Greenspaces (formal and informal) and urban nature

"Greenspace" is a term that was coined to give recognition to the services and benefits derived from particular environmental resources found in and around urban areas such as remnant patches of native vegetation, urban wastelands (i.e., brownfields, vacant lots), gardens and yards, to highly engineered green infrastructure such as bioswales and green roofs.

Such ways of conceptualising nature and the associated benefits have been given global priority through sustainable development goals that aim to improve the liveability and quality of life of urban residents (Haq 2011, Lennon 2015). Greenspaces are increasingly recognised as providing complimentary benefits to ecological, economic and social spheres within cities. To assist with planning, a great deal of work has gone into categorising the different types of green spaces according to function, size and management regimes. In the literature, green spaces are usually classed in two broad categories, namely, formal and informal green spaces. Formal green spaces have often been described as clearly demarcated and managed green spaces within urban areas, whilst informal green spaces are said to be ambiguous, in-between or liminal vegetated surfaces within towns and cities (Rupprecht & Byrne 2014). Examples of formal green spaces include parks, botanical gardens and cemeteries (Schipperijn 2010) whilst informal green spaces include riparian zones, vacant lots and spaces in between buildings (Rupprecht & Byrne 2014). However, such constructs are deeply embedded in a Eurocentric understanding of what is "formally" managed and what is not. To give recognition to the diversity of human-nature interactions within urban areas, we have also included the term urban nature to encapsulate the diversity of experiences that are experienced, rather than just the physical spaces in which nature is presumed to be found, and restricted to.

Structure of this book

This book comprises 13 chapters. The majority (11) are case studies from around the world on the relationships that humans forge with nature in urban settings and are written from a variety of academic backgrounds. The book opens with this introductory chapter providing a general background and rationale for the book. Thereafter follow 11 case studies. Two are from Africa, two are from Asia, two are from Europe and four are from New World.

The first of the case study chapters begins with a detailed account of how Japanese philosophy, worldviews and ecology have influenced the very intimate design, spatial structures, elements and interpretation of Japanese gardens, based on an ethos of working with nature to emphasise particular elements that have immensely deep meaning. Particular elements and their combination (or omission) contribute to the creation of "space in nature" to which different meanings are attributed. Opportunities to restore kinship with nature are seen as crucial to ensure that Japanese gardens remain a part of cityscapes. This is followed by a chapter outlining how Yoruba cosmology has influenced the creation and design of urban natures including royal palace gardens, courtyards and scared groves in the city of Lagos, much of which was lost during the colonial period. A variety of wellbeing strategies are adopted by urban residents through interaction with nature, such as direct worship of spirits in sacred groves and through the gathering and use of indigenous medicinal and food plants. This is followed by a moving and provocative part memoir, part reportage on the cultural meanings given to gardens in Iran. The case study illustrates the significant role that gardens have played in the Persian psyche as a place of refuge, reflection and respite from a harsh environment. Gardens have become increasingly threatened by property development and economic and political control within Iran, which severs not only nature contact but also identity and psyche. Insights into how urban nature in the city and planning approaches have been conceptualised in Finland over different historical time periods is given and how these have influenced the types and forms of urban nature now found in the city of Helsinki. Residents living within the city of Helsinki have given different meanings and developed different forms of attachments to each of these.

A case study from South Africa reveals how urban nature remains colonised and continues to be framed in Eurocentric thinking and designs, which undermine the appreciation and value for such places for many indigenous residents, the majority of whom hold values and meanings of nature that differ to such framings yet continue to remain unacknowledged in the ways in which urban natures continue to be conceptualised and planned for in South African towns and cities. This is followed by a description of how migrant women from Mexico attempt to create "places of refuge" by adapting rural indigenous practices from their country of origin with non-indigenous and urban ones in their new place of residence. The example also highlights how race, gender and class discrimination contribute to the suppression of biocultural relations in the city.

To further illustrate how biocultural relations are transported, a description of how the waterways of Jamaica Bay, New York (USA), and how they have become imbued with spiritual value by the local Indo-Caribbean Hindu community is given. Jamaica Bay is seen as a proxy for the sacred Ganga River in India by this community. Devotees routinely worship and offer sacred items into the bay. The practices have, at times, resulted in tensions between management organisations, residents and worshipers. As a response, a grassroots environmental stewardship movement led by Indo-Caribbean Hindu civic organisations and community members has emerged. The chapter also shows how the community's religious practices have led to ecological connections and deep attachment to the spiritual and physical landscape. A case study from Brazil brings alive the vibrancy, activities and biocultural manifestations of the beachfront in Rio de Janeiro. A diversity of relationships is described, many of which strengthen the sense of wellbeing and belonging amongst different social groups within the city, but which also result in conflict between different social groups in time or space. A case study from Vancouver (Canada) further contextualises how colonialism, multiple waves of immigration, mainstream green city discourses, population growth and densification and recent efforts to engage in a process of reconciliation impact human-nature relationships. The analysis also provides insight into the range of issues that need to be taken into consideration in attempts to understand, support and celebrate biocultural relations in cities.

Attention is then given to how forms of active citizen engagement, namely recreation, community gardening and community-led nature management, can contribute to citizen health and wellbeing in super-diverse cities in Europe. Particular attention is drawn to how the social interactions that take place whilst engaging in such activities contribute to feelings of social cohesion and belonging amongst those involved in such activities. Ways of enhancing stewardship capacity of natural resource professionals and community members through sacred relationships with nature are given. Examples include how engaging in traditional rituals and adapting rituals for new settings, communities and managers can collaboratively work to cultivate and express personal connections to place, which in turn can enhance equitable access to urban nature in diverse cities such as New York City.

The final chapter of the book considers emergent themes from the case studies and discusses them in relation to biocultural theory and practice. The main themes are the

1. embeddedness of emotion in biocultural relationships
2. colonial legacies and the rupture of biocultural relationships
3. migration in shaping and enriching such relationships
4. harvesting nature as an intimate expression of these relationships
5. the importance of having a diversity of nature spaces and species in cities and towns to meet the diversity of needs that humans have for nature and biodiversity

The implications of these for biocultural theory are then presented. The chapter ends with suggestions for urban policy and planning and approaches to recreate or strengthen biocultural relationships and opportunities for human wellbeing and liveable and sustainable cities.

Acknowledgements

The compilation of the chapters presented in this book was made possible through funding provided by the National Research Foundation of South Africa. Any opinion, finding, conclusion or recommendation expressed in this material is that of the authors, and the NRF does not accept any liability in this regard.

References

Agarwala, M., Atkinson, G., Fry, B.P., Homewood, K., Mourato, S., Rowcliffe, J.M. & Milner-Gulland, E.J. 2014. Assessing the relationship between human wellbeing and ecosystem services: A review of frameworks. *Conservation & Society*, 12(4): 437–449.

Agnoletti, M. & Rotherham, I.D. 2015. Landscape and biocultural diversity. *Biodiversity Conservation*, 24(13): 3155–3165.

Agnoletti, M. & Emanueli, F. 2016. *Biocultural Diversity in Europe*. Cham, Switzerland: Springer International Publishing.

Aswani, S., Lemahieu, A., Sauer, W.H. 2018. Global trends of local ecological knowledge and future implications. *Plos One*, 13(4): e0195440.

Barnett, C. & Parnell, S. 2016. Ideas, implementation and indicators: Epistemologies of the post-2015 urban agenda. *Environment & Urbanization*, 28(1): 87–98.

Barthel, S. Folke, C. & Colding, J. 2010. Social–ecological memory in urban garden: Retaining the capacity for management of ecosystem services. *Global Environmental Change*, 20: 255–265.

Bell, P.A., Greene, T., Fisher, J. & Baum, A.S. 2001. *Environmental Psychology* (5th ed.). Mahwah, UK: Lawrence Erlbaum, Wadsworth.

Berdegué, J.A., Proctor, F.J. & Cazzuffi, C. 2014. *Cities in the Rural Transformation, Working Paper Series N° 123, Working Group: Development with Territorial Cohesion, Territorial Cohesion for Development Program*. Santiago, Chile: Rimisp.

Berry, B.J. 2008. Urbanization. In Marzluff, J.M., Shulenberger, E., Endlicher, W., Alberti, M., Bradley, G., Ryan, C., Zumbrunnen, C. & Simon, U. (Eds.), *Urban Ecology: An International Perspective on the Interaction Between Humans and Nature* (pp. 24–38). Boston, MA: Springer.

Botzat, A., Fischer, L.K. & Kowarik, I. 2016. Unexploited opportunities in understanding liveable and biodiverse cities: A review on urban biodiversity perception and valuation. *Global Environmental Change*, 39: 220–233.

Bouteligier, S. 2011. Exploring the agency of global environmental consultancy firms in earth system governance. *International Environmental Agreements: Politics, Law and Economics*, 11(1): 43–61.

Braudel, F. 1982. *Civilization and Capitalism, 15th–18th Century, Vol. II: The Wheels of Commerce*. London, UK: Fontana Press.

Buizer, M., Elands, B. & Vierikko, K. 2016. Governing cities reflexively: The biocultural diversity concept as an alternative to ecosystem services. *Environmental Science & Policy*, 62: 7–13.

Camfield, L. & McGregor, A. 2005. Resilience and wellbeing in developing countries. In Ungar M. (Ed.), *Handbook for Working with Children and Youth: Pathways to Resilience Across Cultures and Contexts* (pp. 189–209). Thousand Oaks, CA: Sage.

Ceuterick, M., Vandebroek, I. & Pieroni, A. 2011. Resilience of Andean urban ethnobotanies: A comparison of medicinal plant use among Bolivian and Peruvian migrants in the United Kingdom and in their countries of origin. *Journal of Ethnopharmacology*, 136: 27–54.

Chan, K., Satterfield, T. & Goldstein, J. 2012. Rethinking ecosystem services to better address and navigate cultural values. *Ecological Economics*, 74: 8–18.

Chandrakanth, M.G., Gilless, J.K., Gowramma, V. & Nagaraja, M.G. 1990. Temple forests in India's forest development. *Agroforestry Systems*, 11: 199–211.

Chiesura, A. 2004. The role of urban parks for the sustainable city. *Landscape & Urban Planning*, 68(1): 129–138.

Chivaura, V.G. 2007. *Hunhu/Ubuntu:* A sustainable approach to endogenous development, bio-cultural diversity and protection of the environment in Africa. In Haverkort, B. & Rist, S. (Eds.), *Endogenous Development and Bio-Cultural Diversity: The Interplay Between Worldviews, Globalization and Locality*, pp. 229–240). Geneva, Switzerland: Compas.

Church, A.T., Katigbak, M.S., Locke, K.D., Zhang, H., Shen, J., de Jesús Vargas-Flores, J. & Mastor, K.A. 2013. Need satisfaction and wellbeing testing self-determination theory in eight cultures. *Journal of Cross-Cultural Psychology*, 44: 507–534.

Cobern, W., Gibson, A. & Underwood, S. 1999. Conceptualization of nature: An interpretive study of 16 ninth graders' everyday thinking. *Journal of Research in Science Teaching*, 36: 541–564.

Cocks, M.L. 2006. Biocultural diversity: Moving beyond the realm of 'indigenous' and 'local' people. *Human Ecology*, 34: 185–200.

Cocks, M.L., Dold, T. & Vetter, S. 2012. 'God is my forest' – Xhosa cultural values provide untapped opportunities for conservation. *South African Journal of Science*, 108(5/6): 1–8.

Cocks, M., Vetter, S. & Wiersum, K.F. 2018. From universal to local: Perspectives on cultural landscape heritage in South Africa. *International Journal of Heritage Studies*, 24(1): 35–52.

Cocks, M.L. & Wiersum, F. 2014. Reappraising the concept of biocultural diversity: A perspective from South Africa. *Human Ecology*, 42: 727–737.

Cohen, B. 2006. Urbanization in developing countries: Current trends, future projections, and key challenges for sustainability. *Trends in Society*, 28: 63–80.

Cunningham, C. & Stanley, F. 2003. Indigenous by definition, experience, or world view: Links between people, their land, and culture need to be acknowledged. *BMJ: British Medical Journal*, 327(7412): 403.

Davidson-Hunt, I.J., Turner, K.L., Te Pareake Mead, A., Cabrera-Lopez, J., Bolton, R., Idrobo, C.J., Mirestski, I., Morrison, A. & Robson, J.P. 2012. Biocultural design: A new conceptual framework for sustainable development in rural indigenous and local communities. *Sapiens*, 5(2): 32–45.

De Lacy, P. & Shackleton, C.M. 2017. Woody plant species richness, composition and structure in urban sacred sites, Grahamstown, South Africa. *Urban Ecosystems*, 20: 1169–1179.

Descola, P. 1994. *In the Society of Nature: A Native Ecology in Amazonia*. Cambridge, UK: Cambridge University Press.

Descola, P. 2006. Beyond nature and culture. Radcliffe-Brown Lecture in Social Anthropology. *Proceedings of the British Academy*, 139: 137–155.

Descola, P. & Palsson, G. 1996. *Nature and Society: Anthropological Perspectives*. London, UK: Routledge.

Diaz, S., Demissew, S., Carabias, J. et al. 2015. The IPBES conceptual framework – Connecting nature and people. *Current Opinion in Environmental Sustainability*, 14: 1–16.

Dickson, D.C. & Hobbs, R.J. 2017. Cultural ecosystem services: Characteristics, challenges and lessons for urban green space research. *Ecosystem Services*, 25: 179–194.

Diener, E., Sapyta, J.J. & Suh, E.M. 1998. Subjective wellbeing is essential to wellbeing. *Psychological Inquiry*, 9: 33–37.

Diener, E. & Suh, E.M. (Eds.). 2003. *Culture and Subjective Wellbeing*. Cambridge, MA: MIT Press.

Dobbs, C. Nitschke, C.R. & Kendal, D. 2014. Global drivers and tradeoffs of three urban vegetation ecosystem services. *PLoS One*, 9(11): e113000.

Dockery, A.M. 2011. *Traditional Culture and the Wellbeing of Indigenous Australians: An Analysis of the 2008 NATSISS*. Perth, Australia: Centre for Labour Market Research, Curtin University.

Elands, B.H., Wiersum, K.F., Buijs, AS.E. & Vierikko, K. 2015. Policy interpretations and manifestations of biocultural diversity in urbanised Europe: Conservation of lived biodiversity. *Biodiversity & Conservation*, 24: 3347–3366.

Ernstson, H. & Sörlin, S. (Eds.). 2019. *Grounding Urban Natures Histories and Futures of Urban Ecologies*. London: The MIT Press.

Ernstson, H., Van der Leeuw, S.E., Redman, C.L., Meffert, D.J., Davis, G., Alfsen, C. & Elmqvist, T. 2010. Urban transitions: On urban resilience and human-dominated ecosystems. *Ambio*, 39(8): 531–545.

Ewing, R. & Hamidi, S. 2015. Compactness versus sprawl: A review of recent evidence form the United States. *Journal of Planning Literature*, 30(4): 413–432.

Ferguson, J. 1990. Modern workers, modernist narratives: A critique of the historiography of transition on the Zambian Copperbelt. *Journal of Southern African Studies*, 16(3): 385–412.

Ferguson, J. 1999. *Expectations of Modernity: Myths and Meanings of Urban Life on the Copperbelt*. Berkeley, CA: University of California Press.

Flint, C.G., Kunze, I., Muhar, A., Yoshida, Y. & Penker, M. 2013. Exploring empirical typologies of human–nature relationships and linkages to the ecosystem services concept. *Landscape & Urban Planning*, 120: 208–217.

Glaz, A. 2017. Worldview as cultural cognitions. *Lumicus*, 1(1): 34–50.

Garnett, S.T., Burgess, N.D., Fa, JE., Fernández-Llamazares, Á., Molnár, Z., Robinson, C.J. & Leiper, I. 2018. A spatial overview of the global importance of indigenous lands for conservation. *Nature Sustainability*, 1(7), 369.

Gopal, D. & Nagendra, H. 2014. Vegetation in Bangalore's slums: Boosting livelihoods, wellbeing and social capital. *Sustainability*, 6: 2459–2473.

Gould, R.K., Ardoin, N.M., Woodside, U., Satterfield, T., Hannahs, N. & Daily, G.C. 2014. The forest has a story: Cultural ecosystem services in Kona, Hawai'i. *Ecology & Society*, 19(3): 55.

Gries, T. & Grundmann, R. 2018. Fertility and modernization: The role of urbanization in developing countries. *Journal of International Development*, 30(3): 493–506.

Haarland, C. & van den Bosch, C.K. 2015. Challenges and strategies for urban green-space planning in cities undergoing densification: A review. *Urban Forestry & Urban Greening*, 14(4): 760–771.

Handley, J., Pauleit, S., Slinn, P., Barber, A., Baker, M., Jones, C. & Lindley, S. 2003. *Accessible Natural Green Space Standards in Towns and Cities: A Review and Toolkit for their Implementation*. English Nature Research Reports, Report Nr. 526, (526).

Haq, S.A. 2011. Urban green spaces and an integrative approach to sustainable environment. *Journal of Environmental Protection*, 2: 601–608.

Hiebert, P.G. 2008. *Transforming Worldviews: An Anthropological Understanding of How People Change*. Grand Rapids, MI: Baker.

Jackson, W. & Ormsby, A. 2017. Urban sacred natural sites: A call for research. *Urban Ecosystems*, 20: 675–681.

Jaganmohan, M., Vailshery, L.S., Mundoli, S. & Nagendra, H. 2018. Biodiversity in sacred urban spaces of Bengaluru, India. *Urban Forestry & Urban Greening*, 32: 64–70.

Jayne, M. 2005. *Cities and Consumption*. London, UK: Routledge.

Kabisch, N. & Haase, D. 2014. Green justice or just green provision of urban greenspaces in Berlin. *Landscape & Urban Planning*, 122: 129–139.

Kabisch, N., Qureshi, S. & Haase, D. 2015. Human–environment interactions in urban green spaces — A systematic review of contemporary issues and prospects for future research. *Environmental Impact Assessment Review,* 50**,** 25-34.

Kabisch, N., Strobach, M. & Haase, D. 2015. Green Surge: Internal project report on inventory of urban green space demand for the two scale levels, ulls and European urban atlas cities. Report MS 24.

Kahn, R. 2014. New communities, new attachments: Planning for diversity in Melbourne's outer-suburbs. *Journal of Intercultural Studies*, 35(3): 295–309.

King, M.F., Renó, V.F. & Novo, E.M. 2014. The concept, dimensions and methods of assessment of human wellbeing within a socioecological context: A literature review. *Social Indicators Research*, 116: 681–698.

Knackmuhs, E., Farmer, J. & Reynolds, H.L. 2017. Student outcomes of eco-restoration service learning experiences in urban woodlands. *Journal of Experimental Education*, 40: 24–38.

Kloek, M.E. 2015. *Colourful Green: Immigrants' and Non-immigrants' Recreational Use of Greenspace and their Perceptions of Nature.* Wageningen, The Netherlands: Wageningen University.

Kull, C.A., de Sartre, X.A. & Castro-Larrañaga, M. 2015. The political ecology of ecosystem services. *Geoforum*, 61: 122–134.

Kuruneri-Chitepo, C. & Shackleton, C.M. 2011. The distribution, abundance and composition of street trees in selected towns of the Eastern Cape, South Africa. *Urban Forestry & Urban Greening*, 10: 247–254.

La Rosa, D., Spyra, M. & Inostroza, L. 2016. Indicators of cultural ecosystem services for urban planning: A review. *Ecological Indicators*, 61: 74–89.

Laird, S.A., Awung, G.L., Lysinge, R.J. & Ndive, L.E. 2011. The interweave of people and place: Biocultural diversity in migrant and indigenous livelihoods around Mount Cameroon. *International Forestry Review*, 13(3): 275–293.

Lawhon, M., Ernstson, H. & Silver, J. 2014. Provincializing urban political ecology: Towards a situated UPE through African urbanism. *Antipode*, 46(2): 497–516.

Lennon, M. 2015. Green infrastructure and planning policy: A critical assessment. *Local Environment*, 20(8): 957–980.

Lohnert, B. & Steinbrink, M. 2005. Rural and urban livelihoods: A translocal perspective in a South African context. *South African Geographical Journal*, 87(2): 95–103.

Lovell, S.T. & Taylor, J.R. 2013. Supplying urban ecosystem services through multifunctional green infrastructure in the United States. *Landscape Ecology*, 28: 1447–1463.

Millennium Ecosystem Assessment (MA) 2005. *Ecosystems and Human Wellbeing: Synthesis.* Washington DC: Island Press.

Maffi, L. 2005. Linguistic, cultural and biological diversity. *Annual Review Anthropology*, 29: 599–617.

Maffi, L. 2008. *An Introduction to Biocultural Diversity*. Salt Spring Island, Canada: Terralingua.

Maffi, L. & Woodley, E. (Eds.). 2010. *Biocultural Diversity Conservation: A Global Sourcebook.* London, UK: Earthscan.

McLain, R.J., Hurley, P.T., Emery, M.R. & Poe, M.R. 2014. Gathering "wild" food in the city: Rethinking the role of foraging in urban ecosystem planning and management. *Local Environment*, 19: 220–240.

McConnachie, M.M. & Shackleton, C.M. 2010. Public green space inequality in small towns in South Africa. *Habitat International*, 34: 244–248.

McConnachie, M.M., Shackleton, C.M. & McGregor, G. 2008. Extent of public green space and alien species in ten small towns in the thicket biome, South Africa. *Urban Forestry & Urban Greening*, 7: 1–13.

Merçon, J., Vetter, S., Tengö, M., Cocks, M., Balvanera, P., Rosell, J.A. & Ayala-Orozco B. 2019. From local landscapes to international policy: Contributions of the biocultural paradigm to global sustainability. *Global Sustainability* 2, 1–11, e7.

Milton, K. 1996. *Environmentalism and Cultural Theory: Exploring the Role of Anthropology in Environmental Discourse.* London, UK: Routledge.

Milton, K. 1998. Nature and the environment in indigenous and traditional cultures. In Cooper, D.E. & Palmer, J.A. (Eds.), *Spirit of the Environment: Religion, Value and Environmental Concern* (pp. 86–99). London: Routledge.

Mowatt, R.A. 2018. Understanding Ifá: Inserting knowledge of an African cosmology in leisure studies and nature-based research. *Leisure Studies*, 37(5): 515–532.

Ngulani, T. & Shackleton, C.M. 2019. Use of public urban green spaces for spiritual services in Bulawayo, Zimbabwe. *Urban Forestry & Urban Greening*, 38: 97–104.

O'Brien, L., De Vreese, R., Atmis, E., Olafsson, A.S., Sievänen, T., Brennan, M., Sánchez, S., Panagopoulos, T., de Vries, S., Kern, M., Gentin, S., Saraiva, G. & Almeida, A. 2017. Social and environmental justice: Diversity in access to and benefits from urban green infrastructure: Examples from Europe. In Pearlmutter, D., Calfapietra, C., Samson, R., O'Brien, L., Krajter Ostoi´c, S., Sanesi, G., Alonso del Amo, R. (Eds.), *The Urban Forest* (pp. 153–190). Cham, Switzerland: Springer.

Orsini, F., Kahane, R., Nono-Womdim, R. & Gianquinto, G. 2013. Urban agriculture in the developing world: A review. *Agronomy & Sustainable Development*, 33: 695–720.

Persica, A. & Martin, G. (eds.) 2008. Links between biological and cultural diversity. Report of International Workshop. Paris: UNESCO.

Peter, K., Elands, B. & Buijs. A. 2010. Social interactions in urban parks: Stimulating social cohension? *Urban Forestry & Urban Greening*, 9(2): 93–100.

Peterson, M.J., Hall, D.M., Parker, F. & Peterson, T.R. 2010. Obscuring ecosystem function with application of the ecosystem services concept. *Conservation Biology*, 24: 113–119.

Posey, D. 1999. *Cultural and Spiritual Values of Biodiversity*. United Nations Environmental Programme. London, UK: UNEP and Intermediate Technology Publications.

Poulsen, M.N., McNab, P.R., Clayton, M.L. & Neff, R.A. 2015. A systematic review of urban agriculture and food security impacts in low-income countries. *Food Policy*, 55, 131–146.

Pretty, J., Adams, B., Berkes, F., Ferreira de Athayde, S., Dudley, N., Hunn, E., Maffi, L., Milton, K., Rapport, D., Robbins, P., Sterling, E. Stolton, S., Tsing, A., Vintinner, E. & Pilgrim, S. 2009. The intersections of biological diversity and cultural diversity: Towards integration. *Conservation & Society*, 7: 100–112.

Rapport, D.J. & Maffi, L. 2010. Eco-cultural health, global health, and sustainability. *Ecological Research*, 26(6): 1039–1049.

Reyes-García, V., Vadez, V., Byron, E., Apaza, L., Leonard, W.R., Pérez, E. & Wilkie, D. 2005. Market economy and the loss of folk knowledge of plant uses: Estimates from the Tsimane' of the Bolivian Amazon. *Current Anthropology*, 46(4): 651–656.

Roy, S., Byrne, J. & Pickering, C. 2012. A systematic quantitative review of urban tree benefits, costs, and assessment methods across cities in different climatic zones. *Urban Forestry & Urban Greening*, 11: 351–363.

Rupprecht, C.D.D. & Byrne, J.A. 2014. Informal urban greenspace: A typology and trilingual systematic review of its role for urban residents and trends in the literature. *Urban Forestry & Urban Greening*, 13: 597–611.

Russell, R., Guerry, A.D., Balvanera, P., Gould, R.K., Basurto, X., Chan, K.M.A., Klain, S., Levine, J. & Tam, J. 2013. Humans and nature: How knowing and experiencing nature affect wellbeing. *Annual Review of Environment & Resources*, 38: 473–502.

Rutt, R.L. & Gulsrud, N.M. 2016. Green justice in the city: A new agenda for urban green space research in Europe. *Urban Forestry & Urban Greening*, 19: 123–127.

Saarikoski, H., Jax, K., Harrison, P.A. Primmer, E., Barton, D.N., Mononen, L., Vihervaara, P. & Furman, E. 2015. Exploring operational ecosystem service definitions: The case of boreal forests. *Ecosystem Services*, 14: 144–157.

Sanesi, G. & Chiarello, F. 2006. Residents and urban green spaces: The case of Bari. *Urban Forestry & Urban Greening*, 4: 125–134.

Santana, C. 2019. Natural diversity: How taking the bio- out of biodiversity aligns with conservation priorities. In Casetta, E., da Silva, J.M. & Vecchi, D. (Eds.), *From Assessing to Conserving Biodiversity* (pp. 401–414). Berlin, Germany: Springer.

Schipperijn, J. 2010. *Use of Urban Green Space*. Copenhagen, Denmark: University of Copenhagen.

Schröter, M., van der Zanden, M.H., van Oudenhoven, A.P., Remme, R.P., Serna-Chavez, H.M., de Groot, R.S. & Opdam, P. 2014. Ecosystem services as a contested concept: A synthesis of critique and counter-arguments. *Conservation Letters*, 7: 514–523.

Shackleton, C.M., Hurley, P.T., Dahlberg, A., Emery, M.R. & Nagendra, H. 2017. Urban foraging: A ubiquitous human practice but overlooked by urban planners, policy and research. *Sustainability*, 9:1884.

Shackleton, C.M., Blair, A., De Lacy, P., Kaoma, H., Mugwagwa, N., Dalu, M.T. & Walton, W. 2018. How important is green infrastructure in small and medium-sized towns? Lessons from South Africa. *Landscape & Urban Planning*, 180: 273–281.

Skandrani, Z. 2016. Considering the socio-ecological co-construction of nature conceptions as a basis for urban environmental governance. *Journal of Geography & Natural Disasters*, 6(1): 1000153.

Sterling, E.J., Filardi, C., Toomey, A., Sigouin, A., Betley, E., Gazit, N., Bergamini, N. 2017. Biocultural approaches to wellbeing and sustainability indicators across scales. *Nature Ecology & Evolution*, 1, 1798–1806.

Summers, J.K., Smith, L.M., Case, J.L. & Linthurst, R.A. 2012. A review of the elements of human wellbeing with an emphasis on the contribution of ecosystem services. *Ambio*, 41: 327–340

Tacoli, C., McGranahan, G. & Satterthwaite, D. 2015. *Urbanisation, Rural-urban Migration and Urban Poverty*. London, UK: Human Settlements Group, International Institute for Environment and Development.

TEEB. 2010. *The Economics of Ecosystems and Biodiversity* (pp. 410). Abingdon, UK: Ecological & Economic Foundations, Routledge.

Thwaites, K. & Simkins, I. 2007. *Experiential Landscape: An Approach to People, Space and Place*. London, UK: Routledge.

Toepfer, G. 2019. On the impossibility and dispensability of defining "biodiversity". In Casetta, E., da Silva, J.M. & Vecchi, D. (Eds.), *From Assessing to Conserving Biodiversity* (pp. 341–351). Berlin, Germany: Springer.

Tov, W. & Diener, E. 2009. Culture and subjective wellbeing. In: Ed Diener (Ed.), *Culture and Wellbeing. The Collected Works of Ed Diener* (pp. 9–41). Dordrecht, The Netherlands: Springer.

UN-Habitat. 2006. *The State of the World's Cities Report 2006/2007: 30 Years of Shaping the Habitat Agenda*. London, UK: Earthscan.

Viveiros de Castro, E. 1998. Cosmological deixis and Amerindian perspectivism. *Journal of the Royal Anthropological Institute (N.S.)* 4: 469–488.

Voeks, R.A. 2018. *The Ethnobotany of Eden* (334 pp.). Chicago: University of Chicago Press.

Wallerstein, I. 1974. *The Modern World System I: Capitalist Agriculture and the Origins of the European World-Economy in the Sixteenth Century*. New York, NY: Academic Press.

Wang, F. & Wang, D. 2016. Place, geographical context and subjective wellbeing: State of the art and future directions. In Wang, D. & He, S. (Eds.), *Mobility, Sociability and Wellbeing of Urban Living* (pp. 189–230). Berlin, Germany: Springer.

White, L. 1967. The historical roots of our ecological crisis. *Science*, 155(3767): 1203–1207.

WinklerPrins, A., & de Oliveira, P.S. de S. 2010. Urban agriculture in Santarém, Para, Brazil: Diversity and circulation of cultivated plants in urban homegardens. *Boletimdo Museu Paraense Emílio Goeldi*, 5: 571–585.

Wolsink, M. 2016. Environmental education excursions and proximity to urban green space: Densification in a 'compact city'. *Environmental Education Research*, 22: 1049–1071.
Young, R.V. 1974. Christianity and ecology. *National Review*, XXVI(51), 1454–1458, 1477, 1479.
Zwiers, M., van Ham, M. & Manley, D. 2018. Trajectories of ethnic neighbourhood change: Spatial patterns of increasing ethnic diversity. *Population, Space & Place*, 24(2): e2094.

2 The veil, the clearing and the flow

New commons of Japanese traditional gardens in Kanazawa city

Juan Pastor-Ivars

> First of all, how do you create a garden? My explanation starts with enclosing a space with something to limit it and then creating a unique world within it ... and there are various ways to reproduce an ideal world ... from realistic reproduction to symbolic.
>
> (Shinji et al. 1999)

In nature, mountains, valleys and rivers contribute to landscape diversity. Cities are born from a relationship established with surrounding nature. Human beings, after centuries of inhabiting nature, are, in many regions in the process of separating from it. However, they retain the original experiences and idealise natural elements based on spatial archetypes. Landscape architecture is usually responsible for converting archetypes into types, and including them in socially established, formal, urban natural settings, such as gardens, parks, sacred enclosures and botanical reserves. In parallel, another urban form of nature, the informal one, develops spontaneously without planning, with some examples being woods, vacant plots and residual green areas. Both the formal and informal types of urban nature indicate the intimate and collective relationship that human beings maintain with nature in cities. This chapter focuses on the formal version of urban nature created in Japan, specifically, the Japanese traditional garden, a space created in between nature.

Japanese traditional gardens are designed without a prior plan from the actual site itself by gardeners who learned the trade from their teachers in lessons received directly at the site. The study of Japanese traditional gardens has been carried out mainly through the characterisation of specific themes and scenes (Kuitert 1988), based on sacred and mythical archetypes (Nitschke 1993), leading to the classification of gardens throughout history. However, not as much research has been devoted to assessing and interpreting the spatial characteristics that comprise the Japanese garden. Nonetheless, there are examples of spatial typologies that classify Japanese traditional gardens as "transparent", "dense" or "strolling" (Nishizawa 1997). This chapter takes up the spatial analysis of Japanese traditional gardens from a model built on the abstraction of "space in nature".

Concretely, the three spatial types of the veil, the clearing and the flow, are advocated as the spatial components structuring Japanese traditional gardens. Through an ambiguous demarcation of the garden space located in between nature, with a sacred, symbolic or realistic intent, the spatial types of the veil, the clearing and the flow, as a combined single image, represent the abstraction of the whole natural landscape of Japan: mountains, valleys and rivers. This abstract concept is adjusted to the ontology of each era, with the three spatial types appearing in different proportions depending on the aspiration of the society. These gardens, and the particular historical form they take, are evidence of the substantial quality of the Japanese relationship with nature.

This chapter investigates the veil, the clearing and the flow: their ancient sacred archetypes, spatial notions, characteristics and techniques, functions and biocultural diversity. To achieve a comprehensive perspective, the subject has been approached from three different but complementary spheres: philosophy, design and ecology. However, the production of the three spatial types is not homogenous; it depends on the degree of socio-economic and cultural transference at local, national and international levels during each historical period. For this reason, the case of Kanazawa city is offered to contextualise the spatial types.

Kanazawa is a city where the snow-clad mountains of the Hakusan range stop right before the Sea of Japan to create three plateaus between which the Sai and Asano rivers run. Kanazawa is a *jōkamachi* (城下町 fortress city) developed by the Maeda Clan at the end of the 16th century. The feudal lord surrounded himself with samurais, merchants and artisans, granting them enough privileges for the creation of a prosperous society and a flourishing, vibrant culture. The castle is located on the central plateau between the two rivers from where an extensive network of more than 50 water canals runs through the entire city. This network supplies water for fire protection and other daily functions, such as washing vegetables, transporting materials, dyeing fabrics and irrigating gardens. Kanazawa is a city of gardens that are inseparable from the abundant water resources within its territory. These gardens add nuances to generic spatial types and are a model of urban biocultural diversity.

Currently, with cities around the world immersed in social, economic and environmental challenges, the production of urban nature through the three Japanese spatial types serves as a tool through which citizens can express their aspirations about how to address these challenges. Lefebvre (1991) points to the political dimension of urban space: "the production of space (i.e., landscape design and architecture) is never politically neutral, nor is it merely 'physical': gardens, monuments and worship places – places with significant symbolic capital – can be employed to convey ideological messages or legitimate particular power configurations" (quoted in Rots 2014). Japanese traditional gardens are not immune to such challenges in the same way that Persian Gardens in Iran have not escaped the forces of broader political and economic interest (Mellegard, Chapter 4). Not only has the number of traditional Japanese gardens diminished, but many gardens have fallen into neglect, largely as a result of the shrinking Japanese population. Strong leadership is needed to address such problems and ensure the permanence of traditional Japanese gardens.

The purpose of this chapter is to deepen awareness about the spatial types of the traditional Japanese garden; to relate these spatial types to the current social, environmental and economic debates about the urban nature of cities; and to develop a suitable framework that enables contemporary reproductions of traditional Japanese gardens. Through reading the three parts: philosophy, design and ecology, a new sustainable and resilient model of the traditional Japanese garden will emerge. The chapter also discusses the need to reinterpret traditional wisdom as the basis for the creation of a new urban nature, and it reveals clues about how to restore kinship with nature.

Philosophy

Landscape and indigenous beliefs

Japan is a country of forested mountains, deep valleys and whitewater rivers. These natural elements exist at a number of different physical scales, including the scale of the human body; this is the "human scale" of the Japanese landscape. Added to the varied topography of the Japanese landscape is a climate of extreme contrasts, characterised by the

> double phenomenon of heavy snowfall and torrential rains … the monsoon blows in the ocean propelling towards the continent an extreme wind current loaded with moisture. … It is often cited as a typical scene of the Japanese landscape, the figure of bamboo, tropical plant, covered with snow.
>
> (Watsuji 2006)

Due to this peculiar climate, seasons are very marked, painting the landscape with delicate tonalities. Nonetheless, natural beauty in Japan is often regarded as fleeting, as countless natural disasters destroy the natural features in which it is manifest, so it must continually be reborn over the same land.

These physical-climatic conditions determined the relationship that the first settlers established with nature. The first inhabitants (Jōmon 14,000–1000 BCE and Yayoi 1000 BCE–300 CE periods) of the islands of East Asia worshipped nature. The natives of that time believed that the *kami* (神 deities) manifest themselves through nature and its phenomena. In this belief system, landscape elements are considered *yorishiro* (依り代natural element appropriable by deities) and can serve as an abode for the *kami* or the soul of the ancestors; nature possesses a conscience. Within this "animistic conception" of the Japanese landscape, natives invoke the presence of the *kami* through rites. The *shimenawa* (注連縄enclosing rope) marks the *yorishiro* where a *kami* manifests, and the *himorogi* (神籬 divine fence) demarcates an unpolluted space around the central *yorishiro*, which acts as the *spiritual antenna for the kami* (Yamakage 2012). The rites, the aim of which is to obtain the energy of the *kami* for benefits in hunting and agriculture, are conducted in forests and along riverbanks, depending on the community's social system. Therefore, Shinto must be seen "as the outcome of the integration of the animism of the forest-dwelling hunter-gatherers with the agricultural fertility cults of the migrant rice-growers living in the valleys" (Rots 2014). Indigenous beliefs are called Shintoism, from the insertion of Buddhism in the Asuka period (592–710 CE). As a result of acculturation, characterised by a selective reassembling (Itō 1971), natives redefined their Shinto philosophy to align with Buddhist cosmology.

Japanese relationship with nature

The relationship that Japanese ancestors established with nature was determined by the physical and climatic conditions, Shinto beliefs and Chinese acculturation. However, defining the essence of this relationship is a matter of controversy among researchers: the commonly held view that the Japanese have a "love of nature" has been developed and repeated literally for centuries by both Japanese and observers of Japan. Closely related to this notion of love is the equally widely-held notion that the Japanese live in harmony with nature, which is frequently contrasted with the quest to "conquer nature" allegedly found among Westerners (Asquith & Kalland 2004).

In general, Japan is characterised by passive dealings with nature, which includes connotations of both affinity and surrender. But, in particular, this relationship depends on a conscious correlation of the human and natural forces within each period. In those periods where human beings have felt superior to nature, the balance is often subverted, bringing the relationship between humans and nature into crisis. Seen from the ontological level, this exchange is currently contradictory; since to most Japanese, human beings can, depending on the context, see themselves either as a part of nature together with spirits, animals, plants and stones (an ecocentric view of nature) or as uniquely superior to other creatures regarding nature as a potential resource to exploit (anthropocentric view). Anthropocentrism and ecocentrism are conflated (Asquith & Kalland 2004).

Seeds of the Japanese traditional garden

The Japanese traditional garden signifies the nature, culture and technology of each place and period. As such, it has its own scale, adapts to the local climate, participates in the animist conception of the landscape, deals subtly with the forces of nature and timeously incorporates foreign influences.

The crucial moment of realisation for the start of the Japanese traditional garden can be found in a change of indigenous mentality in which people ceased to see nature as a divine creation, thus allowing themselves to reproduce it. As stated by Saitō and Shigemori (1976) (quoted by Tschumi, 2004): nature is a world made by the gods for example *iwakura* (磐座 divine rock), … as time passed people started to bring other big stones and added them to an existing *iwakura* … the desire to reproduce nature and create a garden like place, means to first separate the god's existence from nature, people are not permitted to put their hands on it.

The animistic beliefs related to natural elements have been gradually relegated to symbolic and "talismanic representation" (Shirane 2012) in which animism evolves into a polytheism of *kami* embodied in the shrines. Expeditions made by the Japanese to the Chinese continent in the Asuka period put them in contact with this civilisation, including the Chinese traditional garden composed by large ponds and islands. Chinese influence has impacted Japanese cosmology through the introduction and reformulation of myths, reinforcing a symbolic deception of nature. With time, the meaning of specific symbols has been forgotten, giving over to a pure aesthetic representation of nature.

When observing the elements that compose a Japanese traditional garden, this amalgam of animism, symbolism and realism must be considered. For example, stones should be perceived as animistic elements where the gods descend and abide, as well as bearers of good fortune; or as aesthetically refined objects capable of moving the human soul. Similarly, different deities believed to inhabit different natural elements such as certain trees and stones are enshrined in sacred grooves in Nigeria (Adedeji Chapter 3).

Therefore, the early roots of the Japanese traditional garden, broadly understood, are around the elements, rites and sacred spaces of Shinto, as well as in the first interventions on the landscape, such as "stone circles, construction of places for festivals, and tombs" (Ono 2009), requiring a sufficient technological level to later incorporate the influences of Chinese traditional gardens. Moving from this, the Japanese traditional garden, already well developed and constituted along certain lines, begins its urban trajectory through different periods, such as the construction of gardens of large ponds for the palatial court, Nara (710–794 AD) and Heian (794–1185 AD) periods; the dry gardens in Buddhist temples, Kamakura (1185–1333 CE), Kenmu Restoration (1333–1336 CE) and Muromachi (1336–1573 CE) periods; the tea gardens of the Momoyama period (1573–1603 CE); the daimyo gardens of the Edo period (1603–1868 CE); the landscape gardens characteristic of the Meiji period (1868–1912 CE) and the mindscape gardens of the Taishō (1912–1926 CE), Shōwa (1926–1989 CE) and Heisei (1989–2019 CE) periods. To the present day, Reiwa period (from 2019 CE), Japanese traditional gardens are a practice to create formal urban nature, which has been recognised and appreciated beyond Japan and exported overseas.

The evolution of the Japanese traditional garden shows how it adapts to the ontology of each period. However, it originated at a time when humans did not exert domination over nature, and that remains the essence of the concept. The roots of Japanese traditional gardens emerge from a holistic vision of the universe where the humans see themselves as one more object of natural creation. Because traces of the Japanese traditional garden remain in the present as reminiscences of the original worldview, Japanese traditional gardens are an intermediary between current humans, ancient indigenous people and timeless nature. In this sense, time and space for any Japanese traditional garden remain unfinished. Many human societies tend to think that they live only in the dominant ontology of their time, however, unconsciously, they are subject to the diverse ontologies of preceding civilisations which flow through them, being like "sleeping ducks floating".

Design

The veil and "Kanazawa satoyama gardens"

The veil is shade; trees with their branches colonise space and light seeps through the leaves. The veil creates spatial depth through multiple layers, concealing a garden's innermost space (Figure 2.1).

The spatial reference of the veil lies in the sacred archetype of the *chinju no mori* (鎮守の森 sacred forest) that human beings were born from the natural pair of the "mountain-forest".

Chinju no mori are forests that have remained from the ancient age of myths until the present time. These are forests where old trees grow in abundance and where high trees, brushwood and plants growing under the trees are in balance. Many birds, insects and micro-organisms have the space to live here. These are forests with rich ecosystems. Inside, one can find *kiyorakana niwa*

Figure 2.1 The sacred archetype of chinju no mori. Ise Jingū, Naikū shrine, Ise. Source: Pastor-Ivars, Juan.

(清らかな庭 purified gardens), where annual *matsuri* (祭 festival) are organised. (Motegi 2010 quoted in Rots 2017).

The most important *chinju no mori* and one of the oldest is Ise Grand Shrine, founded in the fourth year BCE, when the goddess Amaterasu (天照大神goddess of the sun and the universe) decided to descend, as recorded in the *Kojiki* (古事記 records of ancient matters, Nara period): "The province of Ise, of the divine wind, is the land wither repair the waves from the eternal land, the successive waves. It is a secluded and pleasant land. In this land I wish to dwell" (Aston 2010).

In Japan, the coexistence of primary forests, *satoyama* (里山 secondary nature), *chinju no mori*, Japanese traditional gardens, and other types of woodlands and forests show high complexity between the synergies of nature-made and human-made groves. Both *chinju no mori* and the veil have spatial depth as a common characteristic where sacred functions are performed. However, while *chinju no mori* remains untouched and logging is prohibited, an aesthetic transformation of trees characterises the veil that shapes Japanese traditional gardens. Gardeners embody the space with actions over the trees, making them grow in specific directions and in desired densities intending to transcend the canonic beauty of a tree in nature.

Spatial depth in Japan is determined by the concept of *Oku* (奥 innermost space). Japanese cities are organised based on multiple spatial layers formed by architecture and vegetation. The purpose of this spatial organisation is to hide the core. This spatial organisation is exclusive to Japanese culture and can be described as its "philosopher's stone" (Maki 1979). The Japanese traditional garden is experienced by unveiling the veils and thereby awakening human beings to spatial profundity. Depth is built using the technique of *miegakure* (見え隠れhide and reveal) with the vegetation; trees are overlapped to partially limit the vision of what is hidden behind them. At the end of the path, meanings are revealed. These functions can be sacred, with the discovery of natural elements, temples, shrines, *hokora* (祠 miniature Shinto shrine) etc.; cultural, with the location of the *chashitsu* (茶室tea house) surrounded by the shadows of the veil; or self-referential, with the end presented as a goal in itself, offering a retrospective view of the garden or finding a unique natural element such as a waterfall or a spring. A stroll through the space of the veil is the initiation ritual in preparation for the functions that will be performed inside, assimilating spatial depth with functional depth. Regarding the use of the veil through the history of the Japanese traditional garden, it is scarcely present in gardens of the Heian period, with little vegetation placed over the islands; while it is fully adopted in the tea garden of the Momoyama period, and intermittently utilised through the rest of the periods.

Kanazawa, known as *Mori no Miyako* (森の都 forest city), incorporates each one of the various woods described above. These include the primary forests of Hakusan, *chinju no mori* of the sanctuaries of Ōnominato (大野湊神社, Nara period), Iōzen (醫王山神社, unknown period), *satoyama* areas and veils of Japanese traditional gardens. However, the veil in Kanazawa acquires similar

Figure 2.2 A "satoyama-garden". The old site of Mr. Kurando Terashima's house. Kanazawa. Source: Wakai, Ken.

characteristics to the *satoyama* areas, due to the intense use of the trees. Woods in the Kaga domain were ruled by the "System of the Seven Trees" composed of pines, cedars, oaks, zelkovas, cypresses, hemlocks and tootsik bamboo. This system, implemented by the Maeda Clan to protect trees required for domestic and military uses, exerts a significant influence on the selection and use of the same types of trees in the gardens of samurai residences. Because trees have had not only an ornamental purpose but are also used for materials and food in daily life, the space of the veil in these residences shows a rich biocultural diversity. With the trees, many objects associated with the art of war, such as flagpoles, weapons and torches, have been made (Hasegawa 2019). The garden of the old site of Mr. Kurando Terashima's house (寺島蔵人邸跡, Edo period) (Figure 2.2) is one of the best-preserved examples of Kanazawa *satoyama* gardens. The biocultural diversity historically linked to the space of the veil in Kanazawa suggests that the sustainable way to preserve and maintain a grove is to use it as a constitutive element of daily life, creating strong bonds between the human beings and the trees.

The Clearing and "Kanazawa moon gardens"

Clearing is light; existing emptiness between natural and artificial bodies. Through time, a void is filled by phenomena. On the clearing, relations between the entities of the garden, the exterior and human beings are obtained.

The spatial reference of the clearing is the sacred archetype of the *shiki no himorogi*[1] (磯城の神籬divine fence of pebbles) (Figure 2.3) that human beings conceptualise from the natural pair "valley-plain".

Shiki (no himorogi) is generally the name for a larger area than *iwasaka* (岩坂 sacred rock), which is a square and covered with pebbles. Such an area is reserved for the deities to occupy and move about in, and it serves as a site for rites and festivals in which deities and man participate jointly. This Japanese approach to making space tangible involves defining, marking off and limiting space in nature (Tange & Kawazaoe 1965).

The *Kojiki* records the *shiki no himorogi* as having been built in the year 92 BCE: "Therefore he entrusted *Ama-terasu* to *Oho-kami* [and] to *Toyo-suki-iribime no Mikoto* (deities of the Japanese myths) to be worshipped at the village of Kasanuhi in Yamato (大和Japan), where he established the sacred enclosure of shiki" (Aston 2010). Materialisations of the sacred archetype of *shiki no himorogi* can be found within the precincts of many shrines and temples.

The relation of the space of the clearing to the sacred archetype of *shiki no himorogi* has, first of all, a visual component. Both materialise its surface with pebbles. However, the similarities are not limited to the scope of the visual, but also happen at a deeper level. Spatially, *shiki no himorogi* supposes, through its delimitation, the creation of a space in between nature. In the origin, they are demarcated in riverbanks or intentionally cleared spaces inside *chinju no mori*. In the Japanese traditional garden, opening a clearing is also the way to create a space in between nature. Functionally, *shiki no himorogi* is built as a place to

Figure 2.3 The sacred archetype of shiki no himorogi. Kamigamo shrine. Kyoto. Source: Pastor-Ivars, Juan.

prepare for the arrival of the *kami* during rites and festivals, and for this to happen, the area must always be kept uncontaminated, even if not permanently but temporarily occupied. In the Japanese traditional garden, the clearing is not an empty space; it is waiting for the natural phenomena to appear. In Japan, where nature exuberantly colonises space, action is necessary to maintain an area free of weeds. The use of pebbles, grass and moss resolves this aspect and shows another way in which the Japanese traditional garden passively deals with nature.

The concept of *Ma* (間 space and time in between) determines a unique way of relating place and time to the entire universe. For *Ma* to happen, a proper design of the emptiness and the masses is required for the parts to attract each other. Such spatial configuration produces a climax in the human perception; and when a natural phenomenon occurs, the human being appropriates the sense of place, similar to when indigenous people obtained the energy of the *kami* through the rites. In this sense, *Ma* is equivalent to *kami*. Incorporating features of the landscape is the basis of the *shakkei* (借景 borrowed scenery technique). For *shakkei* to occur, a clearing must be interposed between the internal and external objects to be visually and audibly appropriated.

At first, functions performed in the space of the clearing consisted of the performance of symbolic ceremonies in the Heian period. Later, in the Kamakura and Muromachi periods, these meetings were moved to the interior of the architectural space, and objects were distributed over the clearing to create the garden's aesthetic composition, for example, in the dry gardens. In the Meiji period, social use predominated with the celebration of open-air tea parties in meadows, as a symptom of a climate of social openness.

In the Kanazawa area, there are also materialisations of the sacred archetype of *shiki no himorogi* with the enclosure of natural elements. Examples include the Kinkengū (金劒宮, Yayoi period) sanctuary with the *kameishi* (亀石turtle stone) and *amano shinobu no ishi* (天忍石sacred stone) located in Tsurugi, the temple of Honsen-ji (本泉寺, Muromachi period) with the fencing of the sacred rocks of either *o-setake ishi* (御背丈石), o-suzuri ishi (御硯石) and the sacred bamboo forest *o-chikujō* (御杖竹). The clearing in Kanazawa, a city with undulating topography, is a space from which to get a panoramic view from above, like an observatory. Such is the case in the Kenrokuen (兼六園, Edo period), Rinshōtei (林鐘亭庭園, Edo period) and Tsuji family's (辻家庭園, Meiji period) (Figure 2.4) gardens. In Kanazawa, in most traditional gardens the clearing is surfaced with vegetation, for example, moss in Rinshōtei garden or grass in the Tsuji family's garden; the exception is the Shōfukakutei garden (松風閣庭園, Edo period), which is covered with pebbles. Specifically, the function of Kanazawa panoramic gardens is not so much to introduce the light of the sun but of the moon, as seen, for example, in the Rinshōtei garden (Tsuba 2019). Thus, in Kanazawa, the biocultural diversity of the clearing is characterised by a panoramic space, covered by natural elements and "brushed" by the beams of the moon. It is a space where human beings can communicate with the universe of the night.

Figure 2.4 A "moon-garden". Tsuji family's garden. Kanazawa. Source: Wakai, Ken.

The flow and "Kanazawa thaw gardens"

Flow is mirror and murmur; its backwaters and streams recreate the stillness and the dynamism of water. The flow falls freely between the veil and the clearing, merging all of the garden's spaces.

The spatial reference of the flow is the sacred archetype of *kami-ike-shima*[2] (神池島divine pond and island) (Figure 2.5) that the indigenous peoples conceived from the natural pair "sea-island". To this place they were supposed to sail to find the *Tokoyo no kuni* (常世の国eternal land) in their cult of immortality. The *Kojiki* in the year 663 BCE narrates it as follows:

> my mother and my aunt are both Sea-Goddesses: why do they rise great billows to overwhelm us? So treading upon the waves, he went to the Tokoyo *no kuni* … Urashima of Midzunoye went fishing in a boat. They went down together into the sea and reached Hōrai-san[3] (蓬莱山treasure island), where they saw the genii (Aston 2010).

The relation of the space of the flow and the sacred archetype *kami-ike-shima*, the sea of tranquillity with emerging islands from which everlasting flow sprouts, is given by reproducing this picture through both tranquil and more alive water bodies in the garden. In doing that, the spatial type of the flow, compared to the veil and the clearing, is the one that shows a more significant influence from the Chinese traditional garden and Chinese cosmology.

Figure 2.5 The sacred archetype of kami-ike-shima. Hirosawa pond. Kyoto. Source: Pastor-Ivars, Juan.

Regarding the calm state of the flow, the Chinese myth of the Penglai Island (蓬萊仙島island of the immortals) is later reinterpreted in the Japanese traditional garden on the island of *Hōrai*. Regarding the vigorous state of the flow, the Chinese cosmology of Feng Shui (風水 wind-water) is incorporated in Japanese traditional gardens by transferring its meanings and orientations to the garden's flow. That being said, the Japanese already had abilities in the third century CE to execute diverse waterbodies. Examples of these are the *Kofun* (古墳, Kofun period 300–592 CE) water rings with inner islands where the tombs of eternal rest of nobles are placed or other ponds with detailed margins to perform "*mizu no matsuri*" (水の祭 water festivals), rituals of purification located where the water runs clearest (Ono 2009).

The flow is the indicator that the clearing is waiting for *Ma* to happen in the garden. Moreover, the flow comes from the garden depths, which is *Oku*. In this sense, the flow is the intersection between *Ma* and *Oku*. The *jikoshōji* (自己相似 self-similarity) technique is used to make the flow appear as natural as possible. This technique, close to fractals, matches macro and micro geometries. Specifically, the rock's geometry shaping the flow's perimeter looks like the flow's shape in general.

Several functions have traditionally developed in the space of the flow. For example, many gardens of the Heian period serve as a recreational place, such as to enjoy a boat trip admiring seasonal changes or to celebrate the *kyokusui-no-en* (曲水の宴winding stream poem party). Later, in the Kamakura, Kenmu

Restoration and Muromachi periods, the flow becomes a mental landscape through the execution of dry gardens that materialise non-physical water with stones and pebbles. Finally, in the Meiji period, the water expresses all its naturalness through the different forms and rhythms of natural water to create the pristine atmosphere required for the *sencha* (煎茶full leave tea) ceremony.

In Kanazawa, the type of flow represented by the *kami-ike-shima* archetype can be seen in the Ishiura shrine (石浦神社, Nara period), the Honsen-ji temple, and the Shōfukaku and Kenrokuen gardens, containing sacred ponds and the island of *Hōrai*. Likewise, Kanazawa city is structured on Feng Shui principles and elements; however, in Kanazawa, the model is inverted because of the city's opposite orientations (Kobayashi 2019). The flow is widely incorporated in the city because the abundance of water in the territory makes its production sustainable (Figure 2.6). The flow comes to the surface in the form of either a canal or springs. This natural water originates two types of garden, the canal-water garden and the spring-water garden, with each of them representing both the stillness and dynamism of water.

Kenrokuen is the most famous example of the canal-water garden. This model exerted significant influence in the gardens of samurai residences, which were strategically scattered throughout the city and located next to the canals. Water penetrates from an adjacent canal into the plots and creates a rich atmosphere of water, with waterfalls, streams and ponds, before leaving the garden

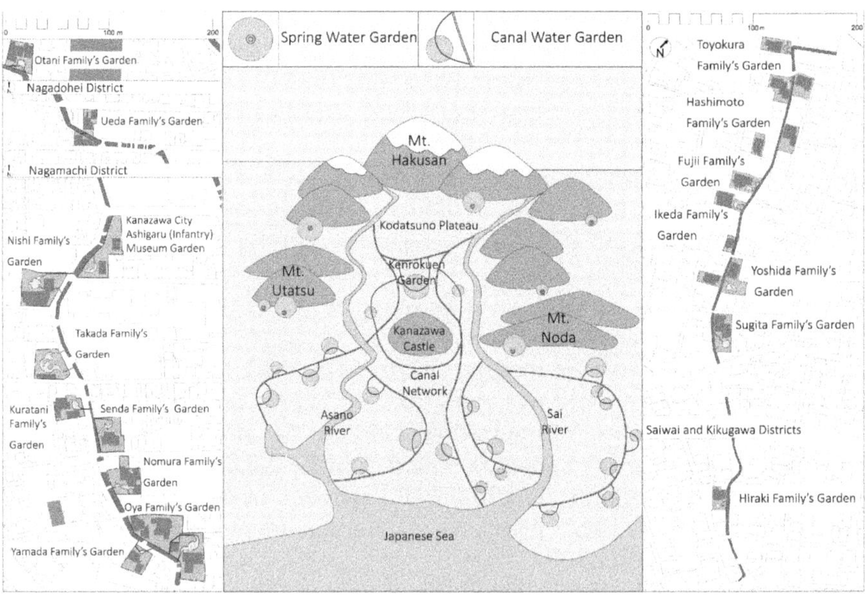

Figure 2.6 Type of Japanese garden depending on the water source in Kanazawa. Historical gardens irrigated by water canals in Nagadohei, Nagamachi, Saiwai and Kikugawa districts. Source: Pastor-Ivars, Juan.

and rejoining the canal. This canal water feeds into a network of gardens as well as rice fields on the outskirts. Therefore, it is "shared water".

In Kanazawa, the well-preserved Senda (千田家庭園, Meiji period) and Nishi (西家庭園、Taishō period) families' gardens are the best examples of canal-water gardens. In addition to the water of the rivers and canals, springs sprout from the subsoil. The most famous of these is the Kinjō Reitaku (金城霊澤, unknown period), which gives its name to the city of Kanazawa. The spring-water gardens are characterised not so much by the dynamism of water but by being backwaters of peace. Ponds are vessels that overflow when water flowing from the bottom exceeds the perimeter of the receptacle. In Kanazawa, the well-preserved Shinrensha temple garden (心蓮社庭園, Edo period) (Figure 2.7) is a good example of a spring-water garden.

Both the canal-water and spring-water garden are characterised by natural water that comes from the melting water from snow on the mountain range of Hakusan. Melted water brings life and the fresh smell of the mountains into the city's residences. Water is appreciated for both the natural elements it contains as well as for the rich cultural elements it embodies. At the biological level, water provides a habitat for numerous fishes, amphibians, reptiles, insects and birds creating a real urban ecosystem within the gardens it flows through. At the cultural level, tea ceremonies are held along a garden's waterbodies, banks and traditional fabrics are dyed, as for example, in the Hiraki family's garden (平木家庭園, Edo period) (Figure 2.8).

Figure 2.7 Shinrensha temple garden. Kanazawa. Source: Wakai, Ken.

Figure 2.8 Mr. Hiraki dyeing in the traditional Kaga style at the canal adjacent to his garden. Hiraki family's garden. Source: Hiraki, Isao.

These examples show the rich biocultural diversity of the type of flow in Kanazawa. The love for pure water coming from the thaw of the landscape has led to the unique creation of gardens in Kanazawa (Figure 2.9).

The veil, the clearing and the flow: understanding the role of these spatial types when shaping urban nature is fundamental. In Japan, it implies a knowledge of their landscape connection, indigenous beliefs, sacred archetypes and the functions traditionally performed in these spaces; these are spiritual practices and social ones such as the tea ceremony.

Knowing the traditional wisdom is necessary for the specialists that dedicate themselves to the design of formal urban nature. The spatial types of Japanese traditional gardens contain many layers, echoes of the past and multiple interpretations, which explain the intensively lived space achieved. If you find yourself taking a walk in a Japanese traditional garden, it's better at 6:00 a.m.

Figure 2.9 A "thaw-garden". Kenrokuen garden. Kanazawa. Source: Wakai, Ken.

on a winter's day when the new light starts to awake. The *Ma* and *Oku* of the garden provide you with a sense of place through a fleeting phenomenon; in this case, the sunrise will literally place you "inside the dawn""

Ecology

Paradigm shift

> My observation is that after one hundred and twenty years of modernisation since the opening of the country, present-day Japan is split between two opposite poles of ambiguity .... The modernisation of Japan has been orientated towards learning from and imitating the West .... Japan was driven into isolation from other Asian countries, not only politically but socially and culturally .... Yet Japan is situated in Asia and has firmly maintained its traditional culture .... What I call Japan's "ambiguity" in my lecture is a kind of chronic disease that has been prevalent throughout the modern age. Japan's economic prosperity is not free from it either, accompanied as it is by all kinds of potential dangers in the light of the structure of world economy and environmental conservation.
>
> (Ōe 1994)

After two centuries of *sakoku* (鎖国closed country), an open attitude to the West characterises the Meiji period. In the field of landscaping, the

European naturalism of the 19th century exerted considerable influence on the Japanese traditional garden, in which a "realism, naturalism and individualism" (Muraoka 2011) was installed following the model of English garden, which fitted better with the Japanese traditional garden than the French one. The main feature of the acculturation of Japan in this period was a greater diversity of plant species (including foreign species) in the veil; for the space of the clearing, a predominance of vast, sinuous prairies crossed by meandering paths; and for the space of the flow the creation of watercourses deployed throughout the garden with picturesque falls and trickles. Garden themes focussed on rural landscapes, in response to a national fear of losing their identity, and at the same time in the creation of ample and luminous spaces influenced by the atmosphere of openness of that period. Functionally, tea ceremonies moved to these prairie-like open spaces following the model of the *enyūkai* (園遊会 garden party) (Pastor-Ivars 2015). Despite some unfortunate examples, the courageous effort made by Meiji designers to update the Japanese traditional garden without losing its essences should be highlighted. The aspiration of this new bourgeoisie, in charge of sponsoring urban nature in the city, has been to obtain the relevant status through beautiful gardens welcoming Westerners. The Japanese garden designer Ogawa Jihei the Seventh was the prime exponent of the Meiji period, a pioneer who created a profound merging of tradition and modernity.

In response to designs of the Meiji period Shigemori Mirei, in the Taishō and Shōwa periods, conducted a historic revision of dry gardens. Mirei understood that excessive naturalism through the use of vast grass areas with less use of stones, meant losing the essence of the Japanese traditional garden (Shigemori & Shigemori 1971). Isamu Noguchi, who designed with carved stones, alludes to the way the modern garden designer wants to consciously make his mark on the forces of nature:

> we live in a modern world. The ancient world, the world of nature confronts. We are in conflict. From that comes creation, the two together. Only one is none. The Japanese garden is made from a collaboration with nature. Man's hands want to be hidden by time and by many effects of nature, moss and so forth, so you are hidden. I don't want to be hidden, I want to show. Therefore, I am modern. I am not a traditional [gardener]. I think it is important for Japan today to know that you have to be involved with nature to be in conflict with it. If you speak to nature you have to be strong, because nature is also very strong. (Alhalel 1989)

This manifesto exemplifies a shift in attitude towards the domination of nature. This paradigm shift consists of moving away from the artisan-gardener, who is an intermediary between nature and the client, to the artist-gardener, who executes human power over nature to satisfy the client's ego.

Currently, the Japanese traditional garden is devalued for various reasons. First, more educated classes do not identify their economic wellness with the opportunity to surround themselves with a model of urban nature that also brings them wellbeing and social recognition. This fact reduces the power of cultural transmission over the rest of society. The Japanese traditional garden has been a way to cultivate urban nature from a model that was propelled by the upper classes, which was then followed and reproduced in the lower levels of society. Second, the decline of *machiya* (町屋traditional townhouses) has had a substantial impact on the production of gardens. While the traditional Japanese house is characterised by spaces that fully relate to the Japanese traditional garden, the new style of home creates an interior that is completely separated from the exterior and thus families no longer live with a Japanese traditional garden as a background. Third, at the educational level, the concept and meanings of the Japanese garden have ceased to be taught in many prestigious institutions. Likewise, the academy does not make the necessary efforts to adapt the wisdom inherent in the Japanese traditional garden to the new trends and environmental challenges.

As a conclusion, the spaces of the veil, the clearing and the flow are much distorted in contemporary Japanese cities. Westernised urban greenspaces, such as stone-paved plazas with aligned trees and the limited presence of water, are rapidly increasing in Japan. However, these Westernised spaces that citizens of the West enjoy through their particular ways of understanding nature do not include the functions that urban nature has usually performed in Japan. The *chinju no mori*, sacred nature, the Japanese traditional garden, sacred-secular nature, or the cherry blossom trees for the celebration of the *hanami* (花見flower viewing) and secular nature space are examples of the formal urban nature forms commonly employed by the Japanese. In addition to the problems arising from the modernisation and Westernisation initiated since the Meiji period, the ageing of the population and population decline in Japan, adds a new challenge for the permanence of formal urban nature, with the ongoing abandonment of Japanese traditional gardens putting their benefits at risk. Specifically, as ecosystems, Japanese traditional gardens provide supporting, provisioning, regulating and cultural services, among which is the strengthening of a sense of belonging to a place and to nature. As sites of biocultural diversity, traditional gardens mitigate environmental degradation, a benefit that is especially in peril. Therefore, strategies aimed at alleviating these risks through initiatives based on both the conservation of Japanese traditional gardens and the creation of new ones must be carried out.

Urban nature conservation initiative – Let's be the gardeners of our cities

The benefits of the Japanese traditional garden transcend the borders in which they are inscribed, whether the property is public or private; or whether the beneficiaries are permanent citizens or temporary visitors. Likewise, the current decline in the human population of the cities of Japan forces consideration of joint management approaches to include the beneficiaries of the urban green spaces as a way to ensure their sustainability. For all of this, both for the democratisation of benefits and collective maintenance, the Japanese traditional garden becomes a new commons (Pastor-Ivars 2019b). The city of Kanazawa crystallises this new vision with the motto "from the city of gardens to citizens as gardeners".

Since 2016, the author has driven a project to conserve existing Japanese gardens in Kanazawa sustainably. This project combines concepts of joint maintenance, ecotourism and sustainable education. The purpose of the project is to assist owners in maintaining their gardens in a proper condition. Due to the ageing population and expensive maintenance costs, it has become necessary to supplement the public subsidies with help from volunteer groups. Small groups of volunteers (<20) are created to reduce the impact on the garden ecosystem. It also prompts better participation and appreciation and learning about the challenges of urban nature conservation, the need to forge deeper human connections with nature to increase community resilience and the uniqueness of learning about and experiencing biocultural diversity directly from a particular place. Three types of "ecotours" have been established for different purposes: voluntary cleaning, learning seminars and a walking workshop. The range of participants is very diverse, including professionals, students, citizens and tourists. In the future, it is also expected to include minority groups and people with disabilities. One of the participants described their experience as follows:

> I participated in the volunteer cleaning at Shinrensha Temple Garden .... Blessed by good weather, we laboured for some time under the instruction of expert staff, cleaning the pond, collecting leaves and pulling weeds. Through this work, I savoured the various beauties of the garden – the soft dappled sunlight, the cold pond water, the fragrance of trees and soil and the frothy eggs of the forest green tree frog (*Rhacophorus arboreus*). These things engendered a feeling of affection towards the garden and even a slight sense of responsibility towards managing it and keeping it alive (Figure 2.10).
>
> (Katō 2019, p. 116)

Within this project, the four fundamental actors involved in the conservation of urban nature – administrators, owners, experts and citizens – work together for the Kanazawa Green Urban Commons. As a result of this experience, several co-created benefits are incorporated into the Japanese traditional garden and Kanazawa New Commons. These include co-creations of a common atmosphere of harmony and improvement of kinship between humans; wellbeing and restoration of kinship with nature (Figure 2.11); and collective knowledge and dissemination of the wisdom inherent in the Japanese traditional garden.

Figure 2.10 Volunteers cleaning at Shinrensha temple garden. Source: Pastor-Ivars, Juan.

Figure 2.11 Restoring kinship with nature. Volunteer cleaning at Senda family's garden. Source: Wakai, Ken.

A future goal is the realisation of a Japanese traditional zero-emission garden. Many owners are now more aware of the need to sustainably recycle the organic materials of gardens, such as leaves and branches. Furthermore, the idea is to test a natural, economic and social model in the gardens, which is capable of synchronising the life cycles of the species that live in the garden, the management cycles that conserve the garden and the cultural cycles for citizen involvement.

Urban nature new creation initiative – Nature reclaims the city

Along with the above-described initiative to conserve formal urban nature through ecotourism in Kanazawa, the author is working to create a new formal nature from the transformation of abandoned grey infrastructure (abandoned houses and empty plots) into green infrastructure. By 2060, the current population of Japan (126.8 million) is expected to drop by approximately one-third as a consequence of the low birth rate and ageing population. This population decline is resulting in a gradual increase in the number of empty houses and abandoned plots (Pastor-Ivars 2019a). This amounts to a revisiting of the population dynamics in the pre-Meiji period characterised by very slow growth. After that period, however, the processes of modernisation and industrialisation in Japan resulted in an exponential increase in the human population. This growth and the excessive consumption of resources underpin the current environmental imbalances of a capitalist system and strongly consumerist society. The demographic shrinking of Japanese cities, if faced positively, should be the starting point of a new urban paradigm: nature reclaims the city. Nature might occupy the place of abandoned grey infrastructure (empty houses and vacant plots) building a green infrastructure capable of correcting these imbalances. For the creation of the formal types of urban nature from which the new green infrastructure will be shaped, the Japanese traditional garden, with its deep philosophy, design and ecology, must serve as a model. Green infrastructure development should not follow current anthropocentric models of urban nature, which are based on colonisation, acculturation and globalisation processes; instead they must be inspired by models premised on the original notions and experiences of kinship with nature. The purpose must be to create holistic green infrastructure, not just for tradable ecosystem services, but for the human spirit and the conservation of nature. This will require interdisciplinary and transdisciplinary integration between the main actors, both in the initial phases of the design and in the later maintenance of the green infrastructure; they must collaborate to create a sustainable and resilient urban nature. It is in this way that the New Commons of the Japanese traditional garden codified through the spatial types of the veil, the clearing and the flow will persist in the future. The author is working on the basis of building a holistic blue-green infrastructure, and a recent survey carried out in neighbourhoods shows that citizens are favourable to the transformation of abandoned grey infrastructure into green infrastructure. The starting place for Kanazawa to

Figure 2.12 Empty houses and vacant plots. Kanazawa. Source: Pastor-Ivars, Juan.

co-create a new urban nature would be the increasing numbers of abandoned houses and vacant plots which are close to the water canal network (Pastor-Ivars 2019a) because the canals have a deep history and significance in the territory and citizens have the necessary tradition to reformulate their historical urban green spaces: the canal-water and the spring-water gardens.

Over two thousand years ago, the indigenous peoples of Japan, with their animistic beliefs, created the spatial archetypes of *chinju no mori*, *shiki no himorogi* and *kami-ike-shima*. It was the place where the *kami* descended, moved and dwelt. Later, their descendants, based on their ancestors, evolved to a new time designing symbols with captivating aesthetics. The future stage must be to add a new layer to these sacred archetypes: the ecological one, always latent but now needed to help address environmental challenges. Kinship between humans and nature will be restored through a wise combination of spirituality and education and recognition of the rights of nature. Today, there are no abandoned plots (Figure 2.12), they reverberate with the Gods of past (Figure 2.13) and are waiting for a future to arrive where the veil, the clearing and the flow will colonise the cities again. However, despite this we are still living "without knowing".

Ukinegamo (浮寝鴨sleeping ducks floating)
Yoake no naka ni (夜明けの中に inside the dawn)
Ite shirazu (居て知らづ without knowing)
Haiku (俳句Japanese poem) by the author

Figure 2.13 Kodenchi (古殿地 empty site beside the shrine where previous shrine and the future shrine will be built). Ise Jingū Naikū shrine. Ise. Source: Pastor-Ivars, Juan.

Notes

1 The *Kojiki* contains the words *shiki no himorogi niwa, niwa* (庭) and *yuniwa* (斎庭). Some authors have referred to the words *niwa* and *yuniwa* appearing in the Kojiki (*niwa* means garden in present Japanese language) to identify them as ancient gardens. In this chapter, the term *shiki no himorogi* is chosen to designate the spatial sacred archetype for this type of clearing. The first step to make a garden is the delimitation of space, but the function of the *shiki no himorogi, niwa* and *yuniwa* that follows from the *Kojiki* is, either to invoke the gods, or to clear a space in front of the house, or a place to grow rice. Therefore, in the author's opinion, it wouldn't be appropriate to refer to the words *shiki no himorogi, niwa* and *yuniwa* contained in the Kojiki as ancient gardens.

2 Nitschke (1993) refers to the same concept as *shinto*, divine islands, and *shinchi*, divine ponds.

3 Mount *Hōrai* is the Peng-Lai-Shan of the Chinese, one of the Three Islands of the Genii, which were believed to lie in the Eastern Sea, opposite to the coast of China. This happy group was the paradise of the Genii, who, there, maintained a sempiternal vigour by quaffing the waters of the fountain of life which flowed for them in a perpetual stream (Aston 2010).

References

Alhalel, R. 1989. Conversations with Isamu Noguchi. *Kyoto Journal*, 10: 32–37.

Asquith J.P. & Kalland, A. 2004. *Japanese Images of Nature: Cultural Perspectives*. Curzon, London, UK: RoutledgeCurzon. p. 304.

Aston, G.W. 2010. *Nihongi: Chronicles of Japan from the Earliest Times to AD 697*. London, UK: Routledge. p. 878.

Hasegawa, T. 2019. *On the Selection and Use of Trees on Kaga Domain Retainer Estates.*(pp. 80-83). In Pastor-Ivars, J. 2019b

Itō, T. 1971. *Nihon no Niwa: An Approach to Nature*. Tokyo: Chuō Kōronsha, p. 201.

Katō, H. 2019. Co-creating new ways to engage with nature. In Pastor-Ivars, J. (ed.), *Restoring Kinship with Nature through Japanese Gardens: The Challenge to Achieve a Sustainable Commons in Kanazawa* (pp. 116–117). Kanazawa, Japan: UNU-IAS OUIK.

Kobayashi, T. 2019. Kanazawa colored by Feng Shui ideas: Concepts of a garden city. In Pastor-Ivars, J. (ed.), *Restoring Kinship with Nature through Japanese Gardens: The Challenge to Achieve a Sustainable Commons in Kanazawa* (pp. 76–79). Kanazawa, Japan: UNU-IAS OUIK.

Kuitert, W. 1988. *Themes, Scenes, and Taste in the History of Japanese Garden Art*. Amsterdam, The Netherlands: JC Gieben. p. 362.

Lefebvre, H. 1991. *The Production of Space*. Oxford, UK: Blackwell Publishing. p. 464.

Maki, F. 1979. *Japanese City Spaces and the Concept of Oku*. Tokyo, Japan: The Japan Architect. N°264. pp. 50–62.

Motegi, S. 2010. *Chinju no Mori in Voice Style Plus: Mori no Baiburu*. Ōmori Kōji. Tokyo, Japan: Ōmori Kōji. pp. 103–111.

Muraoka, K. 2011. Influence of English landscape gardens on Japanese gardens in the Meiji Era. *Nippon Teien Gakkaishi*, 25: 31–39.

Nishizawa, F. 1997. *Tatemono to niwa*. Tokyo: Kenchiku Shiryō Kenkyūsha. p. 123.

Nitschke, G. 1993. *Japanese Gardens: Right Angle and Natural Form*. Taschen. Koln. p. 238.

Ōe, K. 1994. *Japan, the Ambiguous, and Myself*. Nobel Lecture. NobelPrize.org. Nobel Media AB 2019. Fri. 30 Aug 2019. https://www.nobelprize.org/prizes/literature/1994/oe/lecture/.

Ono, K. 2009. *Nihon Teien: Kūkan no Bi no Rekishi*. Iwanami Shoten. Tokyo. p. 235.

Pastor-Ivars, J. 2015. *Ma* and *Oku*, distance and depth, Ogawa Jihei and the 19th century Japanese garden. Doctoral thesis unpublished. RiuNet repository Polytechnical University Valencia. p. 593. http://hdl.handle.net/10251/48539.

Pastor-Ivars, J. 2019a. *Conservation and foundation of urban ecosystem services, the creation of a green blue infrastructure in Kanazawa*. In Kikuchi N. & Ueno Y. (Eds), *Creation of Urban Landscape by Green Infrastructure: Question from Kanazawa*, pp. 51-61. Tokyo: Kōjin No Tomo.

Pastor-Ivars, J. 2019b. *Restoring Kinship with Nature through Japanese Gardens: The Challenge to Achieve a Sustainable Commons in Kanazawa,* Japan: UNU-IAS OUIK, Kanazawa. p. 132.

Rots, P.A. 2014. Shinto, nature and sacred space in contemporary Japan. Doctoral thesis unpublished. University of Oslo. p. 416. http://hdl.handle.net/10852/39760.

Rots, P.A. 2017. Shinto, Nature and Ideology in Contemporary Japan: Making Sacred Forests. New York: Bloomsbury and Academic. p. 260.

Saitō, T. & Geite S. 1976. Mirei *Shigemori's Collection of Works, Gardens: Approach to Gods.* Seibundō Shinkō-sha. Tokyo.

Shirane, H. 2012. *Japan and the Culture of the Four Seasons: Nature, Literature, and the Arts.* New York: Columbia University Press. p. 311.

Shigemori, M. & Shigemori, K. 1971. *Nihon Teien-shi-taikei, Gendai no niwa*. Kyoto: Shakai Shisō-sha.

Shinji, I., Suzuki, H., Nakamura, Y., Uchi, S., Berque, A. 1999. *Shin-sakuteiki: Kokudo to Fūkei-dzukuri no Shisō to Hōhō*. Tokyo: Marumo Pub. p. 309.

Tange, K. & Kawazaoe, N. 1965. *Ise: prototype of Japanese architecture*. Cambridge: MIT Press. pp. 212.

Tschumi, C. 2004. *Mirei Shigemori 1896–1975: Modernizing the Japanese garden*. Zurich: ETH Zürich, p. 326.

Tsuba, T. 2019. Gardens that incorporate their surroundings. In Pastor-Ivars, J. (Ed.). *Restoring Kinship with Nature through Japanese Gardens: The Challenge to Achieve a Sustainable Commons in Kanazawa* (pp. 88–91). Kanazawa: UNU-IAS OUIK.

Watsuji, T. 2006. *Antropología del paisaje: climas, culturas y religiones*. Salamanca: Ediciones Sígueme, p. 254.

Yamakage, M. 2012. *The Essence of Shinto: Japan's Spiritual Heart*. Kodansha USA: New York.

3 Urban biocultural identity of Yorubas

Intersection of philosophy and nature for wellbeing at Osun Sacred Grove UNESCO Site, Osogbo, Nigeria

Joseph Adeniran Adedeji

Introduction

The Yorubas are the largest ethno-linguistic group in southwestern Nigeria and in the diasporic parts of Brazil, Cuba and Benin. Among the 374 ethnic groups in Nigeria the Yorubas have the largest population, 42 million (WPR 2019) and are well-known for their distinct and unique biocultural relationships with nature. All Yorubas, including the many ethnic sub-groups, are the children of *Okanbi*, whose father, *Oduduwa*, was the progenitor of the Yoruba nation (Fadamiro & Adedeji 2016). The nation originated from Ile-Ife town and extended beyond Nigeria to parts of the republics of Dahomey (now Benin) and Togo, with whom the Nigerian Yoruba*s* share a common language, culture and tradition (Gordon 1979). The Yoruba nation, region and cities existed well before colonialism (Parkin 1971). Yoruba cities of the past were characterised by open spaces at the residential, neighbourhood and city level where sacred shrines and natural elements are contained and displayed.

Sacred spaces, both now and in the past, are demarcated by certain trees. Plants are grown for medicinal, gastronomic and condiment purposes. The royal palaces of the past contained multiple courtyards to meet the needs of the royal family, as illustrated below.

> The immediate physical needs of the Yoruba kings were obviously partly met by the purely utilitarian gardens such as kitchen gardens, farm gardens, herb gardens, the wilderness stocked with game, and the grazing fields. Similarly, the king's spiritual needs were met by the religious temples, monoliths and statuettes erected in these gardens where open air worship often took place.
>
> (Falade 1990, p. 54)

At the neighbourhood level, squares were demarcated and trees were planted to provide shade for recreation, such as playing *ayo* (a traditional recreation), holding communal meetings and socialising. At the city level, there were traditional city squares, including the Kings' Squares which automatically doubled up as market squares. The spatial arrangement of each square was compatible with the number of people it was meant to serve. The neighbourhood squares were considerably smaller in size compared to the city squares. The squares were reflections of the cultural, social, political and economic lifestyle of the dwellers. With the advent of colonialism, many of these spaces became severely affected. Over time, the communal living pattern that gave birth to courtyard housing in the core areas of the cities was replaced with single-family dwellings. City groves were encroached upon and in some instances even built upon. These changes describe the situation in all parts of the city and in all the cities. Most of the urban shrine areas became reduced in size to mere foregrounds of the shrines. In 1966, under the instruction of *Oba* (King) of Ile-Ife, the cultural trees in the Ile-Ife palace gardens were replaced with exotic trees and shrubs to resemble the "stereotyped grassed parklands of the English landscape style introduced by the colonial administrators" (Falade 1990, p. 52). Residential and communal open spaces of a Western character were introduced into the administrative cities, especially in the segregated European Government Areas (Adedeji et al. 2016). Such styles continue to dominate as reflected in the establishment of the Agodi Gardens at Ibadan, and the Freedom Park, Abiola Gardens and Muri-Okunola Park in Lagos as part of the post-colonial urban renewal processes.

Despite these colonial overlays and processes of transformations, some traditional open spaces have remained in the cities. Local attachment to such places and legal frameworks have assisted in ensuring that such places still exist. This attachment is deeply embedded in Yoruba's cosmology and ideology, which is based on the traditional belief-system and understanding of the benefits that can accrue to urban dwellers from relationships with elements of nature found in these spaces. This chapter explores this relationship by drawing on the meanings attributed to Osun Sacred Grove, which has become a UNESCO World Heritage Site. I do this by (1) explaining the meaning of Yoruba mythology in relationship with nature, (2) situating it within the wider concept of biocultural diversity and (3) presenting case studies of where wellbeing strategies emerge.

Yoruba philosophies and nature

Yoruba philosophies are based on *Orisas* (gods), being deities whose cosmological considerations demanded that "major routes had an orientation to the palace such that they coincided with the cardinal points, with the east-facing one associated with *Sango*, the west one, *Esu*, the one to the north, *Obatala*, and the south, *Ogun*" (Obateru 2006, p. 223). These are local deities whose shrines are demarcated with trees and located at the traditional core areas of the cities, based on cosmology and mimesis. According to Okedele and Adejumo (2011) "while cosmology refers to mythology and Yoruba aboriginal city foundation, mimesis highlights the principle of mimicking heavenly divinities in this material realm to arrive at 16 radial forms with prominent *four* cardinal points" (p. 522). The 16 radial points are 16 deities including Osun goddess, out of which four are cardinal.

The morphology of the Yoruba cities is therefore based on geosophy and mimetic principles. Okedele and Adejumo (2011) categorised the guiding conceptual framework of this philosophy into four as "cosmological world views, harmonic ideal, seat of power, and nodal symbol", where the first "emerged as the core category and the remaining three constitute sub-categories" (p. 517) in the bio-mimicry of the mimetic theory of art. The theory states that "art is essentially an imitation of nature" (ibid., p. 518) to construct site-specific identity, where every age has its own inimitable view of nature (Verhagen 2008) and nature and the universe use a fastidious set of active associations. According to Adejumo and Adebamowo (2012), this harmonious relationship is based on unity with nature by uninterrupted flows through connections between values and city form and it "accommodates biomimetic principles" that tie the natural world to the people in a definite place (p. 73). Nature in this sense "embraces the aesthetic configuration of available flora, rhythmic sounds of water bodies, the influence of fauna, and the circular movement of celestial bodies as culturally interpreted" and all methods of understanding and relating with the workings of the 'natural' world (ibid., p. 75). Again, Osun Sacred Grove is archetypal of this description (Adedeji & Fadamiro 2018). Its rich primary flora and fauna blends with the naturally flowing, wide and deep River Osun, which is traditionally believed to harbour the sun goddess of fertility as the object of traditional worship for human wellbeing.

These philosophies of cosmology, mimesis and geosophy are the foundation of how Yoruba people relate to nature. The belief-system is that the deities (*Orisas*) are supernatural forces and agents of a supreme, invisible and ubiquitous divine Being (*Olodumare*) who created the universe of nature. In order for humans to have access to the invisible *Olodumare*, these deities are believed to inhabit elements of nature that are visible – rivers, trees, wind, stones, hills, forests, iron, etc. – and these elements are therefore sacred. This personification of the *Orisas* in natural elements means that the Yoruba people perceive the entire body of nature to be *Olodumare*. The belief-system is therefore summed up as a Yoruba philosophical mantra that "man is a tenant on God's earth" (Idowu 1978, p. 206), the earth implying the entire environment of non-human nature inhabited by the *Orisas*. The *Orisas* and the entirety of non-human nature are believed to be created by *Olodumare* for the wellbeing of humans.

The Yoruba philosophies date back to about the 18th century AD and remain reflected in local architecture, palace gardens, courtyard greens and settlement groves (Falade 1984). In line with these philosophical connections between humans and nature, Crowe (1981, cited in Falade 1984, p. 52) asserts that gardens are the link between humans and "the world in which they live". Humans "in every age have felt the need to reconcile themselves with their surroundings and created gardens to satisfy their ideals and aspirations" (Falade

1984: 52). See Pastor-Ivars (Chapter 2) for insights into how Japanese philosophies have influenced the design of Japanese gardens and the elements contained within them.

Myths are central parts of the overarching philosophy and are the storerooms of beliefs, traditions and values, which are in turn ethnocentric views of the actions of humans, living and non-living objects, natural and supernatural, and metaphysics of the factual and fictional. Myths can be located in contexts beyond space and time, capturing the intangibles of life, illuminating the culturally embedded assumptions of daily activities and offering reasons and causes for events from the mundane to the profound. They are generally transmitted orally from one generation to another (Adeleke 2004). According to Moolla (2012), "myth establishes African authenticity" and myths are Yoruba social codes brought into being through mythology (p. 37).

Yoruba biocultural diversity, cosmology, wellbeing and Osun Sacred Grove

The above account of Yoruba cosmology corresponds with the cosmological value systems of biocultural diversity for self-conception, identity and interpretative tradition (National Research Council 1999). These are not only true of the Yoruba nation of Nigeria but also the diaspora. For example, Gordon (1979) describes Yoruba cosmology and culture in Brazil as:

> Yoruba religion is mainly animistic, that is, mysterious, the inexplicable, the unfamiliar, as well as the great forces of nature, are regarded as being indwelt by spirits through a whole agency they possess the peculiar forms of supernatural powers which mark them as objects of veneration.
>
> (p. 234)

This animistic intricacy of the cosmological relationship with nature exhibited at Osun Sacred Grove is a major attribute that qualified it as a world heritage site (Adedeji & Fadamiro 2018). Oseghale et al. (2014) describe the grove as "an organically evolved cultural and landscape site associated with the Yoruba traditional religion and culture" of about "75 hectares of rainforest vegetation" harbouring "over four hundred species of plant of which two hundred are found to be of high medicinal value to the Yoruba people" (p. 2).

Osun Sacred Grove is the largest enduring model of Yoruba urban sacred groves and it is the best manifestation of Yoruba cosmology (Ogundiran 2014). This grove and the annual traditional worship festival that takes place there and in the city of Osogbo is not only a trademark of Yoruba mythology in relation to urban nature, it signifies the sumptuousness of the Yoruba culture. The festival is attended by Yorubas in Nigeria, neighbouring countries and the diaspora, including Cuba and Brazil. The carnival-like festivals are attended by thousands of tourists and promoted by the Nigerian state as a prime tourist event (Probst 2013).

Biocultural identity has to do with biocultural diversity (BCD) as defined in Chapter 1 of this publication. Within the context of the Yorubas considered in this chapter, BCD deals with the cultural values of the natural environment where the notion of culture is multidimensional and concerns specific ways of life and prevailing ways of interfacing with the natural environment, including specific facets of behaviour such as worship in sacred forests (Cocks 2006). In this way, "the relationship between humans and their environments is mediated by culture" (Cocks 2006, p. 191). Culture and nature are not only inseparable; each can be further and more deeply understood by learning from the other (Buergin 2015, Segerstrale & Molnar 2018). The nature-culture interface in BCD is almost imaginary in the traditional and urban settings of the Yorubas (Fadamiro & Adedeji 2012). The concept of BCD can be used to explain the many ways in which people live and identify with green areas in the urban landscapes in developing countries and the world over (Comacho-Benavides et al. 2013, Buergin 2015, Buizera et al. 2016) and the way green areas enhance the wellbeing of urban populations (Elands et al. 2015, Buizera et al. 2016, Vierikko et al. 2016).

To examine the concept of urban nature in Yoruba cities, the following sections explain the socio-cultural and provisioning ecosystem services of Osun Sacred Grove, based on qualitative interviews with 23 people comprising Osung traditional worshipers, administrative staff of the Grove, herbalists, trado-medical specialists and dwellers in other cities. These cities follow or embrace biocultural diversity tenets, even though not labelled as such at the time (Eisler 2013, 2015).

Socio-cultural ecosystem services

Aworo is an Osun traditional worshiper in the Osun Sacred Grove. The interview with him reveals the deep philosophical relationship of the Yorubas with urban nature. He proceeds: "*Olodumare*, here is your servant. It is far deep inside. *Opon ifa, ike ifa*, is here. This is the place of *Ifa* divination. The person to be divined for will enter with the back backwards. Once he enters with the back, there is a seat at the lowest end of [*this*] the *ifa* forest. An ordinary person cannot enter there. The herbalists will be waiting there. As you see it, it is very deep. The instruments of *Ifa* divination are there. We are coming to the river in August for worship. Your prayers will be answered. You will experience promotions."

Ifa forest is a section in the Grove devoted to *Ifa* worship. *Ifa* is one of the deities that are approached for divination about the wellbeing of an individual, a family, a community or the entire city. The forest harbours the shrine where divinations take place. To confirm the archetypal status of the Grove, Aworo added, "there are other *Ifa* shrines allover other cities in Yoruba land". It is believed in Yoruba myths that the deities are intermediaries between a divine Creator, *Olodumare*, and humans. The deities are consulted through divination. The Grove hosts 40 shrines for different deities, including *Esu*, *Sango* and *Obatala*, among others. The worship of Osun goddess and other deities in the Grove is guided with taboos and myths.

Osun goddess worship, the mythical pact between a deity and humans for wellbeing

The Grove has a Welcome Centre with offices in an artistically constructed building where Engineer Yinka Adesiyan's office is located as head of the Education Department at the Grove. He works with the National Commission for Museums and Monuments that oversees the administration of the site in collaboration with UNESCO. In an interview at the Grove office, he unravels the origin of Osogbo city as follows: "That building you are looking at where those men are ahead is the first palace of Osogbo. That is where Osogbo started. The founders are *Larooye* and *Timehin*. They migrated from *ipole*. That shrine is where they first encountered the Osun River goddess that if you continue to do such and such she will continue to protect you and to – a kind of pact among them. If that river spirit did not have the power to protect them, they would have stopped worshipping it. The ritualists that have been coming here have been receiving answers to their requests."

It is on this basis that the worship of Osun goddesses started and has been sustained for more than 500 years. While explaining the global importance of the grove and the mimetic significance of some of the sculptures, Adesiyan said "Presently we have only two World Heritage Sites [in Nigeria] where the first one is in Adamawa State also [*a*] cultural landscape. This is the second one. During the time of Osun festival, we have some traditional local masquerades that come to perform to the audience of the people in the town. These are some of the traditional masquerades. We call this one *Alasoire*. This one represents *Timehin* the hunter who discovered Osun River who was an elephant hunter in those days. We are talking about the history that happened over 500 years ago. You can imagine how this place looked like. Elephants used to be in this area before. We call this one *Egungun-meta*, this one *Egungun Elefuru*, [then] *Egungun Arenigbonbi*. This one represents the Osun River goddess. This one represents the Osun Chief Priest and this one represents *Arugba*, the Osun votary maid. She must be taken from the current ruling family. First, it is a taboo for her to knock her foot against any object as soon as she reaches the Grove. Secondly, there are some rituals which will be performed for her before she would reach the main shrine. We call this entrance the entrance gate of *Arugba* Osun. She will pass through this place together with the Chief Priest and the Chief Priestess. There is no time in the history that she passes here naked. She must wear something".

The worshipping and attendant ceremonies are guided with taboos, intricacies of myths and interwoven procedures involving all the deities believed to be inhabiting the Grove. The *Arugba Osun* "must be a virgin, and she must be chosen from the incumbent royal family. She must be chosen through *ifa* oracle divination. Her own task is to carry the offering calabash all the way from the King's palace down to the main shrine. So when they are coming, this is the entrance she must pass through" as Adesiyan explained.

Hypocrisy, multiple allegiance and religious unification for wellbeing

Aduke Tawa has been working at the Grove as an education officer under the National Commission for Museums and Monuments. She reveals the multiple religious' allegiances in search of wellbeing at the Grove. She emphasises that "some people are hypocrites. You see them go to the church and still come here and worship. You see some in *hijab*, *hijab* sisters: they are here for spiritual needs, not for sightseeing. They hide their religion, they still practice traditional religion. Even they come here with their *hijab*. Some, they park their cars and they go to the river to have their spiritual baths. They consult. Some are hypocrites. They hide their religion and practice Islam. Some believe purely in traditional." This shows that modern religions are practised along with traditional worship for wellbeing by the Yorubas.

Exposing the significance of allegiance for wellbeing at the Grove, she continued

> working here; we've witnessed a lot. Some people will come here, pretend as if they are here for sightseeing. At the end of the day, when you come to the middle of the road, they tell you, Auntie, I came here last year for a child, I pledged. I come to pay my pledge.

Pledges are made by seekers of wellbeing at the Grove. It is also mandatory to pay pledges after receiving the benefits, no matter the time interval. She continued, "No consequence, no side-effect, if you don't pledge, [but if you do pledge] and you don't pay you might face some difficulty. Here, when you are here, if you pledge, fulfill your pledge. If you don't pledge no problem". This illustrates that the biocultural identity and the Yoruba relationship with nature has deep mythological implications, which are metaphysical.

Provisioning and multiple ecosystem services at Osun Sacred Grove

The provision of water was the primary ecosystem service of the Grove on which Osogbo city was founded. Engineer Yinka Adeseun explained: "According to history, the people of Osogbo migrated from *Ipole*, they migrated because of water. *Larooye* was hunting one day and discovered the river and drank from it. They settled there because of water. There was nothing like Christianity or Islam. As they were clearing land for farming and fishing, they heard a voice that said, 'you have broken all my dye pots'. Another voice said '*Oso igbo pele o. Oso igbo rora'*, meaning well-done, forest spirit". "They had to make a pact with the spirit for wellbeing and settled near the river", Adeseun added. To date, the Grove supplies herbs and herbal roots for the wellbeing of the Yoruba urban population. Aside from provisioning, Adeseun emphasised that

> We have quite a number of people that normally come here for different purposes. We have some that normally come for sightseeing purposes, we have some that come here for ritual purposes and we have some that come for research purpose.

These multiple ecosystem services have not only popularised the Grove but also globalised Osogbo city.

David is a middle-aged Christian who works as a member of the Research Planning and Education staff at the Grove. He discussed the economic benefits of the Grove to the city. He emphasised as a tourist during the Osun Osogbo annual festival, "you will spend at least a week, you rent hotel, buy food, buy this" which adds value to the economy. He explained that "the Osun worshipers believe *Ela* is the intermediary between god and man. *Orunmila* is the messenger of *Ela*." David believed that

> this grove has a kind of plus to the environment in the sense that when they are doing their festivals, the economic value to the state increases, because people are coming from different kinds of places in the world to come and see what is going on about the festival. So at least they will spend like, let me say, a week.

However, as observed by Adeseun,

> the animals in the scarce groves are not supposed to be killed because all the animals that are here, we see them are being sacred which is symbolic of endemic sons and daughters. They are also sacred and are to be preserved for future generations.

Material and metaphysical biocultural identity

The worship of deities in traditional religion is characteristic of the Yorubas, not only at the Grove but also in all Yoruba cities. In this regard, herbalists (*babalawo*) are the most traditional actors with respect to urban nature because their wellbeing activities are carried out with the aid of herbal nature. Their operations are carried at the nexus of material and non-material nature. On the physical realm their natural material operational instruments are herbs, while on the metaphysical realm their operational instruments are non-material and spiritual. Accordingly, they operate at both the physical and metaphysical realm, both of which are derived from Yoruba philosophies. The basis of their operational approach, in being both physical and metaphysical, is an understanding that wellbeing is first spiritually induced before manifesting physically.

This case can be illustrated with Lasisi, an herbalist born in 1947, who obtained secondary school and technical college education. He planted a herb garden in his yard, which he uses for herbal procedures. He mentioned that he handles cases of delays in childbearing, sterility and various sicknesses, including hypertension. According to Lasisi, he combines various herbal ingredients to make concoctions which are efficacious. He emphasised the reason for planting the garden was for convenience in sourcing the necessary herbs. He explained "the reason why I plant them is that, instead of searching for them here and there, I will pluck it fast, and use it for what I want to do. That is why I planted them". For many herbalists, the place of consultation is informal with their residences doubling as consultation places for their clientele and addressing wellbeing through the use of herbs.

Because of these dual operational approaches, they are often anachronistically referred to as "native doctors" or "witch doctors" and "*babalawo*" in the Yoruba language. Before they apply herbal remedies, they engage in mythical divination to understand the spiritual root causes of wellbeing challenges. Often their application of material nature (herbs) is not necessarily a symptomatic treatment but a symbolic appeasement of the gods, who are deemed to be responsible as the source, or believed to be "naturally" powerful to influence all events relating to human wellbeing. This accounts for their dealings with a category of herbs (material nature) that have spiritual connotations. The availability of herbs in time and space is believed to be influenced by particular spirits (non-material nature). Their operational mechanisms therefore involve invoking the spirits of the herbs to make them available at the particular time and place. The herbalists also cultivate herbs that are considered to be simply material in the operational courtyards and backyards that are used as their "herbal workshops". The use of these herbs might involve prayers, in the form of incantations, to summon the spirits into action as the herbal preparations are applied for wellbeing, at times also accompanied with ritual incisions (not scarifications) on certain parts of the body.

Metaphysical uses and applications of herbs consider both good and evil dimensions with respect to human wellbeing. A middle-aged male interviewee, who is a Christian and secondary school certificate holder, affirmed that

> *Ewe-ati-egbo* (leaf-and-root) is used for successful fortunes and to ward off evil and chase off evil spirits. There is a particular herbal leaf that one can use to ward off an evil attack of the enemy. Every evil advance of the enemy will return back to them. This is what our forefathers used their intelligence to discover. We are reaping it now. This is our inheritance from our forefathers. The White men have been like us. Western education made them develop their own engines and drugs.

Trado-medical healers situate the biocultural identity of the *Yorubas* within the framework of Western approaches. In this way, they project their knowledge and craft as acceptable to the modern urban governance of healthcare services.

Though still based on the same Yoruba philosophies as traditional concoctions, the ingredients are sourced from nature and formally packaged in both liquid and solid forms. This is exemplified by one of the interviewees, a 50-year-old Muslim female and secondary school certificate holder. She is of the *Awori* ethnic sub-groups and inherited her knowledge from her paternal grandmother between 1971 and 1985 during her training on the uses of herbs for wellbeing. She explained that

> *Tewe-tegbo* (leaf-and-root) is extremely good. So, different types of concoction do different types of things in the body, such as varieties of complaints (ailments), including delays in child bearing. Improvement in sexual performance, headache, leg pains, peppery body, injury from magical charms.

These accounts suggest that the practice of herbal administration for wellbeing in Yoruba cities is being increasingly influenced by modernisation in which some domains of the herbalists are being adopted by herbal medical healers. Currently, traditional herbal healing is practised more in rural areas or it is more secretive "behind closed-doors". In contrast, herbal medical healers have introduced a mobile dimension into their businesses by participating in urban trade fairs from one urban centre to another, like Ibadan, Osogbo, Akure and Ado-Ekiti among others and sometimes in cycles. Sometimes the fairs are planned for only herbal services by their operational association group.

Gastronomic herbal lifestyle for wellbeing

A good proportion of Yorubas in the cities derive wellbeing from plant materials without consultation and divination. They rely on local knowledge of medicinal and gastronomic herbal use. See also Lagunes Gasca & Merçon (Chapter 7) for a description of how migrant women continue to rely on such knowledge and use to maintain their wellbeing in the host cities they have come to occupy.

One of the interviewees, an 82-year-old *Egba* man and a Christian from Ogun State, has a home garden and explains how he gets

> fresh air, fresh food – and the people in the vicinity always come to ask for bitter leaf, fruits, tea leaf and some other things. And I use most of these things as herbs for malaria. You see, when you take *efinrin* or bitter leaves, they purge the body. When you eat them, you go to the toilet easily. Many times, you don't need to consult a doctor. You become healthy.

Another interviewee said

> There's another one we call curry leaves. Curry leaves is meant to be added to a stew for eating. There's a vegetable we call *ugwu*. It's *ergovin*. It's very good in the body. There are some sicknesses you get. Doctors will tell you to get *ugwu*, squeeze it and drink it with milk or malt. It pumps blood into the body. It's also good for vegetable soup with water leaf. Or

> *ugwu* with small bitter leaf. We have a leaf called *okazi*. English people call it *afoms*. It's used with water leaf to cook soup.

It is a common belief in the Yoruba cities that all parts of plants are essential to wellbeing. An interviewee remarked that

> You see, they are medicinal, almost all of them. People use it all now. They cut the bark and boil it to cure rashes on new born children. They are also used to cure malaria. I have *moringa* too. Instead of going to the market, people come here to take them and use.

To reinforce the emphasis on leaves as an important component of nature for wellbeing, an interviewee remarked as follows:

> This is bitter leaf. It is used for soup. There is something we call bitter leaf soup. It's medicinal. It is good against malaria. When you squeeze it and get the water. Actually, it's a very nice leaf. And more or less, during festival people like to cook it because it lowers blood sugar. It's good for diabetes.

He continued

> This is scent leaves. We use it for pepper soup. It's all round leaf. One can wash it, squeeze it and use it. It's also good for soup like melon soup. We can use it to cook black soup with *ugwu*, *oziza* and bitter leaf. We call it black soup and it's very good for health.

On the usefulness of pawpaw leaves, the interviewee explained that

> It's very good for hypertension. The water that comes out when you cut the pawpaw is used to treat wounds. Guava can be used for diabetes miletus. Chewing the stick stops tooth decay. Mango is very good. Everything about it is very good. The leaf is good for eye, it's good for hypertension. One can use the bark. It also works for the skin too. It's also good for inflammation. *Moringa* can also be used as a vegetable. I have *Efinrin*, which is an appetizer. If somebody doesn't have an appetite, one can get the leaves, grind, add a little oil and fish and take. The appetite will return. It also acts as an anti-inflammatory and also works against bacteria. The leaves can be squeezed to cure abdominal upsets. It can also be used to cure diabetes.

Furthermore, one interviewee, a male herbal dealer (*ẹlewe-omo*) Christian aged between 40 and 50 years, declared

> the herbal (health) benefits [of plants] are numerous. I sell herbs and not necessarily a concoction. The ingredients are leaves, nuts, seeds, bark. God made everything complete in nature. It depends on us human beings. Just like the Bible said that we should subdue the world. There are a lot of

> things in the world that we can tap from if we have the knowledge. These things are God's creation. God created them to help mankind. They have the ability to restore, regenerate and replenish the system. They cleanse the system from all toxins from what we eat, the drugs we take and all. They have antibiotic effects in that they kill germs and the rest of them. They help to maintain body sugar and blood pressure.

This assertion reinforces the connection between the wellbeing of urban dwellers and the non-human nature of urban greens. It supports the conception of humans being part of nature. The costs of the nature-based herbal materials are very low compared with Western medicine, which adds to their wide acceptability for wellbeing among the Yorubas. One interviewee pointed out: "Not everyone has financial strength to buy medicine. One day I wanted to buy an anti-malaria drug, I was given a prescription to the sum of N2,500". This interviewee believed that "in Western medicines, there is too much chalk".

Another interviewee described the multiple significance to human wellbeing of growing plants in an urban area: "You see that I have vegetables and trees. For the fact that you are living where these plants are, you are healthy. They give you fresh air and it will work in your body. Secondly, you see the vegetables, they are fresh. I eat them. I get all the nutrients. Like water leaves it gives blood and they contain vitamins. You see that *Aloevera* (a local medicinal plant) it's very good. If you rub it on your body, rashes will go. You see the pawpaw too, pawpaw is very good in vitamins. Even the unripe one is useful". Similar sentiments were shared by gardeners growing vegetables within their home gardens in the South African case study (see Cocks et al. (Chapter 6)). Another interviewee noted: "Our forefathers used to have long life, but presently, humans don't get that old. This period we are in now is different from that period". Part of the reasons attributed to this is modern agricultural practices with fertilizer incentives and poor diets detached from herbal plants".

Herbal natural intuition and wellness

Natural intuition is a metaphysical mechanism in which nature reveals its benefits to the Yorubas through the subconscious realm, including dreams. Many of the interviewees attributed their knowledge and those of many others of the efficacy of nature for wellbeing to God. One of the interviewees noted that "I will say it's natural. God has put this in us. It's natural because all these traditional people, all these herbalists, they use these things". When asked about the ancients with regard to natural intuition of the wellbeing value of urban nature, another interviewee remarked:

> But for the ancients, it is a gift. It's a gift from God, I think. Look at that *ewe* [herb] that you are sighting, last week I was looking at that herb. In this dry season, see how it is very fresh. I now considered it that this herb

> that is like this in this dry season. One spirit came to tell me that if I use this herb for this, for this, that it will stop. Since then I have been observing it although I am yet to use it for anything. Humans used to try something. So, it's a gift. If we patronize a herbalist they will say, that if we use this one like this, this one like that, those ailments will stop. There are some like that but one has to be guided by someone who has the gift, that God gave that gift.

They identified God as the giver of natural plants in the urban environment and cited biblical references. For instance, one interviewee said

> Is it not God that gave them that power? It is because of unbelief. The bible says if we have faith like mustard seed, if we say that a mountain should be removed, it shall be removed – and enter into the sea. It is faith. The pastors say if we believe, we should not take drugs, it is only if one's faith can carry it. That is faith. Do you understand like that? That is more than all we have been discussing since.

Furthermore, local knowledge was a pivotal mechanism in coalescing urban human and non-human nature in Yoruba cities. For instance, an Ijesa man, who was a Christian and a prophet, adduced the discovery of the importance of nature and particular plants to human wellbeing to local knowledge based on trial and error:

> We may say that it is cultural, being handed down to us from our fathers. As the western people do their research so our great grandfathers also do their own research then. They will observe that, since I have eaten this thing, or administer this thing, this particular ailment has left my body. They will then begin to affirm that a particular thing is a cure for a particular ailment. Therefore, it is culturally inherited. Because it is long that we have known that the bark of the mango tree, if it is sourced, and put inside a pot, or bottle, and soaked in palm wine till the following day, and then you drink it, it will cure malaria in the body. Such things are culturally transferred.

Coincidence was also identified as a means of discovering the biocultural significance of nature for wellbeing. For example, one of the female interviewees in Akure narrates a story that summarises such:

> That time, they had not built this church. There were some *ewe* [green leaves] there at that time. I asked what they are useful for. She said they use it to cook soup. That time they are yet to build this church. There was water here. They said it grows naturally where there is water. Do you know, she just said it to my hearing and I did not talk. So one day I

> went to see the leaves. Because at that time, I was not financially buoyant, but we have pepper at home, cooking oil and other ingredients. My junior asked: will it not kill someone? I said she should not worry, that God [*Olodumare*] said that all that He created are very good. So I fetched the herbs. That's how we eat it. That's how I eat it. You see, I had an ailment. I am AA [blood genotype], so I frequently had real malaria. The second day, we could not rise up – we experienced a loss of gravity. It is as if someone takes anti-malaria drug. I now said we should take blood capsules, because that vegetable is anti-malaria. I now said we should take blood capsules.

Urban biocultural identity of Yorubas – realities, reasons and reactions

Various domains of the biocultural aspects of urban Yoruba identity have been presented in this chapter. In particular, the interplay of space, in terms of sacred groves and courtyards, and plants within these spaces which contribute to spiritual, mental and physical wellbeing are essential to Yoruba identity. This identity is mostly based on Yoruba philosophies and enhances the wellbeing of the urban dwellers. Drawing upon a case study of the Osun Sacred Grove UNESCO Heritage Site at Osogbo, Nigeria, it has been shown that nature is paramount to Yoruba physical, mental and spiritual wellbeing. This calls for the need to revisit and reconnect with Yoruba philosophies on nature. The many negative effects of colonialism and urbanisation have threatened and eliminated many urban sacred groves in Yoruba cities, which undermines this biocultural identity. Pastor-Ivars (Chapter 2) and Mellegård (Chapter 4) also describe how processes of urbanisation continue to threaten the existence of Japanese and Persian gardens, respectively.

The push to enlist more cultural sites in Yoruba land by UNESCO, especially the city of Ile-Ife, because of it being the religious headquarters of the Yorubas with more than 400 shrines, is highly desirable. A similar enlisting of an entire city by UNESCO has been achieved in Brazil which is culturally connected with the Yorubas of Southwest Nigeria.

Though the biocultural basis to Yoruba identity is under transition and under threat in many Yoruba cities, it can be fostered and supported through relevant urban planning and regulations that maintain courtyards, sacred groves and access to plants used in everyday life. Promoting and respecting local cultures, knowledge and practices will also be useful. Also, encouraging further scientific research on Yoruba identity can strengthen its recognition and practice. "Natural" gastronomic lifestyles in Yoruba cities can be improved by encouraging the inclusion of both community and private urban green spaces in urban planning policies, especially in new sectors of cities. The cultivation of more urban spaces with local medicinal herbs can be boosted, while the surviving urban groves should be conserved through the subsisting taboos and legal frameworks (Fadamiro & Adedeji 2016). The ongoing urban renewal efforts in the core areas in most of the Yoruba cities would benefit from traditional biocultural identity if these efforts are carried out with urban nature and local wellbeing in mind.

References

Adedeji, J.A. & Fadamiro, J.A. 2018. Urbanisation forces on the landscapes and the changing value-systems of Osun Sacred Grove UNESCO Site, Osogbo, Nigeria. *Landscape Research*, *43*(6): 798–816.

Adedeji, J.A., Fadamiro, J.A. & Adedeji, Y.M.D. 2016. Residential fencing and house gating: An overview of social inequality and urban insecurity in Ilorin, Nigeria. *Cities*, *52*: 123–131.

Adejumo, T. & Adebamowo, M. 2012. Environmental harmony and the architecture of 'place' in Yoruba urbanism. *WIT Transactions on Ecology and The Environment*, 160: 73–86.

Adeleke, D.A. 2004. Lessons from Yoruba mythology. *Journal of Asian and African Studies*, *39*(3): 179–191.

Buergin, R. 2015. Contested rights of local communities and indigenous peoples in conflicts over biocultural diversity: The case of Karen communities in Thung Yai, a world heritage site in Thailand. *Modern Asian Studies*, *49*(6): 2022–2062.

Buizera, M., Elands, B. & Vierikko, K. 2016. Governing cities reflexively – The biocultural diversity concept as an alternative to ecosystem services. *Environmental Sciences & Policy*, *62*: 7–13.

Cocks, M. 2006. Biocultural diversity: Moving beyond the realm of 'indigenous' and 'local' people. *Human Ecology*, *34*(2): 185–200.

Comacho-Benavides, C., Porter-Bolland, L., Ruiz-Mallén, I. & McCandles, S.R. 2013. Introduction: Biocultural diversity and the participation of local communities in national and global conservation. In Porter-Bolland L., Ruiz-Mallén I., Camacho-Benavides C., McCandless S. (eds), *Community Action for Conservation*. (pp. 1–10.) Springer: New York, NY: Springer.

Crowe, S. 1981. *Garden Design*. Chichester, UK: Packard Gibson Publishers.

Eisler, R. 2013. Human possibilities: An integrated systems approach. *World Futures*, *69*(4–6): 269–289.

Eisler, R. 2015. Human possibilities: The interaction of biology and culture. *Interdisciplinary Journal of Partnership Studies*, 1(1), 3.

Elands, B.H., Wiersum, K.F., Buijs, A.E., & Vierikko, K. 2015. Policy interpretations and manifestation of biocultural diversity in urbanized Europe: Conservation of lived biodiversity. *Biodiversity and Conservation*, *24*(13): 3347–3366.

Fadamiro, J.A. & Adedeji, J.A. 2012. *Cultural Morphology of Cityscapes: Semiotic Analysis of Oshogbo, Nigeria*. Saarbrücken, Germany: Lambert Academic Publishing.

Fadamiro, J.A. & Adedeji, J.A. 2016. Cultural landscapes of the Yoruba of South-western Nigeria demystified as solidified time in space. *Space and Culture*, *19*(1): 15–30.

Falade, J.B. 1984. Yoruba sacred groves and squares, *Edinburgh Architectural Research*, *I*(1): 21–49.

Falade, J.B. 1990. Yoruba palace gardens. *Garden History*, *18*(1), 47–56.

Gordon, J.U. 1979. Yoruba cosmology and culture in Brazil: A study of African survivals in the New world. *Journal of Black Studies*, *9*(4), 231–244.

Idowu, E.B. 1978. *African Traditional Religion – A Definition*. London, UK: SCM Press.

Moolla, F.F. 2012. The body unbound. Ritual scarification and autobiographical forms in Wole Soyinka's Aké – the years of childhood. *The Journal of Commonwealth Literature, 47*(1): 27–46.

Obateru, O.I. 2006. *The Yoruba City in History: 11th Century to the Present.* Ibadan, Nigeria: Penthouse.

Ogundiran, A. 2014. The Osun-Osogbo grove as a social common and an uncommon ground: An analysis of patrimonial patronage in postcolonial Nigeria. *International Journal of Cultural Property, 21*(2): 173–198.

Okedele, O. & Adejumo, T. 2011. Geosophic perspective in Yoruba urbanism. In Laryea, S., Leiringer, R. & Hughes, W. (Eds.), *Proceedings of West Africa Built Environment Research (WABER) Conference*, 19–21 July 2011, Accra, Ghana, pp. 517–527.

Oseghale, G.E., Omisore, E.O., & Gbadegesin, J.T. 2014. Exploratory survey on the maintenance of Osun-Osogbo sacred grove, Nigeria. *African Journal of Hospitality, Tourism and Leisure, 3*(2): 1–22.

Parkin, D.J. 1971. *Eva Krapf-Askari: Yoruba Towns and Cities: An Enquiry into the Nature of Urban Social phenomena* (Oxford Monographs on Social Anthropology) xii, 195pp. Oxford: Clarendon Press, 1969. 42s. *Bulletin of the School of Oriental and African Studies, 34*(1): 213.

Probst, P. 2013. From iconoclasm to heritage: The Osogbo art movement and the dynamics of modernism in Nigeria. In Salami, G. & Visona, B.M. (Eds.). *A Companion to Modern African Art* (pp. 294–310). Hoboken, NJ: John Wiley & Sons.

Segerstrale, U. & Molnar, P. (Eds.). 2018. *Nonverbal Communication: Where Nature Meets Culture.* London, UK: Routledge.

Verhagen, F.C. 2008. Worldviews and metaphors in the human-nature relationship: An ecolinguistic exploration through the ages. *Journal of Language & Ecology, 2*(3).

Vierikko, K., Elands, B., Niemela, J., Andersson, E., Buijs, A., Fischer, L.K., Haase, D., Kabisch, N., Kowarik, I., Luz, A.C., & Stahl, A.O. 2016. Considering the ways biocultural diversity helps enforce the urban green infrastructure in times of urban transformation. *Current Opinion in Environmental Sustainability, 22*, 7–12.

World Population Review (WPR). 2019. Accessed from http://worldpopulationreview.com/.

4 Troubled paradise

The Persian garden in Post-Islamic revolution Iran

Viveca Mellegård

I

At 6:00 a.m. the sun casts a swath of hot light over Elahiyeh. From my grandmother's balcony, there is a view of the whole neighbourhood, Tehran's garden suburb, nestled in the purple shadows of the Alborz mountains. Cranes like mechanical trees swing across the skyline. Even now, a woodpeckering of machines – drilling, boring, smashing, scraping – disrupts the early-morning calm. In Elahiyeh, and all across the city, construction is king.

I can see Amin Street, and my family's old plot of land, between the emerging monoliths. The brick wall that surrounds the property is still there, buckling like an old man's creaky knees, but within its perimeter, in place of the villa built by my grandfather, stand two white-stone apartment blocks, three stories high, with dark and dusty balconies. Of the gardens, only the plane trees (*Plantanus orientalis*) along one side of the narrow street remain, flanking a meltwater stream or *joob* that flows down from the mountains – a typical sight in Tehran, one that hasn't yet changed.

The city's population is expected to swell to ten million by 2030; multi-story apartment buildings have replaced the big single-family houses along my old street. Despite increasing pollution, overcrowding and the threat of earthquakes – despite even the threat of economic collapse, and war – skyscraper construction continues. The capital is in the throes of a housing boom.

II

Tehran lies on three active fault lines, the Mosha, the North Tehran and the North Ray (Ritz et al. 2003 as cited in Moradi et al. 2015). Nevertheless, when Agha Mohammad Khan established the Qajar dynasty, he chose Tehran as the capital city in 1786. Before that, the capital had been Isfahan, a city 340 km south of Tehran. According to some sources, during a power struggle between the warring Zand and Qajar tribes, Agha Mohammad Khan united the Qajars, based in the north, and began asserting control of the country as he swept south from the Caspian. When he reached Tehran, he found the city gates closed against him. He was told that the citizens of Tehran followed the rule of whoever held Isfahan. Since his rivals, the Zand princes, held the capital Agha Mohammad Khan rode south to defeat them. In 1785, he deposed the Zand rulers and took control of Isfahan and then promptly returned to Tehran. This time, he was welcomed and Tehran has been the capital city ever since (Axworthy 2016).

Over the next two centuries, political and economic activities became increasingly centralised and attracted people from other cities and regions. Tehran grew from a modest town to a metropolis. At first, the changes were gradual – agriculture was still the dominant form of production and engaged a large, rural workforce (Amirahmadi & Kiafar 1993). Under the Qajar kings, the *bazaar* or marketplace expanded and small-scale trade led to increased commercial and administrative activities that began to draw ever more people. The first official census in 1922 estimated the population of Tehran as 210,000 (Amirahmadi & Kiafar 1993).

Then, like a bolt of lightning, a century of Qajar rule came to a sudden end. A coup d'état in 1921 changed the course of Iran's history and Tehran's urban landscape. Reza Khan, a commander in the Iranian Army, rose to power and within four years he founded a new dynasty, the Pahlavi, and became its first monarch, Reza Shah. His dream of a modern Iran set in motion massive modernisation and the centralisation of commerce, bureaucracy and industry in Tehran. A concentration of state-owned banks and the foreign trade that began to flourish created new urban professionals, technocrats and a new urban, industrial workforce (Amirahmadi & Kiafar 1993).

Waves of immigration to the city continued with the nationalisation of the oil industry in the 1950s. Revenues soared and opportunities to invest in urban development increased rapidly (Habibipourzare 2015). Between 1934 and 1976, Tehran grew from 46 km^2 to 250 km^2 (Madanipour 2006) and now exceeds 700 km^2. Stretching beyond the old gates and walls that set the city's boundaries, the capital continued to spread and engulf what had up until then been separate villages and towns (Amirahmadi & Kiafar 1993).

At the beginning of the 20th century, gardens and agricultural lands sprouted and spread across the earthquake fault lines and supplied the city with food and fresh air (Habibipourzare 2015). However, a gradual decline in the government's investment in agriculture led to some farmers and villagers moving to the cities (Mahdizadeh 2014). And as the numbers of city dwellers increased, the need for housing also grew (idem). Large plots of land and orchards were sold and transformed into residential areas that were densely populated (idem). Mostly concentrated in the southern part of the city, the new dwellings rejected the traditional style of architecture that favoured a central courtyard. As a secluded inner space, the courtyard offered respite from the harsh heat of summer as well as expressing Iranian cultural and social norms (Mahdizadeh 2014).

Reza Shah was fiercely determined to forge a more European-style capital city (Amirahmadi & Kiafar 1993). New neighbourhoods, connected by wide avenues and other urban influences from Europe, rose up in the north of Tehran and became popular with the new class of educated, urban professionals. When the royal family, the Pahlavis, moved to the north of the city in 1967, where Elahiyeh neighbourhood is located, the spatial segregation mirrored social divides. Wealthier households, the professional, educated classes and foreigners working in Tehran lived in the less crowded, prestigious *balay-e-shahr* (uptown), many of them in stand-alone villas surrounded by gardens. The more cramped and less convenient, *payin-e-shahr* (downtown) became increasingly associated with poorer, migrant or rural citizens (Mahdizadeh 2014).

III

Elahiyeh means "paradise" in Farsi. From my grandmother's balcony, some greenery is still visible through the skyscrapers – the remnants of walled gardens that date back to 1921, when the whole of Elahiyeh, once composed of three estates, was broken up after the fall of the last Qajar king. While it's nearly impossible for me to imagine the wheatfields and cherry orchards that once dotted the neighbourhood, my grandmother cherishes such memories. She and her family would make the daylong journey from southern Tehran by donkey, riding towards the coolness of the Alborz foothills and Elahiyeh. All summer long the children raced up and down the hills and soaked themselves under waterfalls, and at night they slept on the roof, under a dizzying canopy of stars. They woke at dawn, sheltered from the mountain chill by thick blankets. Elahiyeh was their Eden.

In the 1930s, my Swedish grandfather came to Tehran and established the construction company Svenska Entreprenad AB (SENTAB). His arrival coincided with Reza Shah's drive to mould a modern nation with all the infrastructure and buildings he envisioned. SENTAB won large contracts from the Shah's government and constructed bridges, hospitals and roads in every corner of the country. It built the University of Tehran and Mehrabad Airport. Whenever I'm there, I think of the stories my parents told me about the city in the 1960s, during the era of the Pahlavis, when an international crowd of partygoers headed to Mehrabad for breakfast after dancing all night. The imprint of that era is very faint now.

He eventually built the house where I spent the first years of my life – a red-brick, one-story villa with a porch and a sloping roof. There was a patio paved with large stones, and a swimming pool and beyond that a garden fringed by rose bushes and tall trees. It was the style of architecture favoured by Tehran's Europeans at the time.

There is a photograph of me on the day I was brought home from the hospital, with my mother and three brothers standing by a garden wall covered with spring climbing roses. During my formative years I played in that garden, making flower-and-leaf stew for the dogs or searching for Oscar, our pet tortoise. Sometimes I would find him in the shaded, hollow space under the weeping willow tree (*Salix babylonica*). My eyes would adjust to the dim, diffuse light, and after a short time I could see the dome of Oscar's armoured shell, camouflaged by leaves that were faded and crispy under my feet. It smelled dusty, and the air felt parched. Peering through the dangling stems, I could see flashes of light darting off the swimming pool and a humming haze hanging over the damascene roses (*Rosa damascene*) – bees.

The love story between Iranians and the rose is over three thousand years old and will never fade. Depicted in miniature paintings, woven into carpets and sung in verse, the love of roses continues to thrive because of their exquisite scent that both soothes the mind and exalts the spirit. The petals are harvested before sunrise, and it takes two thousand of them to make a single drop of rose oil (Nikbakht & Kafi 2004). Rose imbues all the senses – it's even ingested as rose essence in ice cream, and rose vapour is sprayed over the face to cool and refresh.

Sometimes I called over the wall to Reza, the boy who lived next door. If he was home, he climbed over, and we chatted together before deciding where to play. I remember thinking the family next door was odd because Reza's mother wore a *chador* inside the home. My own mother taught aerobics to her friends. Once a week they would arrive in shiny leotards and swing their hips to Donna Summer in the living room at my grandparents' house.

They built that house on an adjoining plot of land, beyond the garden where I played, and my family moved into the original house. A wrought-iron gate separated the properties. When it was closed, I would shake it to attract the attention of the gardener who would open it for me so that I could go to my grandparents. Steps led up to a yard in which my grandmother planted daisies, cornflowers and tall grasses, attempting to make the garden resemble a Swedish meadow. There was a small paddling pool for me; inside the house, my grandfather kept a low-standing bookshelf for a game I liked to play, which involved pulling out his architecture books and piling them into a heap on the floor. The odd trips to the baker at the end of Amin Street were filled with wonder for me: the sight of the bric-a-brac man, the smell of corn on the cob grilled on a brazier and dipped in saltwater and butter.

Occasionally, city life intruded. In spring, a man brought carts piled high with camel dung, which he tipped onto the lawn, creating an unbearable stink. Later, it was evidence of impending revolutionary change that spiked the serenity of the garden. First the muezzin's call to prayer, unheard of until then in that part of the city, and later bullet shells hurled over the walls by marauding young men.

IV

In the old Iranian language of Avestan, according to Mehrdad Fakour in the *Encyclopaedia Iranica*, "the word *pairidaêza*, Old Persian *paridaida*, Medes *paridaiza* ('walled-around,' i.e., a walled garden), was transliterated into Greek *paradeisoi*, then rendered into the Latin *paradisus*, and from there entered into European languages, the French *paradis*, and the English *paradise*".

According to the archaeological record, the garden is a Persian invention (Hobhouse 2004). Emperor Cyrus of the Achaemenid Empire (558–528 BC) created what is thought to be the first garden at his seat in Pasargadae in central Iran (idem). Archaeological remains suggest a quadripartite ground plan with a central axis of canals and basins meeting in a pool (idem). Symbolically, the number four also represents the elements of water, fire, air and earth; the four seasons and the division of the universe into four parts as conceptualised by the ancient Zoroastrian religion (Faghih & Sadeghy 2012). *Chahar bagh*, meaning four gardens, refers to the division into four parts and establishes the basic layout of the Persian garden.

"In a land of harsh light and with a climate too hot for exertion, shade and protection – from wind as well as sun – were essential elements of the Persian garden" (Hobhouse 2004 p.9). Cypress (*Cupressus sempevirens*), poplars (*Populus caspica*) and Asian plane (*Platanus orientalis*) trees cast a shadow over the water rills and protect them from the hot, dry wind that blows across the arid plains. There are pavilions for shade. Cool air, the sound of flowing water and the scent of almond blossoms and roses soothes the senses and invites contemplation. Anahita, the goddess of all the waters of the earth, rides across the universe in a carriage drawn by wind, cloud, rain and snow. She purifies women and men and gives life to all living beings (Taghvaee et al. 2008). Herodotus, the ancient Greek historian, recounts a story about an emperor of the Sassanid Persian dynasty who was so overwhelmed by the beauty of a plane tree that he adorned it with gold and assigned a soldier to watch over and protect the tree (Taghvaee et al. 2008).

In the garden, humans and nature create a perfect place, like an earthly Eden (Alon-Mozes 2004), "the purest of human pleasures and the greatest refreshment to the spirits of man" (Bacon 1883). Symbolic meanings retain their importance through time and even after the Arab invasion that brought Islam to Iran in the seventh century, the Persian garden embodied the Islamic paradise promised to the practising Muslim (Hobhouse 2004). Preparations for prayer start with water as a means of cleansing and purifying. The central pool reflects the mosque, doubling its size and importance to the devotee (Hobhouse 2004).

Fierce summers and searing heat combine with extreme cold in Iran, and it is a wonder that gardens flourish at all. Three thousand years ago, Persian engineers built a system of underground aqueducts, called *qanat*, that carried meltwater from the mountains across arid expanses of desert to cities. The *qanat* system is documented in detail in numerous historical texts which describe the techniques by which water was transported from aquifers at high altitudes. Egyptian inscriptions record that when Darius I of Persia conquered Egypt in 518 BC, his forces introduced the qanat (Wulff 1968). The *qanat* became a popular cultural and technological export across the Middle East, North Africa, Sicily and Spain (Wulff 1968).

Once the surveyor locates the aquifer, the first in a series of shafts is dug and connects to channels that follow a gradient that allows the water to flow slowly so that it doesn't stir up and carry sediment. An extensive network of approximately 22,000 qanats supplied water for consumption, agriculture and orchards, avenues of trees, fountains and pools until hydroelectric schemes began to replace them in the later part of the 20th century (Wulff 1968, Hobhouse 2004). Ranging in length between two and thirty kilometres, the conduits exited into a *joob* or open canal with sluices. A *mirab* or "prince of water" controlled the distribution of water. The title befits the status and power of the person in charge of such a precious commodity.

V

Change came abruptly to Elahiyeh. In 1979, when the embassies closed their gates and the wealthy families abandoned their houses, the new government appropriated many of the abandoned properties. Islamic charities, such as The Foundation of the Dispossessed (Bonyad-e- Mostazafan) invited in by the nascent regime, laid claim to land belonging to the Shah's ministers and the international crowd. Up to 150,000 properties, including houses, hotels, villas and apartments came under the new foundation's control (Mahdizadeh 2014). A different set of neighbours moved in – well-connected supporters of the new government and, guided by the charities, the homeless and the poor.

Many gardens were deliberately destroyed (Mahdizadeh 2014), perhaps as acts of vengeance for the injustice and poverty many had suffered under the Pahlavi's rule. Historical gardens and the private gardens of the wealthy or foreign-influenced elites represented exclusive comforts and an extravagant lifestyle that most could not afford. In the absence of the property owners who'd fled the country, destruction of the gardens went some way in obliterating a past era and condemning the values and norms of a past regime (Mahdizadeh 2014).

When Iraq invaded Iran in 1980, floods of people fled the towns and villages on the southern and southwestern border and sought refuge in and around the capital. By 1982, Tehran's population had mushroomed; three million new residents had arrived in the years since 1976.

The Islamic Revolution and the war's demands for cannon fodder, meanwhile, pushed the birth rate upwards (Hossein 2013); the new government dismantled the Shah's family-planning clinics, and pressed its citizens to help keep their country's numbers strong. Girls as young as nine were allowed to marry. By 1988, Iran's population had grown to 55 million people – double what it was 20 years before. In response, the government found a good use for the empty plots in *Elahiyeh*. The era of skyscrapers had begun.

Following the Iran/Iraq War, Tehran faced acute problems. Many of them, such as unemployment and poverty, resulted from the desperate aftermath of a long and brutal war. Others, like unplanned growth of the city, air pollution, poor public services, a shortage of green spaces, cultural areas and shortcomings in infrastructure including sewage, called for a transformation in urban management (Hossein 2013). In 1990, Gholamhossein Karbaschi, Tehran's newly appointed mayor, devised a series of controversial policies to make the municipality financially independent; the most lucrative policies involved development. A more liberal government was in power and under the new, decentralised system, Tehran Municipality had greater freedom in setting regulations (idem). By 1998, Karbaschi's administration had collected \$6 billion in taxes and fees relating to construction. Karbaschi became known as "the man who sold Tehran's air". For the right price, his administration allowed developers to bypass Tehran's height limits, and turn one-story buildings into structures with as many stories as their builders could afford.

Karbaschi's administration also established public parks, libraries, cultural centres and sports facilities; the mayor himself helped found Iran's first full-colour newspaper while in office. For the first time, environmental, cultural and social aspects of Tehran's development became a focus. Decades of unplanned and uncontrolled growth had created a sprawling city and neglected degraded green spaces. Perhaps, in part because of concerns for the health and wellbeing of the post-war generation, there was more recognition of the value of urban parks as cultural spaces where citizens could literally breathe clean air and replenish body and mind. Composting and recycling services became available. Citizen participation in some decision making led to projects like The Comprehensive Plan of Green Space, Pishgaman *Fazaye Sabz* (idem). Public gardens for children to play in and more resources for maintaining and restoring historical gardens and monuments as sites of cultural significance were also prioritised (Hossein 2013).

VI

An Anglo-French jeweller travelling in Iran in the 17th century noted in his diaries that the Persian garden was a place to rest rather than walk (Chardin 1711). *Badgir*, or wind towers invented by ancient engineers as a kind of air conditioning system, caught the dry air and guided it across pools of water. Islamic architects celebrated the sonorous qualities of water by creating carved slopes over which it flows mellifluously. Fountains produce a different, more playful sound and cool the air, helping to keep insects away. Pools become reflective surfaces which lull the mind towards mysticism. The shrines of two poets, Sa'adi and Hafez are in gardens that are very popular places for families, couples, tourists and anyone looking for a peaceful place. Verses of their poetry are inscribed in decorative screens that let the light flow through in a blending of architecture, poetry and nature.

Carpets also recall the peaceful state of mind that becomes possible to attain in a garden. Designs depict symbols that have become woven into the Persian psyche. Avenues of evergreen cypresses for immortality, the flowering almond tree is regeneration and the plane trees are a reminder of life-giving water. Functional aspects of the garden appear too. An edible landscape with an abundance of pomegranate, wild sour cherry, plum, apricot, mulberry, almond, saffron and fig. Iranian garden design incorporates a multipurpose diversity of species, both ornamental and productive that provide food, pollination, sound and colour (Khalilnezhad & Tobias 2016). Vegetables and herbs commonly used in cooking appear in the same garden as clover, tulip, iris, poppy and roses that attract birds, butterflies and insects.

VII

In 1998, as the national political climate swung in a more conservative direction, Karbaschi was imprisoned on corruption charges. His swift departure appeared to inaugurate a pattern. Outnumbered by conservatives in parliament, reformists like Karbaschi were found guilty either of embezzlement or mismanagement.

Elahiyeh's skyline had already been thoroughly reshaped by the time the next mayor, Morteza Alviri, came to office; the problem was that no one could decide how high the buildings should be allowed to go. Alviri attempted to re-establish a five-story limit; when he resigned in 2002, under his own cloud of accusations, the next mayor, Mohammad-Hassan Malekmadani, lowered the limit to two.

Meanwhile, land prices skyrocketed. More and more of *Elahiyeh's* garden estates were transformed to construction sites, and again attempts at regulation, and preservation, did little to slow the pace. Since the 1980s, Tehran's Municipality had been responsible for protecting green spaces larger than 500 m^2, public or private (Mahdizadeh 2014). The Constitution also stipulated that damaging natural resources for the sake of development purposes was prohibited (Mahdizadeh 2014). Around the same time, there had been calls to safeguard the quality of the city's air; at one point the Revolutionary Council asked the municipality to levy taxes on the felling of trees to protect them. Walnut and mulberry trees were valued at up to 5 million *tomans* each (roughly $1,200 in 2019).

Reaction to the new law was swift. The tax applied only to healthy trees, not those that were diseased or dead; so, people stopped watering their trees or poisoned them, and fires consumed many of *Elahiyeh's* remaining gardens. Tehran's mayors struggled to keep to the city's guidelines. In January 2003, Malekmadani was given a five-month jail sentence for approving construction permits that threatened the preservation of green spaces and ignored the minimum standards for the distance between buildings.

VIII

Whenever I come home to Elahiyeh now, I like to wander through our old neighbourhood and take in the changes. I do this despite the fact, as a woman walking alone, I tend to attract unwanted attention. Men stare and mutter under their breath.

At first I was rattled by this; if there were any benefits to the compulsory Islamic garb that has been enforced inconsistently since the Revolution, it was the occasional equality between men and women that it seemed to establish. By observing the dress code, women from conservative families could participate in activities that, before the Revolution, would have been deemed unsuitable. A woman who might have expected a future as a wife, mother and homemaker had the option to study at a university and pursue a career, all thanks to the headscarf.

In more recent years, however, an ambiguous space has opened up between modesty and the subversion of it. Prostitution lies at one end of the spectrum; there are parts of the city well known for being pick-up points. For everyone else, the gaze has usurped any possibility of sexual temperance.

As the only public contact between men and women, the eyes have become weapons of mass desire. Nowadays, most people prefer to drive.

Two women walking together is still somewhat socially acceptable. Samaneh Ghadarkhan, a city reporter, agreed to accompany me on my last visit. Samaneh, who was born after the Revolution, is fascinated by Tehran's past – one that she, too, believes is on the verge of disappearing completely. She emigrated to Canada since the trip described here, having found it too difficult to continue her work uncensored.

On that walk, I took us on a shortcut that avoids the ever-present traffic, leading us from my grandmother's apartment block down the hills to my old street. Huge, greenstone steps have been cut into one hill; few people use the path anymore, but I can remember my grandmother and I walking it, and the challenge it posed for a three-year-old. Going down the massive steps now, I see they are beginning to fall away on one side, where the foundation for a new building has been laid.

Directly ahead is Amin Street. Ten metres wide, it seems narrower with the cars racing along with it, hugging the stream on one side and our old brick wall on the other. At the end of the street, alongside the apartment blocks, the smell of fresh bread has attracted a breakfast queue – mostly Afghans and construction workers waiting for sheets of stone-baked *lavash* to be folded into steaming piles once they've been pulled from the *tanoor*. A young man named Nasari sits outside the last remaining brick villa on the street, tearing into the hot, thin bread and rolling strips of it around a bit of cheese.

Some Swedish neighbours of ours lived in that house before the Revolution. When they left, an Islamic charity took it over, and as far as I know no one has lived there since. A glimpse through the garden door reveals that the roof has fallen through; branches are growing into the windows. Nasari, who looks after the property, speaks too quietly. He came to Tehran from Kabul and quickly found that his nationality gave him an advantage. Afghan migrant labourers are preferred for caretaker jobs, as Afghans are thought to be generally more sceptical than other ethnicities, and less likely to allow snoopers onto a property. When I explain that I used to live on the street, though, he agrees to show us the house and garden.

Samaneh and I walk back up the street, to the lot where my old house once stood. The address on the wall is the same, but now there is a buzzer for the flats occupying what used to be our garden. Mr. Sadekhi, a distinguished, grey-haired man in his sixties, comes to the door when we ring. When I introduce myself, he immediately recognises my surname. He tells me that the electricity metre is still registered to the Mellegards, and I joke that I hope we won't have to pay 26 years' worth of bills.

Mr. Sadekhi bought the plot from the woman to whom my mother sold the property, and for six or seven years he lived in our former house. The only thing he neglected was the swimming pool; he had been afraid he might be reported for having decadent Western habits, so it lay empty and withered. While his children were earning their degrees in Germany, he built apartments on the property for each of them. By 1999, the villa, garden and pool had given way to the flats. My grandparents' house next door had already been built over.

That transition coincided with a period before the 1997 presidential elections, when journalists, writers and political activists were murdered in a brutal purge. The thread of suspicion that weaves through most societies in a state of flux knotted itself into the fabric of everyday Iranian life; people who lived on their own, in houses, became fearful that they were exposed to prying eyes and the attention of the secret police. Many moved into what they considered to be the relative safety of the new apartment blocks. There were economic incentives, too; interest rates abroad were at a high, and profits from Tehran property sales could be tucked away in foreign bank accounts.

Mr. Sadekhi asks how it feels to stand where my old house used to be. In many ways, I tell him, I'm glad that it no longer exists; the fact that there's no physical trace of the world I knew means that there's nothing that could challenge my memories of it.

A faint breath of wind lifts up the dust; traffic crawls along. There is never enough parking. A few of the old single-family dwellings have been replaced by 20-story buildings, with their concomitant demands on the public space around them.

Samaneh and I make our way down the street, dusk encircling the neighbourhood, and I wonder again: how did the largest and most obviously illegal tower blocks come to be built in the first place? If the law prevented anyone from building more than two stories, or five, how was it ignored? A 20-story building isn't exactly easy to hide. We're walking past an estate agent's office; I figure someone in the property business should be able to tell us.

"Construction is more powerful than the law of the land", explains the agent, who asks to remain anonymous. He takes us a few streets away, to the site of the old Swedish embassy, a 70-thousand-square-foot property where a giant skyscraper is under construction. A notice has been placed on the gate: in October 2003, a waiver was granted allowing the developer to build to a height of 20 stories. When I ask the agent what happened, he shrugs.

"Property, power, and politics – they're intertwined", he says.

Elahiyeh has the most expensive real estate in Tehran, if not the country. It is no surprise that high-ranking government insiders and their families long ago snapped up a significant portion of the area. It's prestigious to live in the few stand-alone villas set in the centre of a garden and an indication of one's social status. During the Khatami years between 1997 and 2005, cultural heritage was put on the agenda. The reformist president engaged in a "Dialogue among Civilisations", and a resurgence in pride for Iran's pre-Islamic culture grew. More funding was allocated to restoring palaces, historical buildings and royal gardens, which also became the sites of diplomatic meetings (Mahdizadeh 2014). Tree-lined avenues where royal feet had walked now hosted officials of the Islamic Republic. Any awkward reminders of a monarchist past were deftly avoided by a policy of "rehabilitation" – by changing the function of a place, its history faded into the background.

IX

As the streets start to darken, the neighbourhood draws a different crowd; Kurds selling binoculars appear and enjoy a roaring trade. Because the buildings are packed so close together, a new pastime has emerged: peering into other people's flats.

One summer I accompanied a cousin to the eye hospital in downtown Tehran and realised that it was one of my grandfather's buildings. The manager, recognising my grandfather's name, told us that a new wing had been added a few years ago. "During the last earthquake", he continued, "that's the wing that suffered damage. You can see the cracks in the walls. But your grandfather's building is as solid as rock".

Official statistics estimate that about 14 thousand hectares of unstable buildings in the city are vulnerable to earthquakes (Tehran, Iran: Disaster Risk Management Profile 2006, p. 2). Japanese seismologists who made an assessment of *Elahiyeh* concluded that a small tremor, like the ones commonplace in the country, would not put the neighbourhood at risk. A larger quake, though, above 6.5 on the Richter Scale, could reduce 50 per cent of the neighbourhood to rubble. Because of the narrow streets, reaching survivors would be a struggle (Moradi et al. 2015).

You would have thought that the possibility of such a catastrophe would lead the government to make the nature of the risk more transparent to the public. But whenever Samaneh calls various borough offices for statistics, she is always told – very courteously – that someone will have to call her back. They never do. Both of us wonder if comprehensive information exists at all.

Eventually, we reach a Mr. Zahedi, a member of Tehran's city council. He is in his eighties and seems focused on other problems. In the past, he tells us, water management was a primary concern; years ago, some of the big houses used to release their sewage into the streams running alongside the streets in the dead of night. That doesn't happen anymore, but the sewage system that was built to solve the issue is now being overwhelmed itself. In some places, sewage burning though the old pipes has seeped into the drinking water.

Next to the bakery on Amin Street is a corner shop that has been in business for more than 50 years. It's crowded with packets of biscuits, cartons of sour cherry juice, squares of white cheese floating in water and boxes of tissues with smiling, brashly coloured cartoon faces on them. From behind his counter, the owner, Mr. Heyderi, has spent decades observing the changes in the neighbourhood; when Elahiyeh was dominated by the embassies, he says, Americans were his best customers, and Europeans his worst. Now his store has become a place where the nouveau riche buy milk alongside the Afghan workmen, a few old-timers, and gangs of chic youngsters from the nearby cafés.

Business is much better for Mr. Heydari now; he prefers the social melting pot. In the past, he says, the only contact he had with the wealthy residents was through their servants, drivers and gardeners. The people in the big houses were from a higher social class, and thought a lot of themselves, he says. He sums it up using an expression I've never heard before: *Haser naboodan ba shah faloodeh boghoran.* "They wouldn't have deigned to eat ice cream with the Shah".

X

Iran leaves the traveller "haunted by those plains of amber, those peaks of amethyst, the dignity of that crumbled magnificence, that silence of two thousand years" (Harold Nicolson in Hobhouse 2004). Nicolson's description finds its echo in the purple peaks of the Alborz mountains to the north of Tehran which have the dusty look of a distant past that is simultaneously very present. A majority of Tehran's population was born after the Islamic Revolution and only know about the lifestyle that existed before through their parents or grandparents. Forty years after the revolution, a popular cable television channel called Manoto screens documentaries about life during the era of the Shah. Many of the locations featured in the films are recognisable in the city today, but the kodachrome colour of the archive film accentuates a growing nostalgia for a lost way of being that many young Iranians both inside and outside the country report feeling – a stinging sensation of regret and yearning for a place that you already occupy but arrived 40 years too late to experience.

Incessant traffic, pollution and the constant thumping of construction in Tehran wears down people's nerves. Tabiat Bridge is a walkway and garden suspended above one of Tehran's busiest highways. Three pathways are stacked on top of each other and supported by steel trunks that look like massive trees. Tulips, climbing vines and meandering wooden decks encourage a slowing of one's pace and pulse. Investments to improve the city's environment have focused on creating more green areas, and Tabiat connects two urban parks large enough that you can lose oneself in the cool shade of a sycamore tree in.

XI

Culture and nature meander in and out of one another along the spectrum of Iran's distant origins and the urban landscape of Tehran today. The Persian garden fulfils the dream of an earthly Eden and the genius it takes to create such a lush paradise, in spite of the harsh and hostile landscapes, is like a promise that the impossible is possible. Persian New Year, *Nowruz*, epitomises just how intertwined Iranian culture and identity are with the ideal of the Persian garden and the hope it breathes into the sometimes-frazzled psyche.

Over a millennium before Islam, the ancient Zoroastrian religion established the festival of *Nowruz*, meaning new day. Nowruz celebrated the vernal equinox in March and marked the start of the New Year. Today, Iranians gather around the *haftsin*, a decorative display of seven items beginning with the Persian letter "*sin*" or "s". Collectively, the items symbolise rebirth in nature and embody the fulfilment of human wishes for a new season, a new year. *Sabze* (lentil or wheat sprouts) for growth, *sib* (apple) is beauty, *sir* (garlic) for good health, *serke* (vinegar) symbolises patience, *somagh* (sumac), a spice that's deep red like the sunrise, *sombol* (hyacinth) represents spring and hope, *senjed* (dried fruit of the Lotus tree) is for love. Families often have their own traditions and choose different items which they display on a cloth in a prominent place in the home. Sometimes, they add candles and a mirror to reflect light into the future.

Even in the densely populated megacity of Tehran, gardens and parks, even small balconies in apartment blocks, blaze with small bonfires on the eve of the final Wednesday before Nowruz. As people leap over the flames, they invoke the element of fire, the symbol of light in Zoroastrianism, to exchange its glow and health for their sickly pallor. In the early years just after the Islamic Revolution, some hardline factions tried to stop Iranians from celebrating Nowruz, possibly because of its pre-Islamic roots and emphasis on the freedoms of joy, light and love. High-spirited gatherings of young and old Iranians who gathered outside and leaped together might have been regarded by the Revolutionary Guard as dangerous solidarity. Any suggestion of stamping out a tradition that's been intertwined with Iranian culture and identity for over 3,000 years was roundly defeated (Mahdizadeh 2014). Nowruz is registered as an intangible cultural heritage by UNESCO (Mahdizadeh 2014).

On the 13th day of the new year, the tradition is to spend the day outside in gardens and nature. Picnic blankets are spread out, and the *sabze* is thrown into flowing water, carrying away all the household's bad luck from the previous year. Whether in public or private, millions of Iranians perform the ancient rituals and entrust their prayers, sometimes ebullient, sometimes urgent, to the potent and transformative alchemy of the elements. During the 13 days of Nowruz celebrations, earth, fire, water and air are the conduits by which humans transform and renew themselves. Bad luck, poor health and hopelessness are offered up as gifts to the natural cycles in an exchange that pays homage to the continual and reciprocal relationship between humans and nature.

> It might be said that the gifts we give at times of transformation are meant to make visible the giving up we do invisibly. And of course we hope that there will be an exchange, that something will come toward us if we abandon our old lives.
>
> (Hyde 2007)

XII

In the fizzy orange streetlight glare, Elahiyeh's streets thicken with exhaust fumes; cars pass by carrying rosy-lipped beauties and sultry-eyed young men. Jammed side by side and inching along, there's enough time and plenty of inclination for them to flirt through their windows. Even though the cafés here still display signs reminding women to remember their proper *hejab*, there's more freedom in this part of town.

On our way home, Samaneh and I pause in front of a tiny, pistachio-green van parked outside the last house on Amin Street. It's filled with keys and key-cutting tools; Hassan, who minds it, says it has been here for eight years. His uncle, the owner, makes a very good living. A string of coloured bulbs like fairy lights hangs across the open side. Hassan taps into the bakery for electricity.

Back again on my grandmother's balcony, I see the pallor of skyscrapers in the making. Construction is like a death rattle in this city. Yet whispers of something else still float over the stream running down my old street, where my house and garden once were; even for someone like Samaneh, who has only known life after the Revolution, that garden remains an ideal. She confided to me that she, too, would one day like to have one, "if any are left".

In 1926, when the British writer Vita Sackville-West visited Tehran, she wrote, "I call it a city, but it's more like an enormous village". Perhaps Tehran has come full circle – more a cluster of villages, once again, than a cohesive capital. Those who can afford it live, shop and entertain themselves locally, while others wake up in the early hours to beat the traffic and get themselves to work. Places like Karaj, to the west, which used to be on the outskirts of the city, are now connected to it by a continuous stream of buildings. It's possible that Tehran will reach its limits only when it overtakes those towns entirely. Another future is suggested by recent government incentives offered to those willing to resettle elsewhere in the country, in order to ease the strain on Tehran's infrastructure – or, perhaps, to scatter the activist base that congregates in the capital.

To my eye, the city's skyline paints a picture of some of the stresses playing out in the country as a whole. The towering buildings, far exceeding legal limits, suggest a separate reality for those with the money and political clout to live above the law. The restive bustle below suggests that such an arrangement may be unsustainable.

At night, I look to the north, and the mountains, away from the glittering sprawl. Even there I can see threads of light delineating the seams of the Alborz. Construction creeps upward, toward cleaner, cooler air, and toward some degree of privacy. It's easier to escape the latest round of worries over an uncertain future and extended economic sanctions when surrounded by nature, peace, and freedom. These are not romantic, sentimental meanderings. A young Tehrani woman, when I asked her what dreams she had for the years to come, told me that "The day I can feel the breeze on my naked neck, that's when I'll feel really free".

References

Alon-Mozes, T. 2004. Gardens and the emergence of culture: The Meir Garden: The first Hebrew garden for the first Hebrew city – Tel Aviv. *Studies in the History of Gardens & Designed Landscapes*, 24(1): 55–64.

Amirahmadi, H. & Kiafar, A. 1993. The transformation of Tehran from a garrison town to a primate city. Urban development in the Muslim world. In Amirahmadi, H. (Ed.), *Urban Development in the Muslim World*, (pp. 109–137). New York, NY: Routledge.

Axworthy, M. 2016. *A History of Iran: Empire of the Mind*. New York, NY: Basic Books.

Bacon, F. 1883. Of gardens (1625). In Dixon Hunt, J. & Willis, W.P. (Eds.), *The Genius of the Place – The English Landscape Garden 1620–1820*, (pp. 51–56). London, UK: MIT Press.

Chardin, J. 1711. *Voyages de Monsieur le Chevalier Chardin en Perse et autres lieux de l'Orient*. Vol. 3. Amsterdam: JL de Lorme.

Faghih, N. and Sadeghy, A. 2012. Persian gardens and landscapes. *Architectural Design*, *82*(3): 38–51.

Habibipourzare, A. 2015. Pushing Tehran across a threshold: Applying the social-ecological resilience approach to a vulnerable city. *Spaces & Flows: An International Journal of Urban & Extra Urban Studies*, 6(2): 75–87.

Hobhouse, P. 2004. *The Gardens of Persia*. Carlsbad, USA: Kales Press.

Hyde, L. 2007. *The Gift: How the Creative Spirit Transforms the World*. Edinburgh: Canongate Books.

Khalilnezhad, S.M. & Tobias, K. 2016. The productive landscape in Persian gardens, foundations and features. *Bagh-e Nazar*, 13: 3–16.

Madanipour, A. 2006. Urban planning and development in Tehran. *Cities*, 23(6): 433–438.

Mahdizadeh, S. 2014. Historical gardens in transition in 20th century Iran: A critical analysis of garden conservation. Unpublished PhD thesis, University of Sheffield, Sheffield.

Moradi, M., Delavar, M.R. & Moshiri, B. 2015. A GIS-based multi-criteria decision-making approach for seismic vulnerability assessment using quantifier-guided OWA operator: A case study of Tehran, Iran. *Annals of GIS*, 21(3): 209–222.

Nikbakht, A. & Kafi, M. 2004. A study on the relationships between Iranian people and Damask rose (*Rosa damascena*) and its therapeutic and healing properties. In *VIII International People-Plant Symposium on Exploring Therapeutic Powers of Flowers, Greenery and Nature 790*, pp. 251–254.

Ritz, J.F., Balescu, S., Soleymani, S., Abbassi, M., Nazari, H., Feghhi, K., Shabanian, E., Tabassi, H., Farbod, Y., Lamothe, M. and Michelot, J.L. 2003. Determining the long-term slip rate along the Mosha Fault, Central Alborz, Iran. Implications in terms of seismic activity. *Proceeding of the 4th International Conference on Seismology and Earthquake Engineering, Tehran,* 12–14 May.

Taghvaee, A.A., Ansari, M. & Nejad, H.M. 2008. Cultural beliefs regarding Persian gardens with the emphasis on water and trees. *African & Asian Studies*, 7(1): 101–124.

Tehran, Iran: Disaster Risk Management Profile 2006, 2.

Wulff, H.E. 1968. The qanats of Iran. *Scientific American*, 218(4): 94–105.

5 Designed and engaged urban nature

Biocultural values in Helsinki, Finland

Kati Vierikko and Mia Jaatsi

Introduction

Compared to Central European countries, urbanisation in Finland, one of the most northern countries in Europe, is a relatively modern phenomenon (Eurostat 2016). The establishment and maintenance of different forms of urban nature, termed urban green space (UGS), has played a significant part in the urbanisation process fulfilling multiple roles during the history of urban development. Today, the capital of Finland emphasises the role of urban green spaces in its City Vision for 2050 (City of Helsinki 2013):

> Helsinki is a green network city. Recreational areas, the sea and recreational services are easily available. Parks and other public outdoor spaces embody high quality and diversity. The increasingly dense city has also been able to retain large unfragmented green spaces. City forests and, on the other hand, cultural environments are strengths for Helsinki. The densifying urban structure benefits from the green spaces and recreational areas nearby. On the other hand, the urban environment may include lots of greenery, in the form of urban gardening and green roofs, for instance.

For centuries in Helsinki nature has fed, protected, attracted and supported the health of urban citizens. The design and management of urban nature have involved many temporal, spatial and social factors, such as geographical location, biophysical preconditions, land-use development, socially shared norms and rules and the citizens themselves (Vierikko et al. 2014; Wolch et al. 2014; Elands et al. 2015). The role of nature in city planning and among citizens is dynamic, due to changes in the socio-political situation, migration flows (shrinking vs. growing cities), climate change, paradigm shifts in urban design and changes in the lifestyles of urban residents (Haase & Gläser 2009). The ideologies, norms, rules and management practices adopted in city planning further shape relationships between cultures and natures (Buizer et al. 2016, Vierikko et al. 2016). The biocultural diversity (BCD) approach accentuates the dynamic, constantly evolving role of nature, relying on the idea that nature is socially constructed and the interrelationship between humans and nature is articulated through practices and relational values towards nature (Buizer et al. 2016).

We discuss how biocultural values and manifestations of nature have been shaped by official urban planning and consider whether these values and manifestations are reflected in the relationships between people and nature. The BCD approach provides the means to investigate these local relationships. Our case study in Helsinki provides an example of how nature within cities has been conceptualised and designed over different time periods. We discuss whether the local residents' perceptions of and engagements with nature respond to these changes. The term *local people* refers to residents living in cities, so we are not emphasising indigenous people (e.g., Sami people), minorities or immigrants in Finland as many other authors do in this book (e.g., Adedeji Chapter 3, Cocks et al. Chapter 6, Nesbitt et al. Chapter 10).

Our chapter is divided into three parts looking at biocultural values from three different ontological perspectives. First, we present materialistic values of urban nature and how the biophysical landscapes of the city have been modified and valued by urban design from the 19th century to modern city landscapes. Second, we look at subjective values of urban nature in terms of what kind of green spaces local residents have identified as unique and typical urban nature in Helsinki, and where citizens' favourite and most meaningful places are situated. Finally, we provide a short view of how urban parks act as places for social cohesion (Comstock et al. 2010, Elands et al. 2018).

Knowledge used in assessing biocultural values in Helsinki

The results presented in the first section of the chapter are based on published online documents and articles (Table 5.1). We show how urban nature is manifested through formal land-use planning and green space design in different historical periods in Helsinki, focusing on introduced (urban parks) and indigenous types of UGS (forests). In the second part, we use two Internet-based public participation surveys, conducted by the City of Helsinki in 2013 and in 2018. The earlier survey is related to the 2013 Helsinki City Plan and the 2018 one to the updated work of the Biodiversity Action Plan which aims to safeguard local biological diversity by adopting different management practices (e.g., preserving dead wood in forests). Internet-based public participation mapping tools (e.g., public participation geographic information system – PPGIS) have recently been developed to record and collect socio-cultural values and experimental knowledge of citizens attached to specific places (Kyttä et al. 2013). PPGIS surveys asked citizens to mark places as "unique urban nature" and "typical nature in Helsinki" on a digital map. The third part of the chapter is based on results collected from a place-based survey, which was administered to 600 park visitors across 12 different parks in Helsinki during the summer of 2015. These results were collected as part of the EU FP7 funded project GREEN SURGE. The aim was to collect park users' perceptions about the biological diversity found within the parks as well as the types of human-nature

Table 5.1 Sources of Knowledge Used in Assessing Biocultural Values

Source of knowledge	*Short description*	*Significance for this article*	*Availability*
Official websites of Helsinki; official documents and reports.	Websites, documents and reports related to the history, socio-economic statistics and land-use planning of the city of Helsinki.	Materialistic values of urban nature; conceptualising nature in city planning.	Open access data. www.visithelsinki.fi; City of Helsinki & Ministry of Environment 2005; City of Helsinki 2013, 2014, 2019: Statistical Yearbook of Helsinki 2017; Vierikko et al. 2014.
Internet-based PPGIS survey related to the Master Plan of the city of Helsinki (2013).	The city of Helsinki conducted a public participation geographic information system (GIS) survey to collect views of city residents for the preparation of the new city plan. The survey allowed respondents to mark locations on the map, e.g., areas for unique urban nature.	Subjective values of urban nature; perceptions on urban nature and biodiversity; meaningful places.	Open access data. Publications: Mapita 2014; Vierikko et al. 2017; Wang et al. 2019.
Internet-based PPGIS survey related to the updating process of Biodiversity Action Plan (LUMO) in Helsinki (Biodiversity Action Plan 2008).	The city of Helsinki conducted a survey to collect opinions about local biodiversity values and "typical urban nature". The survey allowed respondents to mark locations on the map to indicate, e.g., typical urban nature, areas with high biodiversity values, nature areas they are concerned with and that need protection, and what they are ready to do to protect nature in cities.	Subjective values of urban nature; perceptions on urban nature and biodiversity.	Open access data. Publications: Mapita 2019.
Place-based qualitative research to study park visitors' perceptions and experiences was conducted in Helsinki, Lisbon, Berlin and Bucharest as a part of the GREEN SURGE project.	In Helsinki, a total of 600 park visitors were interviewed in summer 2015. The survey covered questions about motivations, experiences and perceptions about biological and cultural diversity of the park the respondent visited. The visitors represented diverse socio-demographic characteristics.	Subjective values of urban nature; favourite places, social cohesion in urban parks.	Available on request. Publications: Puttonen 2017; Vierikko et al. 2017; Vierikko et al. 2020.

interactions they participated in. A detailed description of this method is presented in the final deliverable of the WP2 of the GREEN SURGE project (Vierikko et al. 2017).

Describing people and biodiversity of the city

Helsinki, the capital of Finland, has long been an arena of political conflict between Western and Eastern powers. Founded at the mouth of River Vantaanjoki in 1550 by the king of Sweden, it was declared the capital in 1812, during the autonomous time under Russian power (www.visithelsinki.fi). Initially, the country was part of Sweden (1157–1809) and then Russia (1809–1917), gaining independence in 1917. The Helsinki metropolitan region is the economic centre of Finland. Bordering on the Baltic Sea it has always been an important international port and passageway between Western and Eastern Europe. The location has shaped the biological and cultural diversity found in the city area. Helsinki is Finland's largest city with a population of 640,000 (Statistical Yearbook of Helsinki 2017) which is expected to increase to 900,000 by 2050. More than half of the population (60 per cent) has a migration background and 14 per cent are foreign work-based immigrants or refugees representing 170 nationalities with the majority being from Russia (City of Helsinki 2014).

Green areas cover a great deal of the city's surface (47 per cent), with 20 per cent covered by forests (Vierikko et al. 2014). The original landscape is a mixture of granite rock hills, forests and open wetlands (reeds) along the coast. The extent of indigenous nature, such as forests, is one of the highest of all the European capitals. Helsinki's urban forests, defined as "semi-natural woodland patches where the ground layer vegetation is natural, and tree structure is multi-layered, similar to that of natural forests and extensively managed" (Lehvävirta & Rita 2002), do not differ much from those in rural areas. Owned and managed by the city, the urban forests are managed primarily to meet recreational needs while taking biodiversity into account (Saukkonen 2011). The forest network, which totals approximately 4,000 ha, is strongly fragmented, featuring over 800 separate patches, the majority of them being smaller than one hectare (Wang et al. 2019).

An urban park is a publicly accessible green area within the city that is intended for recreational use by urban dwellers and includes a wide range of biophysical and cultural elements, such as trees, shrubs, flowering plants, mown lawns, playgrounds, water bodies and sculptures (Cvejić et al. 2015). In Helsinki, most of the high-quality parks with diverse vegetation and facilities are situated in the city centre, the oldest part of the city, while indigenous types of UGS (e.g., forests) are the most common types of UGS in suburban and peripheral areas.

Ahern et al. (2014) have revealed that urban areas can be biologically rich, offering habitats for many native species. The cultural history of cities is the primary factor contributing to species richness, seen, for example, as high vascular plant richness in Helsinki (Kurtto & Helynranta 1998, Pyšek 1998,

Kühn et al. 2004). There are over 1,600 species found in Helsinki, 70 per cent of all of the vascular plants and 80 per cent of all of the amphibians and reptiles recorded in Finland (Vierikko et al. 2014, Hyvärinen et al. 2019). In Helsinki, the conservation value of forests is estimated to be high even at the national level, which makes its forests potential habitats for native species that could not survive in intensively managed commercial forests in rural landscapes (Kowarik 2011, Vierikko et al. 2014). There is a strong will among city authorities and local decision-makers to protect urban biodiversity. The Biodiversity Action Plan (2008) makes recommendations for biodiversity-friendly actions and the Nature Conservation Programme (2015) to establish new conservation areas.

Bringing nature to the city – or vice versa? Conceptualising and designing nature

Parks, one of the oldest types of urban green space, are in most cases open to the public providing space for biocultural interaction and opportunities for residents to exercise, relax, meet others and enjoy nature (Chiesura 2004, Kabisch et al. 2015). Historically, urban parks have been considered as instruments of ecological modernisation, reflecting diverse ideologies of "socio-natural projects" (Byrne & Wolch 2009). These ideologies have affected the types of urban nature that are perceived to represent the "right kind of nature", which contributes towards citizens becoming healthy, moral and socially responsible (Figure 5.1) (Byrne & Wolch 2009). However, urban parks have typically

Figure 5.1 Urban parks have traditionally represented the "right kind of nature", which contributes towards citizens becoming healthy, moral and socially responsible. Alppipuisto Park is one of the first public parks in Helsinki and was founded in 1851. Source: Jaatsi, Mia.

had strict socio-cultural norms determining who and how these areas should be used (Byrne & Wolch 2009). See Elands et al. (Chapter 11) and Cocks & Shackleton (Chapter 13) for further discussion on this. In late 19th-century Helsinki, urban parks were introduced as a form of nature intended to uplift the "brutal" nature of the young capital city. Initially, forests, rocks, mires and other forms of "wild nature" were perceived as unattractive and not belonging in a cityscape:

> In the 1800s the southern tip of Helsinki was undeveloped land covered by rocks, forest and a few marshy areas. Locals would come here for long walks and picnics. The story of the park Kaivopuisto began in 1834, when a businessman by the name of Henrik Borgström (1799–1883) leased the undeveloped land from the city for 50 years. The 50-year lease agreement expired at the end of 1885, and the City of Helsinki took over the park again. Kaivopuisto became a city park open to everyone. Visitors to the park were expected to be gentlefolk who behaved with propriety. Aristocrats were replaced by refined bourgeoisie. Kaivopuisto was not yet a people's park, as uncultivated people who hung hammocks between the trees, for example, were frowned upon.
>
> (Vihreät sylit, City of Helsinki 2019)

As the short narrative above indicates, public parks have had strict socio-cultural norms for who should use them and how they should be used (Cranz 1982, Byrne & Wolch 2009). The growing working class had restricted possibilities to enjoy the beauty and recreational benefits of Helsinki's parks. As revealed by Schönach (2015), there was markedly unequal access to parks in the early 20th century with the extent of green space in Helsinki's low-income neighbourhoods much lower (9 per cent of land surface) than in higher-class subdistricts in the western part of the city (49 per cent of land surface). When the needs of the working class were finally taken into account, the city developed special national parks (such as the island of Seurasaari) and also allotment gardens for food production primarily for use by working-class people. However, these green areas were located on the city periphery, far from working-class housing. The next large migration from the rural areas happened after the Second World War. Urban planning and the role of green space underwent a significant change, which included attempts by architects, such as Alvar Aalto (1898–1976), to ensure that "every man should have the right to walk through forest on their way to work".

The "garden city" ideology, developed at the end of the 19th century to improve living standards and the quality of urban space, originated in Great Britain during the Industrial Revolution (Hirvensalo 2006, pp. 80–81) and was later adopted by Finnish architects. Within Finland, an emphasis was placed on mimicking nature and biological organisms, and on assimilating buildings with the surrounding environment. This led to the rise of "organic architecture" (Hirvensalo 2006, pp. 80–81), the impacts of which are reflected in the incorporation of original biotopes – pine, spruce-dominated forests and granite rocks – into the city plans and the design of green space in the second half of the 20th century. This is in contrast to other parts of Europe, which saw the incorporation into city planning of English-style urban parks along the lines of Kaivopuisto Park, as a proxy for the "right kind of nature". In Finland, the move towards organic architecture derived from a fear that modernisation and technological innovations would sever human-nature connections:

> Primary nature, old growth forest, granite rock covered by lichens are much more valuable than planted shrubs constructed sideways … water, plants and vegetation try to connect us back to nature and decrease the cold and unattractive environment caused by technology and mechanization.
>
> (Hirvensalo 2006, p. 81)

To mitigate such threats, an ideology of bringing nature (back) to the city was adopted (Hurme 1991). Finnish architects introduced the "forest neighbourhood" concept, developing residential areas that were surrounded by indigenous nature and which had everyday services, such as nurseries, schools and grocery stores, within walking distance. There was a determination to develop neighbourhoods where residents could derive readily accessible health benefits by walking through natural environments, as well as to provide newly urbanised residents with places to connect to their neighbourhoods and to remind them of their home rural landscapes (Hirvensalo 2006, p. 66). The forest neighbourhood ideology strongly supported direct daily engagement with nature to enhance human-nature relationships and promote multisensory natural experiences.

In the 1970s, the role of indigenous nature was still central in green space planning, while at the same time the idea of developing green spaces for specific functions was growing in popularity. A green space (playground, recreation forest, lawn) would have a specific function, according to which it was categorised and its use standardised (Hirvensalo 2006, p. 161). In the 1980s, a more utilitarian attitude to nature (to supply human needs) strengthened in urban planning discourse, and the ideology behind the symbolic role of close-to-home nature virtually disappeared. However, the experiential role made a comeback in the 1990s (Hirvensalo 2006, p. 237) when ecological modernisation (Hajer 1996) influenced urban policies globally through emerging ideologies such as the "sustainable city" and "eco city" (Jong et al. 2015). In Helsinki, an example of ecological urban planning can be found in the "Eco Viikki" district where nature was again brought near to citizens' everyday life, but now in the form of edible, rich and diverse gardens, which would support local community identity through urban farming (City of Helsinki and Ministry of Environment 2005).

Today, modern and smart cities conceptualise nature as part of city infrastructure, which can be constructed and controlled by humans (Hirvensalo 2006, p. 64). In addition, nature is commonly seen as an instrument to increase property value (Votsis 2017) giving little attention to the reciprocal relationship between culture and nature. The concept of urban green infrastructure (Wright 2011) together with the concept of ecosystem services (Millennium Ecosystem Assessment 2005) has strengthened the ideology of urban nature as being a set of "common goods" offering (economically) valuable services and fulfilling the needs of citizens (Buizer et al. 2016). Nature in these contexts is thus perceived to fulfil a strong utilitarian, consumerist and service-providing role in urban planning and green space design.

Subjective values of urban nature

Besides the strong role forests have played in city planning, they have had a key role in Finnish society, providing economic growth and developing cultural identity in the 19th and 20th centuries. The archetypical image of lakes and primal forests remains very strong among Finns, who like to call themselves a "forest nation". Every Finn has the right to the recreational use of commercial and privately owned forests, including collection of wild berries and mushrooms without the landowner's permission or charge (Ministry of Environment 2019). See Cocks & Shackleton (Chapter 13) for implications of such acts. Forests, due to their strong cultural and personal significance, are a common place for nature-based conflict. In the following sections, we consider a few questions: Are forests with high biodiversity value important or meaningful for urban residents? What kind of urban nature do residents value and consider meaningful? Do urban green spaces enhance interaction between people and attachment to place?

Mapping biocultural values in Helsinki

In the Internet-based PPGIS survey conducted by the city of Helsinki in 2013 (Table 5.1), 1,404 respondents marked locations as "unique urban nature" on a map of the city. A majority of respondents (64 per cent) marked forests as locations for unique urban nature. Respondents mostly valued forests as unique places to relax, enjoy nature and feel a real connection to wildlife, as well as offering a place for environmental education and a playground for children. Urban forests were also valued for traditional practices, such as collecting berries and mushrooms: "This forest patch offers real nature experiences. You can find forest berries to collect". However, only 30 per cent of all forested patches in the city received marks by respondents.

Just over a quarter of respondents (26 per cent) recognised urban parks as "unique urban nature" (Wang et al. 2019), with citizens living in the city centre (Vierikko et al. 2017) placing particular value on this aspect. Reasons given by respondents included that parks provided important meeting points and places for cultural events. Cafés were seen as essential elements of the urban park experience. Many residents indicated that parks represent "true urban nature", which included open lawns, large trees and flowerbeds: "a true English-style park with intensively managed and cultivated vegetation".

Our results reveal a positive correlation between high biodiversity and citizens' perception of uniqueness in urban forests (Wang et al. 2019). Citizens – at least those who replied to the two surveys – seem to be well aware of the biodiversity values of forests and what threatens local biodiversity. Residents' relatively good knowledge (Yli-Pelkonen & Kohl 2005) about local biodiversity values could be explained by the easy access to and regular use of forests. Green areas cover 47 per cent of the city's land surface, and 20 per cent of that consists of forests (Vierikko et al. 2014). Almost half of the population has good access to semi-natural urban forests that are generally valued and visited all year round for nature experiences and outdoor recreation, such as cross-country skiing (Neuvonen et al. 2007, Vierikko et al. 2017, Harlio & Tuhkanen 2018, Korpilo et al. 2018). Tyrväinen et al. (2007) and Hauru et al. (2015) reveal that close-to-home forests are highly valued by local residents for recreational purposes in Finnish cities.

In the second Internet-based PPGIS survey conducted by the city of Helsinki in 2018, respondents were asked to identify "typical nature in Helsinki". A total of 883 respondents gave 1,213 locations as the most "typical nature in Helsinki". Most respondents indicated forests and sea shores as typical nature with their diverse biological and structural characteristics (Figure 5.2). One respondent commented: "Old growth forests growing in rocks, covered by lichens, mosses and decaying wood. True oases!"

Perceptions of urban parks were similar to those in the 2013 survey, with only a few respondents identifying urban parks as "typical nature". In the second survey, citizens indicated an enjoyment of easy access to urban green spaces and the mixture of urban and rural atmosphere: "while swimming in the sea, you can feel like being in your summer cottage in the countryside, and at the same time have a view of a metro train". The most prominent concern expressed related to the existence of forests in the future. Residents were concerned about the disappearance or decrease of forest patches, and harm to sensitive vegetation from trampling and the intensification of forest management, as typified by this response: "Forests are biologically diverse places, their micro-level functions support life, they offer habitats for many species. Forests are the lungs of the city".

In the on-site survey conducted in 12 public parks, 600 park users were asked to indicate their favourite places in the city. Previous studies in Finland showed that residents' favourite places were located in indigenous green space, such as forest areas (Tyrväinen et al. 2007, Korpela et al. 2010), whereas our 2015 survey found that many park visitors preferred constructed parks over indigenous green spaces. Respondents mentioned not using any other green

Figure 5.2 Urban forests represent indigenous nature in Helsinki. They harbour several endangered species at the national level and are highly valued by residents. Keskuspuisto forest area in Helsinki. Source: Pitkänen, Susanna, used with personal permission.

space except the park they were visiting when the survey was conducted. This finding could imply that despite the relatively high availability of green spaces, the use of green spaces in Helsinki is fairly restricted (Harlio & Tuhkanen 2018, Neuvonen et al. 2007).

A study of interlinkages between uses, experiences and structural characteristics in 33 parks in Berlin, Bucharest, Helsinki and Lisbon revealed that nature was highly enjoyed during park visits, and the structural diversity of a park – large trees, different kinds of biotopes – increased the number of visitors who enjoyed nature (Vierikko et al. 2020), mirroring other studies on the higher appreciation of more diverse green spaces (Fuller et al. 2007, Southon et al. 2017). The study revealed that the structural characteristics of parks, especially vegetation structure, was dependent on the park's location and the socio-economic status of the surrounding neighbourhood (Vierikko et al. 2020). Previous studies have shown that parks that are high in greenness enhance residents' connectedness with local nature (Nisbet et al. 2009, Wood et al. 2018). Large trees and diverse vegetation offer residents nature experiences to which restorative benefits are attributed (Berto 2014). This could be especially significant for urban residents who do not have the opportunity or the will to visit other natural places and prefer constructed parks as places to visit. One respondent said: "Even rabbits enjoy it here, along with dogs, Frisbee throwers and sunbathers. There are trees and land here, hillocks, and a lot of space. … We refer to this by the nickname 'Paradise'".

Based on the 2013 PPGIS survey, we identified the six green spaces most recognised as unique urban nature. Respondents assigned diverse values (use and experience, existence, symbolic, bequest) to each green space, indicating the multi-layered meanings attached to these places (More et al. 1996, Gustafsson 2001, Kumar & Kumar 2008, Horlings 2015). The two green spaces, which received marks from over 100 respondents, were large forest areas. Three public parks located in the city centre received marks from over 50 respondents. The meanings attributed to these areas reflect a deep attachment to these places.

The Meri-Rastila forest area is highly valued by local residents and the biodiversity values of the area are significant. Forests in Meri-Rastila have high conservation values and are important for several bird species, polypores (bracket fungi) and vascular plants (Vierikko et al. 2014). See Cocks & Shackleton (Chapter 13) for further discussion on this. Many respondents mentioned multiple historical and biological values of the area and a concern over the possible disappearance of special characteristics. One respondent remarked:

> Meri-Rastila outdoor park combines diverse nature and ancient shoreline. These special characteristics will be lost if the recreation area will be shrinking due to the construction of a new residential area. Recreational areas need to be sufficiently large in size.

The Kivinokka area is popular for recreation and has a strong history of social inclusion as summer cottages were built in the 1920s so that working-class citizens could receive health benefits from the park (Figure 5.3). The cottages, used by urban residents, create a village-like community. The old-growth forest has a high level of biodiversity and has been included in the Nature Conservation Programme (2015). Local citizens and several NGOs have actively participated in the city planning process and campaigned against city authority plans to develop Kivinokka into a residential area. Typical remarks are: "Kivinokka should definitely be maintained; it is a gorgeous place as it is, an oasis in the middle of the city for those who do not have their own cottage in the area", and "Important recreation area for local residents, especially when city structure become denser in the future".

Southern Keskuspuisto (Central Park) is part of a larger forest corridor from the city centre to the peri-urban regions. Well patronised by residents, it is used as a thoroughfare to walk to work, to exercise in – jogging, mountain biking and cross-country skiing during winter – and to pick mushrooms and berries (Korpilo et al. 2018). The park attracts over a million visits annually (Saukkonen, pers. comm.). The forest corridor was incorporated into the city plan in 1920 and currently city authorities have plans to build new residential areas along the park edge. Commenting on these plans, residents said: "Do not touch Keskuspuisto!"; "Important lungs and veins of the city beginning

Figure 5.3 The Kivinokka area is a popular village-like recreational area in Helsinki. Source: Helsinki City Museum.

from the central area"; "The size of Keskuspuisto has already been diminished enough by previous land-use plans. The rest of the park should be protected"; "Keskuspuisto is a wonderful forest area, not like public parks, but like a true forest!"

Situated by the sea, **Kaivopuisto** Park dates back to the 19th century and is a landmark visited by tourists and citizens (Vierikko et al. 2020). It contains geologically valuable rock formations and attracts many bird species. As a representative of an English-style park it is highly valued: "Kaivopuisto park is a great park among other European parks, it is a true English-style park, where you can find rocky nature and managed, planted vegetation".

Töölönlahti, a close-to-home green space, is a popular tourist destination due to its location in the city core and along the attractive Töölö Bay. It is intensively used for jogging, working out in the park gym, dog walking, sunbathing, picnicking and playing chess. It provides habitats for a variety of bird species and one for bats. Elements of "wild nature" within Töölönlahti are appreciated, as is the year-round use of the area. Park users remarked: "Please no more building of Töölönlahti, it should be conserved as a Finnish place, not an English park"; "Töölönlahti is unique urban nature that is very nice to walk around throughout the year. It would be nice to have open cafes in wintertime as well".

Alppipuisto Park is also located in a close-to-home neighbourhood, mostly populated by young adults and students. Containing an amphitheatre and two pavilions, it is a favourite place for cultural and social events, including public concerts, which are held nearly every summer weekend. The urban forest surrounding the park provides breeding grounds for many bird species and there is a geological site within the park (City of Helsinki 2019). Typical respondent comments were: "Please don't change Alppipuisto"; "The park needs to be maintained!"; "The landscape and characteristics of Alppipuisto park resemble a European park more than other parks in Helsinki. There are some negative user groups, but the park should be developed as a green oasis and meeting point for citizens!"

Perceived values of urban parks

While forests have an important role in Helsinki (most common type of UGS, within reach of the majority of residents, valued for unique nature and restoration), parks might play a more significant role in promoting social cohesion. Local studies (e.g., Korpela et al. 2010, Silfverberg 2010) have found that residents perceive urban parks as more social environments than forests. The place-based survey conducted in urban parks in Helsinki revealed that in many cases, the use of parks resembled the use of other kinds of urban public space, such as city squares, where socio-cultural actions take place among residents (Puttonen 2017).

In the on-site park survey, we studied human-nature interactions in terms of place attachment. Our place-based study in 12 urban parks revealed that almost two-thirds (65 per cent) of the 600 park users interviewed felt that the particular park they visited was especially important to them (Puttonen 2017). The reasons given varied, but the most common factor was the close-to-home location. Many urban residents referred to "my park'" indicating a sense of ownership over their neighbourhood park. However, personal attachment was not a prerequisite, and many young adults were simply using the space to enjoy the moment, for reasons such as warm weather or good company, as typified by this statement: "now this park is important because of my breakfast, sunny weather and good feeling, but maybe it is not important tomorrow". Temporality and the dynamic use of place are thus important notions to take into consideration when discussing engagement with any specific (green) space.

The study paid special attention to the figures of speech used by park users to describe the park. According to Lakoff & Johnson (1980), figures of speech are a tool to conceptualise the experienced environment and a sign of a deeper connection to a place. We categorised the figures of speech into three groups reflecting (1) self, (2) others, or (3) the environment, and attempted to explain their meanings (Table 5.2). As most of the metaphors had to be translated from Finnish, readers must understand the loss of linguistic connotations.

The survey results and on-site observations indicated that Meurman Park had an especially strong social character. Serving as an insightful example of the "Others" category in Table 5.2, we describe Meurman Park here in detail.

Meurman Park is a small pocket-sized park located in the Käpylä neighbourhood, famous for its historical wooden houses and neighbourhood spirit. The park hosts a wooden coffee kiosk, a signature element of Helsinki, where neighbours gather and socialise and music is often played: "Everybody gathers here, it's THE place to hang out". It was the only park among the twelve surveyed with a commercial service, which clearly set it apart in terms of its socio-spatial characteristics (i.e., the atmosphere brought by the kiosk and its users) (Figure 5.4). Social reasons were the most commonly mentioned motivation to use the park, with good company and atmosphere highly valued. Users felt that the park was the heart of the neighbourhood and a place to meet others, for which they were ready to fight. While it might be accurate to compare this intimate park to a small city square from the point of view of its functionality, the greenery is essential for providing enjoyment, comfort, a place to sit, and aesthetic and restorative value. Meurman Park is therefore an interesting example of collective human-human interactions within nature.

As mentioned, urban parks are considered to be the "right kind of nature" and use of parks is strictly controlled and guided by authorities. However, there has been a marked change in the use of urban parks since the 1960s, when stepping on the lawn was forbidden. The vision for 2050 (City of Helsinki 2013) sees urban parks as multifunctional spaces that residents can use as their own "living rooms" for individual and collective action. In our place-based park observations, we observed that besides individual park use, there were social gatherings, music performances, art installations, games, exercise classes and even company staff summer days. These actions suggest that parks are valued as socially inclusive meeting places. As one respondent described: "This is a meeting place, we call it 'the five corners'. We call our friends on the phone and meet at five corners [in the park] to hang out".

In recent years, urban parks increasingly function as core locations for citizen-led events, such as Restaurant Day or Cleaning Day. The parks' public character, openness and accessibility, with no restrictive opening hours or fencing, unlike in many European parks, promote such uses. The events have not raised any notable objections among the general public or city officials, which might be due to relaxed public policies, new types of civic activism movements and a growing sense of collectivity in the urbanising cities (Mäenpää & Faehnle 2017). As Helsinki has only recently diversified in terms of urban lifestyles, services and cultures, new urban generations who are comfortable using public spaces for a range of activities are only emerging. This might suggest a stronger role for urban parks as places for social cohesion in the future.

Table 5.2 *A* List of Collected Figures of Speech Park Users Mentioned When Asked Why the Urban Park Is Important for Them

Category	*Figure of speech*	*Meanings*
Self	Living room Backyard Substitute for yard Extension of home Home park GardenUrban office "My park"	Private space and private actions extend to public space (e.g., eating, drinking, working, reading, playing board games) Sense of ownership and belonging
Others	Heart of the neighbourhoodMeeting place "Five corners" (place where people cross)Gathering spotPlace to hang outLove for home district	• Interactions with other people (both friends and strangers) • Source of community spirit • Neighbourhood centre • Inclusiveness
Environment	OasisParadiseCathedral-likeLungsRespiration spotPlace to breatheSpirit lifter	• Nature and its sacredness • Place to be calm and recover • Aesthetics

The last column provides possible explanations for their perceived meanings.

Figure 5.4 The wooden coffee kiosk in Meurman Park is a popular place for local people to gather and socialise. Source: Jaatsi, Mia.

Conclusion

> The city is first and foremost for people. There is space for diverse nature in Helsinki: for primary, constructed and culturally dominated nature. We have a long cultural history that has had a positive influence on nature. The human-nature interaction should not be in conflict.
>
> Respondent, survey in 2018

The role of nature in urban planning has shifted from a mediator of sensitive human-nature relationships towards a more human-produced, functionalised and commercialised service within cities. Citizens still have deep bonds with forests in Helsinki and are quite aware of the levels of biodiversity of urban forests. At the same time, urban parks have become a place for citizen activism and provide "living room" for the growing population.

With Helsinki growing rapidly through national and international migration, the questions raised are: What kinds of urban green space should there be in the future city? Should the emphasis be to maintain a deep bond with the original types of urban green space to enhance citizen's stewardship towards nature, or should urban green spaces be designed and managed primarily for people as places for active and socially inclusive place-making? We argue that it's not just a question of having the "right kind of UGS" for promoting human-nature relationships or social cohesion. The biocultural diversity approach strongly argues that development of urban green spaces should be defined in the local context, so as to strengthen the rights, knowledge, identities and resources of local residents and also of migrants who have not yet settled down in the city (Maffi 2010, Davidson-Hunt et al. 2012,

Njwambe et al. 2019). We also argue that non-human urban dwellers have a right to inhabit urban green spaces.

Although urban nature can be controlled and designed by planners, designers or engineers, citizens will develop their own perceptions, meanings and values towards nature. Urban green space that is too controlled or neutralised may leave people emotionally empty, lacking a sense of place or bonds towards the surrounding environment (Lewicka 2011). Forest areas can be understood as "non-human-controlled" spaces free from social norms or national definitions, and thus as universal places for all (Jay & Schraml 2009). Moreover, young urban generations are challenging the traditional uses of forests and the values associated with them, such as peace, silence and respiration. One curious example can be witnessed during the full moon in the summer, when a hidden, grassroots-initiated, full-moon party takes place in a large forest area in Helsinki.

In wild nature, such as urban forests, individuals, especially children, have an opportunity to feel a connection with nature, to be innovative and impulsive and to create by themselves. Can different generations and user groups develop their own, co-created human-nature interaction, or human-human interaction within nature? Attention must be paid to different types of bonds with and needs of nature, as all new inhabitants or emerging urban generations may not share the legacy of a "forest nation" (Buijs et al. 2009).

In turn, planning and design of urban green space play a key role in stimulating (or hindering) interactions between people and nature in constructed environments (Jennings & Bamkole 2019). The humanistic approach in architecture has yielded successful stories which illustrate how to make places more social, alive and inviting for residents to interact with the place itself and with other people (Gehl 2010). In cities, constructed urban parks have the potential to be important places for residents, which promote social cohesion, active place-making and a sense of nature, without jeopardising the deep bond with urban forests or seashores. Our work shows that in addition to the traditional emphasis on the wellbeing and health benefits arising from the physical environment of urban green space, it is also essential to consider the benefits emerging from the social environment as part of the human-nature interaction and to develop deep bonds with nature, places and people.

Acknowledgements

We want to thank Professor Christopher Raymond for his valuable comments on the first version of the chapter.

References

Ahern, J., Cilliers, S. & Niemelä, J. 2014. The concept of ecosystem services in adaptive urban planning and design: A framework for supporting innovation. *Landscape & Urban Planning*, 125: 254–259.

Berto, R. 2014. The role of nature in coping with psycho-physiological stress: A literature review on restorativeness. *Behavioral Sciences*, 4(4): 394–409.

Biodiversity Action Plan. 2008. Helsingin luonnon monimuotoisuuden turvaaminen. Toimintaohjelma 2008–2017. Available at: https://www.hel.fi/helsinki/fi/asuminen-ja-ymparisto/luonto-ja-viheralueet/monimuotoisuus/monimuotoisuuden-turvaaminen/.

Buijs, A.E., Elands, B.H. & Langers, F. 2009. No wilderness for immigrants: Cultural differences in images of nature and landscape preferences. *Landscape & Urban Planning*, 91(3): 113–123.

Buizer, M., Elands, B. & Vierikko, K. 2016. Governing cities reflexively – The biocultural diversity as an alternative for ecosystem services. *Environmental Science & Policy*, 62: 7–13.

Byrne, J. & Wolch, J. 2009. Nature, race, and parks: Past research and future directions for geographic research. *Progress in Human Geography*, 33(6): 743–765.

Chiesura, A. 2004. The role of urban parks for the sustainable city. *Landscape & Urban Planning*, 68: 129–138.

City of Helsinki. 2013. Helsinki City Plan, Vision for 2050, Urban Plan – The New Helsinki City Plan. Reports by the Helsinki City Planning Department general planning unit 2013: 23. 84.

City of Helsinki. 2014. Helsingin ulkomaalaisväestö vuonna 2014. Helsingin tilastoja 2014:36. 60. (Foreign residents in Helsinki in 2014).

City of Helsinki. 2019. Vihreät sylit – internet page. Available at: https://vihreatsylit.fi/en/kaivopuisto/.

City of Helsinki and Ministry of Environment. 2005. *Eco-Viikki – Aims, Implementation and Results*. Dark Ltd, Vantaa. 50.

Comstock, N., Dickinson, L., Marshall, J., Soobader, M.-J., Turbin, M., Buchenau, M. & Litt, J.S. 2010. Neighborhood attachment and its correlates: Exploring neighborhood conditions, collective efficacy, and gardening. *Journal of Environmental Psychology*, 30: 435–442.

Cranz, G. 1982: *The Politics of Park Design: A History of Urban Parks in America*. Cambridge, MA: Harvard University Press.

Cvejić, R., Eler, K., Pintar, M., Železnikar, Š., Haase, D., Kabisch, N., et al. 2015. *A Typology of Urban Green Spaces, Ecosystem Provisioning Services, and Demands*. EU FP7 project GREEN SURGE, deliverable D3.1. Retrieved from: www.greensurge.eu

Davidson-Hunt, I., Turner, K., Te Pareake Mead, A., Cabrera-Lopez, J., Bolton, R., Idrobo, J., Miretksi, I., Morrison, A. & Robson, J. 2012. *Biocultural design: a new conceptual framework for sustainable development in rural indigenous and local communities. Sapiens* [Online], 5, 2, 2012, Online since 12 August 2012, connection on 30 April 2019. Retrieved from: http://journals.openedition.org/sapiens/1382.

Elands, B., Wiersum, K., Buijs, A. & Vierikko, K. 2015. Policy interpretations and manifestations of biocultural diversity in urbanized Europe: Conservation of lived biodiversity. *Biodiversity & Conservation*, 24: 3347–3366.

Elands, B.H., Peters, K.B. & de Vries, S. 2018. Promoting social cohesion – Increasing wellbeing. In van den Bosch, M. & Bird, M. (Eds.), *Oxford Textbook of Nature and Public Health* (pp. 116–121). Oxford, UK: Oxford University Press.

Eurostat. 2016. *Urban Europe – Statistics on Cities, Towns and Suburbs*. European Commission, 282. doi: 10.2785/91120.

Fuller, R.A., Irvine, K.N., Devine-Wright, P., Warren, P.H. & Gaston, K.J. 2007. Psychological benefits of greenspace increase with biodiversity. *Biology Letters*, 3(4): 390–394.

Gehl, J. 2010. *Cities for People*. Washington, DC: Island Press.

Gustafsson, P. 2001. Meanings of place: Everyday experience and theoretical conceptualisations. *Journal of Environmental Psychology*, 21: 5–16.

Haase D. & Gläser, J. 2009. Determinants of floodplain forest development illustrated by the example of the floodplain forest in the District of Leipzig. *Forest Ecology & Management*, 258: 887–894.

Hajer, M.A. 1996. *The Politics of Environmental Discourse: Ecological Modernization and the Policy Process*. New York, NY: Oxford University Press.

Harlio, A. & Tuhkanen, H. 2018. NATOURS 2016–2018 kaupunkiluontohanke Helsinki & Tallinna (Urban nature project in Helsinki and Tallinn). Oral presentation 03.05.2018.

Hauru, K., Eskelinen, H., Yli-Pelkonen, V., Kuoppamäki, K. & Setälä, H. 2015. Resident's perceived benefits and the use of urban nearby forests. *International Journal of Applied Forestry*, 2: 1–23.

Hirvensalo, V. 2006. Modernin kaupungin luonto muutoksessa – kahdeksan esimerkkiä suomalaisesta asuinalueesta. Turun yliopiston julkaisuja, sarja C osa 245. Academic dissertation (in Finnish with English abstract: The role of nature in the modern city – eight examples). 263.

Horlings, L.G. 2015. Values in place: A value-oriented approach toward sustainable place-shaping. *Regional Studies & Regional Sciences*, 2(1): 256–273.

Hurme, R. 1991. Suomalainen lähiö Tapiolasta Pihlajanmäkeen (in Finnish). *Suomen tiedeseura* 142.

Hyvärinen, E., Juslén, A., Kemppainen, E., Uddström, A. & Liukko, U.-M. (Eds.). 2019. *The 2019 Red List of Finnish Species*. Helsinki: Ympäristöministeriö & Suomen ympäristökeskus. 704 p.

Jay, M. & Schraml, U. 2009. Understanding the role of urban forests for migrants: Uses, perception and integrative potential. *Urban Forestry & Urban Greening*, 8(4): 283–294.

Jennings, V. & Bamkole, O. 2019. The relationship between social cohesion and urban green space: An avenue for health promotion. *International Journal of Environmental Research & Public Health*, 16(3): 452.

Jong, M., Joss, S., Schraven, D., Zhan, C. & Weijnen, M. 2015. Sustainable–smart–resilient–low carbon–eco–knowledge cities: Making sense of multitude of concepts promoting sustainable urbanization. *Journal of Cleaner Production*, 109: 25–38.

Kabisch, N., Qureshi, S. & Haase, D. 2015. Urban nature: Human-environment interactions in urban green spaces – contemporary issues and future prospects. *Environmental Impact Assessment Review*, 50: 25–34.

Korpela, K., Ylén, M., Tyrväinen, L. & Silvennoinen, H. 2010. Favorite green, waterside and urban environments, restorative experiences and perceived health in Finland. *Health Promotion International*, 25(2): 200–209.

Korpilo, S., Jalkanen, J., Virtanen, T. & Lehvävirta, S. 2018. Where are the hotspots and coldspots of landscape values, visitor use and biodiversity in an urban forest? *PLoS One*, 13: 9–19.

Kowarik, I. 2011. Novel urban ecosystems: biodiversity and conservation. *Environmental Pollution*, 159: 1974–1983.

Kühn, I., Brandl, R. & Klotz, S. 2004. The flora of German cities is naturally species rich. *Evolutionary Ecology Research*, 6: 749–764.

Kumar, M. & Kumar, P. 2008. Valuation of the ecosystem services: A psycho-cultural perspective. *Ecological Economics*, 64(4): 808–819.

Kurtto, A. & Helynranta, L. 1998. Helsingin kasvit: Kukkivilta kiviltä metsän syliin. *Helsingin kaupungin ympäristökeskus*, Yliopistopaino, Helsinki, 400.

Kyttä, M., Broberg, A., Tzoulas, T. & Snabb, K. 2013. Towards contextually sensitive urban densification: Location-based soft GIS knowledge revealing perceived residential environmental quality. *Landscape & Urban Planning*, 113: 30–46.

Lakoff, G. & Johnson, M. 1980. *Metaphors We Live by*. Chicago, IL: University of Chicago Press, pp. 256.

Lehvävirta, S. and Rita, H. 2002. Natural Regeneration of Trees in Urban Woodlands. *Journal of Vegetation Science* 13(1): 57–66.

Lewicka, M. 2011. Place attachment: How far we have come in the last 40 years? *Journal of Environmental Psychology*, 31: 207–230.

Mäenpää, P.A. & Faehnle, M.E. 2017. *Civic activism as a resource for cities*. Kvartti: Helsingin kaupungin tietokeskuksen neljännesvuosijulkaisu 1/2017, pp. 68–81.

Maffi, L. 2010. What is biocultural diversity? In Mafi, L. & E. Woodley (Eds.), *Biocultural Diversity Conservation: A Global Source Book*, (pp. 3–11). London, UK: Earthscan.

Mapita. 2014. Helsinki 2050 -kysely. TULOKSET (PPGIS survey in 2013 in Finnish). Available at: http://www.yleiskaava.fi/wp-content/uploads/2014/02/mapita_ksv_raportti_10022014.pdf.

Mapita. 2019. Rikas ja rakas Helsingin luonto. Kysely Helsingin luonnon monimuotoisuudesta 2018. Yhteenveto kyselyn tuloksista (Survey about urban biodiversity in 2018 in Finnish). Available at: https://www.hel.fi/static/ymk/lumo/tulokset-LUMO-asukaskysely-2019.pdf.

Millennium Ecosystem Assessment (MA). 2005. Ecosystems and human well-being: synthesis. Washington: Island Press. 155 pp.

Ministry of Environment 2005. Kysely suomalaisten luontosuhteesta. Kyselyn tulosten koonti 28.6.2018 (in Finnish). Available at: https://www.ym.fi/fi-FI/Luonto/Iakkaammat_kantavat_nuoria_enemman_huolt(47263).

Ministry of Environment. 2019. Everyman's Rights. Available at: https://www.ymparisto.fi/en-US/Nature/Everymans_rights(27721).

More, T., Averill, J.R. & Stevens, T.H. 1996. Values and economics in environmental management: A perspective and critique. *Journal of Environmental Management*, 48: 397–409.

Nature Conservation Programme. 2015. Helsingin luonnonsuojeluohjelma 2015–2024. Ympäristökeskus, Helsingin kaupunki. Available at: https://www.hel.fi/static/ymk/lso.html.

Neuvonen, M., Sievänen, T., Tönnes, S. & Koskela, T. 2007. Access to green areas and the frequency of visits: A case study in Helsinki. *Urban Forestry & Urban Greening*, 6(4): 235–247.

Nisbet, E.K., Zelenski, J.M. & Murphy, S.A. 2009. The nature relatedness scale: Linking individuals' connection with nature to environmental concern and behaviour. *Environment & Behavior*, 41: 715–740.

Njwambe, A., Cocks, M. & Vetter, S. 2019. *Ekhayeni*: Rural–urban migration, belonging and landscapes of home in South Africa. *Journal of Southern African Studies*, 45(2): 413–431.

Puttonen, M. 2017. Helsinkiläisten yhteiset olohuoneet: Kaupunginosapuistojen merkityksiä sosiaalisena julkisena tilana (in Finnish with English abstract). Master thesis, University of Helsinki.

Pyšek, P. 1998. Alien and native species in Central European urban floras: A quantitative comparison. *Journal of Biogeography*, 25: 155–163.

Saukkonen, T. 2011. Helsingin kaupungin luonnonhoidon linjaus (Helsinki Nature Management Strategy). Helsinki, Finland: Public Works Department (in Finish).

Schönach, P. 2015. Historical paths of environmental injustice: A century of placing industrial facilities in Helsinki, Finland. *The International Journal of Justice & Sustainability*, 21(4): 397–413.

Silfverberg, O. 2010. Kaikkien oma tila: Kaupunkiluonnon merkityksiä itähelsinkiläisten asukkaiden näkökulmasta, Master thesis, University of Helsinki.

Southon, G.E., Jorgensen, A., Dunnett, N., Hoyle, H. & Evans, K.L. 2017. Biodiverse perennial meadows have aesthetic value and increase residents' perceptions of site quality in urban green-space. *Landscape & Urban Planning*, 158: 105–118.

Statistical Yearbook of Helsinki 2017. Helsingin kaupunki, kaupunginkanslia, kaupunkitutkimus ja -tilastot. 315 sivua. Available at: https://www.hel.fi/hel2/tietokeskus/julkaisut/pdf/18_01_02_tilastollinen_vuosikirja2017.pdf.

Tyrväinen, L., Mäkinen, K. & Schipperijn, J. 2007. Tools for mapping social values of urban woodlands and other green areas. *Landscape & Urban Planning*, 79: 5–19.

Vierikko, K., Niemelä, J., Salminen, J., Jalkanen, J. & Tamminen, N. 2014. *Sustainable green infrastructure of Helsinki – urban ecological research report and recommendations for the Helsinki master plan 2014*. Helsingin kaupunkisuunnitteluviraston yleissuunnitteluosaston selvityksiä 2014. 113 s. (In Finnish with English summary) Available at: https://www.hel.fi/hel2/ksv/julkaisut/yos_2014-27.pdf.

Vierikko, K., Niemelä, J., Elands, B., Buijs, A., Haase, D., Kabisch, N., Kowarik, I., Fischer, L., Luz, A., Olafsson Stahl, A., Andersson, E., Száraz, L. & Konijnendijk van den Bosch, C. 2016. Considering the ways biocultural diversity helps enforcing the urban green infrastructure in times of urban transformation. *Current Opinion in Environmental Sciences*, 22: 7–12

Vierikko, K., Andersson, E., Branquinho, C., Elands, B., Fischer, L.K., Goncalves, P., Grilo, F., Haase, D., Ioja, C., Kowarik, I., Lindgren, J. M., Mendes, R., Niemelä, J. K., Pieniniemi, M., Principe, A., Puttonen, M., Santos-Reis, M., Teixeira, D., Vieira, J. & Yli-Pelkonen, V.J. 2017. Identifying, quantifying and qualifying biocultural diversity: Assessment of biocultural diversity. GREEN SURGE Technical Report; vol. September 2017. Available at: http://greensurge.eu/working-packages/wp2/.

Vierikko, K., Goncalves, P., Haase, D., Elands, B., Ioja, C., Jaatsi, M., Lindgren, J., Pieniniemi, M., Grilo, F, Santos-Reis, M., Niemelä, J. and Yli-Pelkonen, V. 2020. Biocultural diversity (BCD) in European cities – Interactions between motivations, experiences and environment in public parks. *Urban Forestry & Urban Greening*, 48: 126501. doi: 10.1016/j.ufug.2019.126501.

Votsis, A. 2017. Planning for green infrastructure: The spatial effects of parks, forests, and fields on Helsinki's apartment prices. *Ecological Economics*, 132: 279–289.

Wang, Y., Kotze, J., Vierikko, K. & Niemelä, J. 2019. What makes urban greenspace unique – Relationships between citizens' perceptions on unique urban nature, biodiversity and environmental factors. *Urban Forestry & Urban Greening*, 52: 1–9.

Wolch, J., Byrne, J. & Newell, P. 2014. Urban green space, public health, and environmental justice: The challenge of making cities 'just green enough'. *Landscape & Urban Planning*, 125: 234–244.

Wood, E., Harsant, A., Dallimer, M., Cronin de Chavez, A., McEachan, R. & Hassall, C. 2018. Not all green space is created equal: Biodiversity predicts psychological restorative benefits from urban green space. *Frontiers in Psychology*, 9: 2320.

Wright, H. 2011. Understanding green infrastructure: The development of a contested concept in England. *Local Environment*, 16: 1003–1019.

Yli-Pelkonen, V. & Kohl, J. 2005. The role of local ecological knowledge in sustainable urban planning: Perspectives from Finland. *Sustainability: Science, Practice & Policy*, 1(1): 3–14.

6 Decolonisation of nature in towns and cities of South Africa

Incorporation of biocultural values

Michelle Cocks, Charlie Shackleton, Lindsey Walsh, Duncan Haynes Amanda Manyani and Dennis Radebe

Introduction

Ways of conceptualising the world around us and being in this world are defined by an ontological understanding. Within a Eurocentric ideological understanding, nature is positioned to be opposite to culture, i.e., nature is considered as "other" of which humans are not a part. Modernity is perceived as the antithesis of nature as processes of production, metabolism and expansion of modern cities represent attempts to tame and control nature. In turn, cities have become viewed as agents of development and change, promoting ideals of progress, thinking and innovation (Jayne 2005). Eurocentric ideals are framed as the forerunners of these processes and have come to influence international policies, global governance, alliances and networks which have in turn informed the design and governance of cities and influenced all aspects of urban liveability (Bouteligier 2011), including how urban natures are defined and constructed and the wellbeing benefits derived from them. These ideals continue to be enforced through formal and social media which produces a constant stream of information not only about what nature is but also about the appropriate ways to use, control or alter it (Castree 2013, p. 12). These constructs of nature and the associated values have been given global priority through sustainable development goals that include improving the liveability and quality of life of urban residents (Haq 2011). The Council of Europe at the 2000 European Landscape Convention in Florence stated that to achieve sustainable development, cities and towns should be "based on a balanced and harmonious relationship between social needs, economic activity and the environment" (Loures et al. 2007, p. 172). Within this context, public green spaces have been conceptualised as spaces that are designed, managed and maintained to provide opportunities to appreciate nature, as well as for recreation and sport (Zhang et al. 2013, Bertram & Rehdanz 2015). The benefits have been outlined as contributing to urban residents' physical and mental health, increasing biodiversity and improving the air quality of cities (Loures et al. 2007, p. 172).

Urban areas well endowed with various types of greenspaces are considered to be indicative of a healthy environment for its residents and wildlife (Swanwick et al. 2003). This has resulted in such spaces becoming legislated for in many countries. The association of an open space with an urban park dates back to the Industrial Revolution in Britain. Living conditions in urban areas during this period for poor and working-class families were particularly dire. The creation of urban open spaces become a priority to improve the quality of air in urban areas (Walker & Duffield 1983). Around the same time, the "myth of the rural arcadia" began to emerge among the middle and upper classes of Europe fuelling a desire to reconstruct the countryside in town (Walker & Duffield 1983).

An approach which is not grounded in a Eurocentric construct is biocultural diversity. Biocultural diversity is based on the premise that nature and culture are not in opposition with each other but are inextricably linked (Posey 1999). Within such a context, nature is perceived as being "that which surrounds [the individual], [making] all relations [and therefore knowledge] ecological" (Milton 2002, pp. 115–116). Such a perspective acknowledges that individuals and nature "embrace an irreducible system; that is, the person is part of the environment and in the same way, the environment is part of the person" (Descola & Palsson 1996, p. 18). Such insights have highlighted the need to explore more nuanced understandings of the ways that humans live in, understand and shape their environment and the nonhuman species with which they share their world (Coates et al. 2006) and how these experiences contribute to cultural ways of being and meaning-making (Feld & Basso 1999; Ingold 2000). Since the mid-1990s, social scientists have challenged modern intellectual traditions working from the dualistic premise (e.g., Descola & Pálsson 1996, Milton 2002, Ingold 2000) and in response calls have recently been made, particularly from the Global South, to reflect other ways of being in this world.

The concept of biocultural diversity was originally proposed to denote the intimate link between nature and culture in relation to indigenous communities (Posey 1999), especially those living in areas of high biodiversity (Maffi 2005, Maffi & Woodley 2010). Within this frame, cultural diversity became portrayed as being threatened by environmental degradation and exploitation resulting from unbalanced tenure and governance relations and economic development-related processes of acculturation and socio-economic change. Changing people's livelihoods, worldviews and value systems became portrayed as leading to a breakdown in local knowledge, practices and languages which would ultimately lead to a loss in cultural diversity (Posey 1999, Woodley 2010, pp. 131–132). The identification of people and places as endangered led to the creation of platforms to address the global decline in biocultural diversity (Merçon et al. 2019). Many indigenous and environmental movements, civil society organisations and engaged scholars have and continue to be involved in protecting indigenous rights in the struggle for land and access to natural resources in attempts to combat the loss of biocultural diversity (Merçon et al. 2019). Local knowledge systems have in turn became increasingly recognised as dynamic and relevant to endogenous processes of development (Posey et al. 1984, Davidson-Hunt et al. 2012).

Since its inception, the concept of biocultural diversity has been modified and broadened in scope. It is now understood to include cultural dynamics as a vital component in the way in which communities and societies live and associate with biodiversity (Pretty et al. 2009). It gives recognition to "the interweave of biological and cultural diversity, people and places, and the continuing adaptation and co-evolution between landscapes and ways of life" (Laird et al. 2011, p. 276) and to humankind's accumulated reserve of learned responses to the environment that make co-existence and self-recognition possible (Posey 1999). Such learned responses do not only include value–practice systems of traditional and indigenous societies in respect to "wilderness" areas, but also the value–practice systems of "modernised" societies who live with biodiversity in human-modified landscapes that reflect a co-evolution between nature and society (Cocks 2006, Cocks & Wiersum 2014, Cocks et al. 2017). The concept has also become increasingly used to acknowledge the diversity of manifestations that are reflected in cultural heritage sites and cultural landscapes (Taylor & Lennon 2011, Rössler 2012, Pungetti 2013). There is now a growing awareness, even in Europe and other economically advanced regions, that a great diversity of manifestations of biocultural diversity exists within both rural and urban landscapes. This recognition resulted in the 2014 Florence Declaration, explicitly indicating the need to acknowledge "the vital importance of cultural and biological diversity for present and future generations and the wellbeing of contemporary societies in urban and rural settings" (Elands et al. 2015). There is now far more acceptance of recognising the ways in which humans live with biodiversity within a diversity of contexts and settings. The term "biocultural creatives" has been introduced to give recognition to "groups of people who, driven by an engagement with society and nature, create new cultural models and practices for interaction with biodiversity" (Elands & van Kloppen 2012, Buizer et al. 2016). As highlighted by Buizer et al. (2015), to understand the diversity of local practices it is important to recognise different cultural orientations, related knowledges and value frameworks of people. Such understandings have emerged in response to the increase in migrant communities living in cities in the Global North, many of whom have different values and needs of nature to the various forms that are offered and provided for within the cities in which they now live. To accommodate this diversity of needs, calls have increasingly been made to incorporate and reflect culturally appropriate designs, elements and features of all ethnic groups within natural areas. For example, in the Netherlands, the Dutch government is attempting to facilitate inter-ethnic interactions by funding more public green areas and parks and by including designs and elements that reflect the needs of all local inhabitants (Peter et al. 2010). Such attempts have been used as a tool to enhance place attachment and wellbeing among migrant communities as design elements have included aspects that reflect components of their heritage (Peter et al. 2010).

Expressions of nature and bioculture in Qonce and Komani, South Africa

The framing of biocultural diversity from an ontological premise provides a platform to draw attention to how other ontological understandings of the human–nature relationship exist and how many of these continue to remain invisible due to the hegemony of certain knowledge and value systems (Merçon et al. 2019). This chapter attempts to highlight how local indigenous understandings of nature remain largely hidden and unaccommodated for within the Eurocentric constructs of urban nature in two former colonial South African towns. Such understandings have and continue to remain unaccommodated in past and present constructs of urban nature. To demonstrate this disregard, this chapter provides an account of the types of urban nature found within the two towns until recently known as King William's Town (now Qonce) and Queenstown (now Komani), and how the region's colonial, apartheid and post-apartheid history has had an impact on the interpretation and manifestations of the urban natures represented. This is followed by a descriptive account of the lived experiences of nature by urban indigenous Xhosa residents living in the two towns. Attention will be given to how the types of nature found in the towns of Qonce and Komani are misaligned to the values and understanding of nature that isiXhosa-speaking residents have. Finally, a call is made for urban nature planning and management to be more inclusive of the ways that urban residents engage and interact with nature. To facilitate how urban natures can be re-imagined within cities and towns, particularly in the Global South, an argument is made for biocultural diversity to be used as an approach.

To gain perspective and understanding of the local associations and meanings attributed to urban nature and the types of urban nature found within the towns of Qonce and Komani, material collected by the authors is reflected on. Qonce and Komani both represent two former colonial towns in the Eastern Cape Province of South Africa. Material reflected on included field data and data sets collected through in-depth interviews, household surveys and recordings of observations made at the various types of urban natures found in the two towns. Within each of the towns the different types of urban natures were visited to record their characteristics. In-depth interviews were carried out among household members living in townships (townships are areas of towns that used to be reserved for Black South Africans during the racist Apartheid era, and were, and remain, poorly serviced with regard to basic needs such as housing, water, electricity, green spaces and economic opportunities), to collect stories, meanings and associations related to these different types of urban nature. Additional surveys made it possible to determine how widespread such associations and meanings are. The recordings and observations have been contextualised in the formal and informal literature. A summary of the different types of urban nature that are available in the public domain in each of the two towns is provided below.

Public urban nature spaces in Qonce and Komani

Types of urban nature

Fifteen different types of urban nature spaces were recorded in Qonce and Komani. Specific attention is drawn to those found in the public domain and fall under the jurisdiction of the local municipality, including botanical and memorial gardens and local commonages. The areas covered by each of these make up the largest portion of the amount of public green space in each of the two towns (Radebe 2018). Each of these types of nature is deeply embedded within Eurocentric constructs and have been heavily influenced by the region's colonial and apartheid history.

Public formal urban nature(s) – gardens and parks

The establishment of public gardens within Qonce and Komani dates back to the mid-1800s when both towns were established. Qonce, originally a mission station, was established in 1847 on both sides of the Buffalo River and served as the military headquarters for German settlers in what was known as British Kaffraria in 1857 (Zituta 1997). It later served as an administrative and wool processing hub for many of the surrounding farming districts. Komani was proclaimed a colonial settlement by George Cathcart in 1853 and named Queenstown after Queen Victoria. The town was originally designed to be a military outpost to protect the British settlers during the Frontier Wars. The town's central business district (CBD) is planned around a central hexagonal structure which was built for European settlers to use as a defensive encampment (laager) during times of trouble. In 1880, a railway line between the port in East London and Komani was built, which enhanced its economic growth (Burman 1984). Komani became a regional centre of industry, commerce and education that boasted sophisticated infrastructure in its CBD.

To elevate the towns' status and importance, local residents and authorities invested in the replication of public amenities found in the Europe. For example, in 1861, Sir George Grey donated a piece of land for the purposes of a botanical garden in Qonce. In 1865, a Deed of Grant was officially issued by the Colonial Government to create a "public pleasure garden". Further funds were raised by local residents of European settler descent. Plants and seeds were obtained from various parts of the British Empire and Central and South America, to populate the 1.6-ha garden. The garden contained ponds, an ornamental stream, a bandstand and the town's first public swimming pool. The Borough Band performed regularly, and it was a popular picnic spot for settler descendants and home to an assortment of birds and large trees (pers. comm. Stephanie Victor 2019). The Walter Everitt Sunken Gardens were established for the same purposes in Komani, in the early 1900s. The hexagon in Komani became a public space and was used as a marketplace. A fountain and garden were later added (Voss 2012). Today, the area is designated as a park and includes a central pathway with tightly packed rows of plants and smaller walkways, a garden style typically found in England (Favretti 1971). Still today, plants grown in the garden represent only European ornamental species. Scattered throughout both towns are several memorial gardens. A memorial was erected in Qonce cemetery to commemorate the European soldiers who died during the 1877–1878 Frontier Wars.

The 1930s in South Africa saw the removal of Black urban residents from towns into designated peripheral township areas; such acts were enforced through the racially discriminatory legislation motivated by the ideologies of the apartheid government (Wentzel & Tlabela 2006). Housing for workers initially only included single men's hostels in township areas. Later four-roomed houses were built for families (Zituta 1997). Throughout the apartheid period (1940s–mid-1990s) the government failed to provide sufficient housing for the demand that existed. By the 1990s 75 per cent of household residential plots in Qonce townships contained informal backyard shacks, doubling the amount planned for in terms of service delivery (Zituta 1997). Mlungisi township was established in Komani in 1959 to accommodate the growing number of Black urban residents. The number increased drastically when White-owned farms were bought by the apartheid state to create the ethically defined homelands of Transkei and Ciskei (Bank 2010). Housing to accommodate the expanding population remained inadequate, and Mlungisi became a slum. During the apartheid period, little attempt was made to provide Black urban residents access to public nature. By law, Black residents could not access the public gardens located in the CBD or suburbs reserved for whites.

Both towns featured prominently in political revolts against apartheid and played a role in the liberation struggle. Qonce is renowned for being the birthplace of the Black consciousness leader Steve Biko. Protests for improved education were held in Qonce, the most famous being the Bisho Massacre, which resulted in 28 people being killed and an unconfirmed number wounded at the Bisho stadium on the 7th of September 1992 (Bond & Mottiar 2013). The revolt became a pivotal spark of unrest portending the eventual demise of the apartheid government. Political revolts were also initiated in Mlungisi as part of the anti-apartheid struggle. In 1985, local Black residents demonstrated their unhappiness by boycotting White-owned shops to illustrate how much White-owned businesses relied on the Black working class.

Tension against the Apartheid government grew (Skinner 2017), culminating in a protest later in the same year against the atrocious living conditions in townships. The apartheid police counterattacked, and 11 unarmed residents were killed in what became known as the Queenstown Massacre.

Post-1994 saw the new democratically elected government wanting to pay homage to residents of Qonce and Komani who fought and died in the liberation struggle. New memorial gardens were established within both the township and former White CBD areas. For example, to commemorate the role that Steve Biko played in the liberation struggle, the Qonce local municipality erected a tombstone in Ginsberg cemetery. To pay homage to all who had died and contributed to the struggle, Steve Biko's family requested that the entire graveyard be upgraded and be maintained as a garden of remembrance. The garden remains maintained by the Steve Biko Trust and is open to the general public (Figure 6.1). To pay homage to residents who died in the Queenstown Massacre, remembrance plaques listing all their names have been erected in Nonzwakazi Methodist churchyard (which used to be a meeting place to initiate protest action). To commemorate the many Xhosa people who starved to death after the infamous Cattle Killing in 1857 (Peires 1990), a memorial was erected in 2006 by the Amathole District Municipality in Qonce (Figure 6.2). The memorial depicts the horns of a bull and is situated in an open space behind the War Memorial of the British soldiers that died during the Frontier Wars. These memorial sites are marketed as tourist attractions, and they appear on municipal websites as places to visit to appreciate the history of each town.

All parks are currently managed by the Department of Sports, Recreation, Arts and Culture. Facilities include children's play structures, tables, benches, braai (barbecue) and ablution facilities In 2015, an upgrade of the Botanical Garden in Qonce included improving its recreational facilities (Buffalo City Metro 2018) (Figure 6.3). The garden has subsequently been fenced, and access is only provided for formal functions. Public parks found in township areas of Qonce and Komani are prone to vandalism and are poorly serviced and maintained. Refuse is not regularly collected.

Figure 6.1 Steven Biko Memorial Garden. Source: von Tonder, Ella.

Figure 6.2 Cattle Killing Memorial erected by the local municipality. Source: von Tonder, Ella.

Local commonages

Local municipal commonages were established during the colonial period on the outskirts of most towns and cities in the Eastern and Western Cape provinces of South Africa (Kepe et al. 2015) including Qonce and Komani. They were established as pasture areas to keep transport and draught animals (Atkinson 2005). Once mechanised transport became the norm, poor and working-class White families were allowed to keep their livestock on the commonage (Anderson & Pienaar 2003). The regulatory use of the commonage later changed from providing a space for holding transport and domestic animals to providing a source of revenue for local municipalities, through leasing commonage land to commercial farmers to generate an income (Anderson & Pienaar 2003; Atkinson 2005). With the demise of apartheid in 1994, attempts to reform urban nature included increasing Black urban residents' access rights to commonages by terminating farmers' leases on commonages. This shift was officially recognised in the *White Paper on South African Land Policy of 1997*, which stipulated that "poor people need to gain access to grazing land and small arable or garden areas to supplement their income and to enhance household food security" in urban areas (Department of Land Affairs 1997). As part of the reddress initiative, 7 000 hectares of commonage land were redistributed to Komani residents (Wotshela 2009). Currently, commonage spans hundreds to thousands of hectares per town and is used for grazing, collection of wild resources, sports and cultural observances (such as entering manhood

Figure 6.3 Botanical Garden in Qonce. Source: von Tonder, Ella.

initiation). They are significant contributors to the household economy of many Black, urban residents (Davenport et al. 2012). Typically, there is little or no active management by urban municipalities, and many commonages have undergone fundamental changes in plant species composition and productivity (Puttick et al. 2011).

Meanings and associations attributed to urban nature by isiXhosa-speaking residents

Through the in-depth interviews, narratives and stories were collected on meanings and associations attributed to different types of urban nature, the majority of which relate to or reflect a particular way of being which is informed by a particular ontological understanding. A significant portion also reflect a sense of pride in reaping produce from nature and being in nature to escape and socialise (also see Nesbitt et al. (Chapter 10) and Elands et al. (Chapter 11)). The meanings and associations collected have been described and discussed under the following three themes: (1) the way of being "umXhosa", (2) a place to escape and socialise and (3) benefits.

The way of being "*umXhosa*"

Urban residents attributed certain kinds of activities and certain places within nature as facilitating a particular way of being "umXhosa". Each of these activities is also linked to a particular knowledge set that, in turn, evoked emotions of feeling a sense of pride, being linked to one's roots, being with ones' ancestors, and a sense of inner peace and wellbeing (*impilo*). To attain *impilo* implies having fulfilling and healthy relationships with God and ancestors, as well as with other people and the natural world (Edwards 2011) and being ritually pure and free of metaphysical pollution (*umlaza*) (Dold & Cocks 2012). For many residents being in nature was also described as bringing about a particular state of wellbeing.

Bonsile, a 67-year-old man, explains

> "who I am as a Xhosa man is because of nature, if I didn't do my initiation I would not be a man, I still go to the *ihlathi* [forest]... even the *ihlathi* in the township, because sometimes it reminds me of my ancestor".

Directly translated *ihlathi lesiXhosa* means "the Xhosa forest". It refers to local indigenous vegetation types found within the Thicket Biome in the region, which is characterised by short, dense and mostly spinescent trees and shrubs, quite different from images or tropical or temperate forest. *Ihlathi lesiXhosa* is considered to be a sacred place where the ancestors communicate with their living descendants by means of messengers (*izithunywa*) in the form of birds, mammals, insects or even the wind (Dold & Cocks 2012; Cocks et al. 2012, 2016). Similarly, Kwanele, a 26-year-old, explains that he likes going to the forest as he associates it with the place where he learnt how to carry out certain cultural practices. His uncle taught him how to slaughter the animal that is sacrificed at rituals. He now feels a great sense of pride in knowing how to perform this duty for his family.

The forest is considered to be a sacred place as it is believed that one's ancestors reside within the forest. The forest is therefore understood to be a place that bestows spiritual health and *impilo* and a place to communicate with ones' ancestors (Dold & Cocks 2012; Cocks et al. 2012, 2016), also see Nesbitt et al. (Chapter 10) for a similar example. Notinu, a 37-year-old woman, explains how she needs to go to nature, particularly the forest on the commonage, because "we can be shown visions from our ancestors [when we are] there". Notinu identified such experiences as being important to her and her family. Others alluded to the significance of certain sounds within such places, also associated with the ancestors, for example, Eunice, a middle-aged woman, who has lived most of her life in an informal township shack where she has had very limited opportunities to visit and interact with nature. The only time she has to experience nature is on her way to work. She explained how each morning she walks to work and will often appreciate the birds singing as she passes through an informal green area. Hearing their calls makes her feel relaxed and positive about life. Others alluded to how they appreciated the sounds of animals and the sound of the wind as it moves through trees or rustled leaves. For some, such sounds are considered an important trigger to contemplate their ancestors.

Growing Xhosa medicinal plants (*amayaza*) in the home garden was also found to be linked to a particular way of being for many urban residents. Many of the medicines identified as *amayeza* are used in a specific way to treat illnesses that relate to *impilo*. To achieve a state of healthy body and mind certain medicines are used when the need arises. This includes cleansing oneself internally by washing, purging, smudging and steaming (Cocks & Møller 2002, Dold & Cocks 2012). Kwanele explains how after his mother's passing he started growing medicinal plants in a garden he established in the yard of his new, low-cost, house built by the government under the Reconstruction and Development Programme (RDP) to address housing backlogs generated during the apartheid era. He got the plants he needed from family members in a rural village. Kwanele elaborates: "Cultural practices are a big part of who I am, when I use *imphepho* or walk in my garden, I get that free feeling of peace of mind. All these things are important to me [gardening and medicinal plants], it would not feel right, to me [if I did not do them] it is important that everyone experiences this [the peace of mind from nature]."

Of 362 households surveyed in *Qonce*, just over half (n = 163) had grown plants that are linked to a specifically embedded meaning. These included *impepho* (everlasting – *Helichrysum odoratissimum*), a medicinal herb which is used as incense to protect and purify the body spiritually; *umnquma* (wild olive – *Olea europaea* subsp. *cuspidata*), planted for its use in religious rituals. The wild olive branches are used as a platter for the consecrated meat (*intsonyama*) of ritually sacrificed animals (Dold and Cocks 2012). *Itswele lomlambo* (wild garlic – *Tulbaghia violacea*) is planted to deter *umamlambo* (the snake of men; a python-like serpent with magical abilities to transform itself which leaves a harmful trail behind it; *intelezi* (air-plant, *Tillandsia* sp., is one of the most powerful and effective charms used in times of crisis and emergency) is grown to protect households from lightning and malevolent forces. A range of medicinal plants (*amayeza*) are grown as home remedies made from recipes passed down from elders. Common examples are *umhlonyne* (wormwood; *Artemisia afra*) which is used as a treatment for colds and coughs and *wendrit* (rue; *Ruta graveolens*) for upset stomachs (Figure 6.4).

Yard spaces were also found to be important places for some urban residents to connect to nature and their ancestors as they provide a place to bring one's ancestors "home". BruVista explains how, when he moved into his new RDP house, he brought his ancestors into his new house. He carried out an opening of the home ceremony, *ukumisa ixhanti*, to do so. This meant sacrificing a goat and brewing traditional beer (*umqomboti*) for his ancestors. A handful of soil from their original family kraal (traditional livestock pen) was brought into his new yard and placed at the foot of the new tethering pole, thus maintaining the link with the ancestors in the new home. He explains further "My house is [now] protected by my ancestors, I brought them here and said this is where I live now". He has also planted two protective plants on either side of the gate to protect his home. Near the entrance to his kraal he has placed a large stone which he uses to crush and prepare medicines he makes to treat himself and his

Figure 6.4 A display of the range of *amayeza* plants grown to prepare home remedies. Source: von Tonder, Ella.

family. The stone also functions as a place "*to knock*" and explain to his ancestors why he is coming to speak to them at the tethering pole. BruVista has also planted trees around his designated kraal area to provide shade for guests when he hosts traditional ceremonies. He explains this is important because "our ancestors are pleased when we have many guests". When Aphiwe, a 24-year-old man in Cape Town, returns to his family home in Komani, the first thing he does is visit the kraal. He explains, "before I can even put my foot in the door, or be welcomed by my grandmother, it is important for me to let the ancestors know that I am home". Aphiwe also enjoys being in his family kraal as he "feels safe" there and is "close to my ancestors".

Yards also represent places that residents could display their connection to their cultural heritage. This is achieved by erecting and or maintaining a kraal (*ubuhlanti*) and a tethering pole (*ixhanti*) within one's yard (Figure 6.5). An *ubuhlanti* is considered a sacred place in Xhosa culture, explained as a temple in which ancestral spirits reside and watch over family members. It is also the place where animals are sacrificed in during traditional ceremonies (Dold & Cocks 2012, Cocks et al. 2008). The tethering pole is generally located in the centre of the *ubuhlanti*, serving as an anchor for the sacrificial animal, and is a symbolic point of contact with the ancestors. The horns of the sacrificial beast are attached to it in reverence to them (Dold & Cocks 2012). *Ubuhlanti* walls are either made of living plants or dry branches harvested from local commonages. Those who did not have the means or space to build and or maintain

Figure 6.5 Mr. Mila sits in front of his *ubuhlanti* which gives him a great sense of pride. Source: van Tonder, Ella.

an *ubuhlanti* erect only a symbolic tethering pole. Preferred species for the tethering pole are sneezewood (*umthathi*; *Ptaeroxylon obliquum*) and wild olive (*umnquma*) trees (Dold & Cocks 2012).

A place to escape and socialise

Places in urban nature also provide township residents with places to "escape" from daily drudgery and places to socialise with friends and family (also see Vierikko & Jaatsi (Chapter 5), Lagunes Gasca & Merçon (Chapter 7), Elands et al. (Chapter 11), for similar examples (Figure 6.6). Brenda, a middle-aged woman, described how she has many fond memories of going to the park with her friends when she was younger. She explains further "When I was younger I had a naughty streak, we would go to the local park and that's where I was introduced to 'the world of cans' [alcohol] … we had a lot of fun there".

Unati, a young township resident, shared similar experiences. He enjoys going to a nearby park with his friends to smoke tik (local name for crystal methamphetamine) and play soccer."I am most happy when I am relaxing with my friends, sometimes we chill and smoke tik …. I love playing soccer at the park. It clears my mind". For others going to nature included going with friends and family to local dams during the holidays. Xola, a middle-aged man, described how when relatives visited during the Christmas holidays, they would go to the local dam to socialise and drink.

Figure 6.6 Friends getting together at local municipal dams to enjoy each other's company. Source: van Tonder, Ella.

Benefits reaped – growing vegetables and fruit trees

Growing vegetables and fruit trees in the home garden evoke a strong sense of pride for many urban residents (see Nesbitt et al. (Chapter 10) and Elands et al. (Chapter 11), for similar examples. MamSukweni, who has lived in Komani township for over 30 years, remembers gardening as a young girl with her father. She learnt everything she knows about plants and their uses from him. In her own garden, she plants spring onions and lettuce. She shares the produce from her garden with her neighbours and grandchildren and explains that "It feels good to look after myself and those around me and it is very economical, I like to mix all my fresh veg". Similar sentiments were shared by Shepherd, a middle-aged man who grew up in the village of Mdingi. Some of his fondest childhood memories are of working in their garden and ploughing their fields with his mother and siblings. They planted maize and pumpkins. For most of his adult life Shepherd worked for a large industrial company doing manual labour. Ten years ago, he fell and injured himself. His injuries were so severe that he was unable to continue working. He moved to Qonce to support himself by making dog boxes to sell and doing basic house repairs. He also planted fruit trees, medicinal plants and vegetables in his garden. He grows lettuce and chillies and has planted fig and plum trees. He also harvests leaves from his *ivendrithi* (*Ruta graveolens*) plants to make a medicinal remedy. Gardening has become a central part of his life. His garden not only helps him financially through hardships but also gives him meaning when he is in a "dark place" and a great sense of pride. He shared the following sentiments about his garden: "I am very happy with my garden, it is important to have a garden because it feeds us and keeps me happy and energetic, especially when the plants start to sprout, it makes me very proud … it has always been important for me to have a garden, even growing up, we survived and ate from that garden".

The benefits are not just what is grown in home spaces, but also what is collected from the wild, usually the local commonage. For example, the wood needed to build the kraals for ceremonies typically comes from the commonage, as do the sacred, spiritual and traditional plant medicines and foods which are part of Xhosa identity. It also includes utilitarian products such as firewood, building timber, thatch grass and wild fruits (Davenport et al. 2011).

Nature of significance to urban residents

The preceding accounts reveal a multifaceted relationship that local urban Xhosa-speaking residents have with urban nature. As demonstrated by a survey distributed among over 200 households living in township areas of Qonce and Komani, these sentiments were shared by an overwhelming majority of urban Xhosa-speaking residents. Of the surveyed sample, 95 per cent of the household members agreed with the statement that "nature is a part of who I am" and that "nature is an important aspect of my culture/traditions" (Table 6.1). (See also Lagunes Gasca & Merçon (Chapter 7) for similar expressed sentiments. Eighty-five per cent agreed with the statement that they "felt connected to their ancestors when they were in a natural space". Ninety per cent found the sounds of nature relaxing. Going to nature to socialise with my friends was shared by 56 per cent of the surveyed sample. Forty-five per cent of households surveyed in Qonce (n = 590) showed signs of yard modification which related to planting.

Of households that engaged in plant-related yard modification (n = 310), such activities included planting vegetable gardens (34 per cent), fruit trees (46 per cent) and or cultural and medicinal plants (26 per cent) and decorative plants (30 per cent) and trees (18 per cent). Sixty-eight per cent of the household members interviewed acknowledged interacting with nature (outside of

Table 6.1 Percentage of 214 Respondents Who Agreed with the Following Statements about Nature

	% respondents (of 214)
Nature is an important aspect of my culture/traditions	95
Nature is a part of who I am	95
I feel good about myself when I'm in nature	95
Being in nature makes me feel free	95
The sounds of nature are relaxing	96
I feel connected to my ancestors when I'm in a natural space	83

their own personal home space, more or less equally between genders. From the in-depth interviews, urban nature areas included local commonages, forested areas (*ihlathi*) found on the commonage and private home gardens. The survey findings to a large extent overlapped with these as 50 per cent of the households indicated that they visited local commonages, of whom 42 per cent made specific mention of also visiting the local forest on the commonage. Yet, 14 per cent acknowledged visiting a park. Ten and eleven per cent visited dams and rivers, respectively. These findings reveal that the majority prefer a more natural state of nature as 69 per cent of the natures visited included such areas. Almost half (46 per cent) of urban residents visited nature at least once or more per week.

Bioculture and decolonising urban nature

The findings reveal regular engagement with urban nature by urban Xhosa-speaking residents. The types of interactions and engagements were diverse. For some this included going to dams and local parks to socialise with family and friends or to keep fit by playing sports. For many young people going to nature included opportunities to escape from parents and be "naughty". The majority (>90 per cent) of urban residents' interactions and experiences with urban natures are linked to culturally specific understandings that are based on particular histories and genealogies. These include initiation ceremonies, growing medicinal plants and culturally important plants, maintaining a kraal or a tethering pole or both and performing traditional rituals and ceremonies. Each of these activities is linked to a particular kind of nature category such as forest, home garden and yard.

The possibility to engage in more natural states of nature such as those found on the commonage and informal green spaces as well as nature found within the family yard was also acknowledged as contributing to a state of holistic health called *impilo*. In such instances, residents placed emphasis on how being in the forest contributed to a sense of wellbeing. Others felt that they were closer to their ancestors in the forest (on the commonage) and within their kraals. Within both of these places, residents felt that they could communicate with their ancestors. Authors have drawn attention to how our very breathing, thinking and moving through our surrounding landscapes contribute to our very existence (Schama 1995; Hinchliffe 2002), our cultural identity and a shared heritage (Williams et al. 1992), thus highlighting the importance of such places for urban residents.

Cocks et al. (2016) have recorded similar findings among urban Xhosa-speaking residents living in Makhanda (formerly Grahamstown), which revealed that 80 per cent believed that ancestors are present in the forest. Eight-four per cent felt that it is necessary to communicate with ancestors who reside in nature. Engagements and benefits derived from nature therefore cannot be understood in isolation. These associations and meanings remain largely unrecognised by urban planners, and the associated needs remain largely uncatered for in the types of urban natures found in the towns of Qonce and Komani. Furthermore, the urban natures that exist and the new ones that have been created, post-1994, remain deeply embedded within a Eurocentric understanding of what nature is and how it should be experienced. No attempts have been made by local authorities to re-imagine what local urban natures should comprise by incorporating local interpretations, meanings, needs and values. This is reflected in the maintenance of botanical gardens, memorial gardens and public parks in township areas. New parks created in the townships post-1994 remain defined solely as places for recreation. Few attempts have been made to include indigenous plant species which have meaning to local residents. Nature found in memorial gardens remains Eurocentric in both the older established memorial parks and more recently established memorial gardens which pay homage to local liberation heroes and struggles. Immortalisation of heritage within natural environments (Batten & Batten 2008) also remains deeply embedded within a Eurocentric construct, and promoting such areas as tourist designations reinforces the regions' colonial history (Marschall 2010, Martin 2004). Tellingly, of the 214 urban residents surveyed not a single respondent cited visiting either the botanical or memorial gardens in either Qonce or Komani.

Meskell and Scheermeyer (2008) and Smith and Waterton (2011) provide a critical account of how "heritages" have been defined, resourced and showcased and reveal patterns of priority and exclusion which are influenced by economic, social and political factors. In the case of Qonce and Komani very few places have been allocated for local Xhosa-speaking residents to engage in practices associated with their heritage or belief systems. Authors such as Basso (1996), Dominy (2001) and Sampson & Goodrich (2009) have identified how engagements in narrative practices promote a sense of attachment and sense of belonging. Antonsich (2010) identifies five factors contributing to an individual's sense of place-belongingness emotional attachment to a particular place. He also speaks to how such processes contribute to the ability of an individual to lead "a life that is meaningful and worth living and how a lack of place-belongingness is experienced as a sense of loneliness, isolation, alienation and displacement" (Antonsich 2010). Failure to integrate concepts of place-making and place identity into the planning and management of urban natures within urban areas by local authorities has the potential to alienate urban residents further as very few spaces have been provided in township areas for local residents to attach meaning to (Williams-Bruinders 2013).

To begin to facilitate transformative ways in which urban natures are conceptualised and experienced, there is a need to acknowledge that there is a diversity of ways of being in nature. This requires embracement of other ontological understandings of human–nature relationships which include recognising humans as being a part of nature (Merçon et al. 2019). Biocultural diversity offers such a starting point as it recognises that nature and culture are inextricably linked (Posey 1999). A person is recognised as being a part of the environment, and in the same way, the environment is recognised as being a part of the person (Descola & Palsson 1996). Such understandings are often maintained despite an adjustment to changing socio-economic and environmental conditions (Cocks 2006, Ianni et al. 2014). The narratives collected from Xhosa-speaking urban residents demonstrate the enduring nature of such understandings. To begin aligning locally held values and meanings, there is a need to unlock the alternative ways of interacting with nature which are held in stories, knowledge systems and experiences of living with nature. Attention needs to be given to recording such accounts and providing appropriate platforms for community members to engage in such dimensions of their lives. Such platforms have been created in Hawaii whereby the teachings carried in stories of elders have been used to help communities understand the way their ancestors related to the environment while contextualising the role they can play in the care of nature (Kealiikanakaoleohaililani et al. 2018, Chang et al. 2019).

The above findings also provide insight into challenges that face many Global South towns and cities within which many urban natures remain colonised. The exclusivity of such places often becomes further entrenched by local authorities marketing such places as key destination points for the tourist market, further alienating local residents. An urgent call is made to embrace the diversity of ways of being in and of nature to ensure that local residents of towns and cities across all town and cityscapes can derive benefit from past and present creation of urban natures.

Acknowledgements

Our thanks are extended to Dubula Maxam and Mluleki Nkosi who have assisted students doing the collection of the field data. Through their own personal stories, valuable insight into the meaning associated with biocultural relations over the years has been provided.

The work was supported by the National Research Foundation (NRF) of South Africa. Any opinion, finding, conclusion or recommendation expressed in this material is that of the authors, and the NRF does not accept any liability in this regard.

References

Anderson, M. & Pienaar, K. 2003. Municipal commonage. Evaluating land and agrarian reform in South Africa. Occasional Paper Series, No. 5. University of the Western Cape, Cape Town, South Africa, Programme for Land and Agrarian Studies, School of Government.

Antonsich, A. 2010. Searching for belonging: An analytical framework. *Geography Compass*, 4(6): 644–659.

Atkinson, D. 2005. Commons southern Africa: People-centred environmental management and municipal commonage in the Nama Karoo. CBNRM Occasional paper series: Commons Southern Africa, No. 11 University of the Western Cape, Cape Town, PLAAS, School of Government.

Bank, L. 2010. Shifting out of gear: Households, livelihoods and public policy on the South African wild coast. In Bryceson, D.F. (Ed.), *How Africa Works: Occupational Change, Identity and Morality*. Practical Action Publishing, London, UK. pp. 82–103.

Batten, B. & Batten, P. 2008. Memorialising the past: Is there an Aboriginal way? *Public History Review*, 15: 92–116.

Basso, K.H. 1996. *Wisdom Sits in Places: Landscape and Language among the Western Apache*. UNM Press, Albuquerque, NM.

Bertram, C. & Rehdanz, K. 2015. The role of urban green space for human well-being. *Ecological Economics*, 120: 139–152.

Bond, P. & Mottiar, S. 2013. Movements, protests and a massacre in South Africa. *Journal of Contemporary African Studies*, 31(2): 283–302. doi: 10.1080/02589001.2013.789727.

Bouteligier, S. 2011. Exploring the agency of global environmental consultancy firms in earth system governance. *International Environmental Agreements: Politics, Law and Economics*, 11(1): 43–61.

Buizer, M., Elands, B. & Vierikko, K. 2016. Governing cities reflexively: The biocultural diversity concept as an alternative to ecosystem services. *Environmental Science and Policy*, 62: 7–13.

Burman, J. 1984. Early railways at the Cape. Human & Rousseau, Cape Town.

Buffalo City 2015. Informal tenders. Retrieved from https://www.buffalocity.gov.za/CM/uploads/awarded/1620389199.pdf.

Buffalo City Metropolitan Municipality. 2018. Community services. Retrieved from https://www.buffalocity.gov.za/communityservices.php.

Castree, N. 2013. *Making Sense of Nature*. London, UK: Routledge.

Chang, K., Winter, K.B. & Linco, N.K. 2019. Hawai'i in focus: Navigating pathways in global biocultural leadership. *Sustainability*, 11, 283: 2–9. doi: 10.3390/su11010283.

Coates, J., Gray, M. & Hetheringten, T. 2006. An eco-spiritual perspective: Finally, a place for indigenous approaches. *British Journal of Social Work*, 36: 381–399.

Cocks, M. 2006. Biocultural diversity: Moving beyond the realm of 'indigenous' and 'local' people. *Human Ecology*, 34: 185–200.

Cocks, M., Alexander, J., Mogano, L. & Vetter, S. 2016. Ways of belonging: Meanings of 'nature' among Xhosa-speaking township residents in South Africa. *Journal of Ethnobiology*, 36(4): 820–841.

Cocks, M.L., Bangay, L. Shackleton, C.M. & Wiersum, K.F. 2008. 'Rich man poor man' – Inter household and community factors influencing the use of wild plant resources amongst rural households in South Africa. *International Journal of Sustainable Development of World Ecology*, 15(3): 198–211.

Cocks, M.L., Dold, T. & Vetter, S. 2012. 'God is my forest': Xhosa cultural values provide untapped opportunities for conservation. *South African Journal of Science*, 108(5–6): 52–59.

Cocks, M. & Møller, V. 2002. Use of indigenous and indigenised medicines to enhance personal wellbeing: A South African case study. *Social Science & Medicine*, 54(3): 387–397.

Cocks, M., Vetter, S. & Wiersum, K.F. 2017. From universal to local: Perspectives on cultural landscape heritage in South Africa. *International Journal of Heritage Studies*, 24(1): 35–52.

Cocks, M.L. & Wiersum, K.F. 2014. Reappraising the concept of biocultural diversity: A perspective from South Africa. *Human Ecology*, 42: 727–737.

Davenport, N., Gambiza, J. & Shackleton, C.M. 2011. Use and users of municipal commonage around three small towns in the Eastern Cape, South Africa. *Journal of Environmental Management*, 92: 1149–1460.

Davenport, N.A., Shackleton, C.M. & Gambiza, J. 2012. The direct use value of municipal commonage goods and services to urban households in the Eastern Cape, South Africa. *Land Use Policy*, 29: 548–557.

Davidson-Hunter, I.J., Turner, K.L., Mead, A.T.P., Cabrera-Lopez, J., Bolton, R. Idrobo, C.J., Miretski, I., Morrison, A. & Robson, J.P. 2012. Biocultural design: A new conceptual framework for sustainable development in rural indigenous and local communities. *Sapiens*, 5(2): 33–45.

Department of Land Affairs. 1997. White Paper on Land Policy. Pretoria, South Africa: Government of South Africa.

Descola, P. & Palsson, G. 1996. *Nature and Society: Anthropological Perspectives*. London, UK: Routledge.

Dold, T. & Cocks, M. 2012. *Voices From the Forest: Celebrating Nature and Culture in Xhosaland*. Johannesburg, South Africa: Jacana Media.

Dominy, M.D. 2001. *Calling the Station Home: Place and Identity in New Zealand's High Country*. Rowman & Littlefield, Maryland, MD.

Elands, B.H. & van Koppen, C.S. 2012. Biocultural diversity in the Netherlands: From ecologically noble savages towards biocultural creatives. In Arts, B. van, S., Ros-Tonen, M & Verschoor, G. (Eds.), *Forest-People Interface: Understandings Community Forestry and Biocultural Diversity*. Wageningin, The Netherlands: Wageningen Academic Publishers. pp. 181–193.

Elands, B.H., Wiersum, K.F., Buijs, A., Buizer, M. & Vierikko, K. 2015. Policy interpretations and manifestations of biocultural diversity in urbanized Europe: Conservation of lived biodiversity. *Biodiversity & Conservation*, 24(13): 3347–3366.

Edwards, S.D. 2011. A psychology of indigenous healing in Southern Africa. *Journal of Psychology in Africa*, 21(3): 335–347.

Favretti, R.J. 1971. Colonial gardens. *Arnoldia*, 31(4): 145–171.

Feld, S. & Basso, K. 1999. *Senses of Place*. Santa Fe, New Mexico: School of American Research Press.

Haq, S.A. 2011. Urban green spaces and an integrative approach to sustainable environment. *Journal of Environmental Protection*, 2: 601–608.

Hinchliffe, S. 2002. Inhabiting – Landscapes and natures. In Anderson, K., Domosh, M., Pile, S. & Thrift, N. (Eds.), *Handbook of Cultural Geography*. London, UK: Sage. pp. 207–226.

Ianni, E., Rivera, E.S. & Geneletti, D. 2014. Sustaining cultural and biological diversity in rapidly changing communities: The revitalization of the voladores ritual in northern Veracruz (Mexico). *Environmental Development Sustainability*, 16: 1197–1208.

Ingold, T. 2000. Ancestry, generation, substance, memory, land. In Ingold, T.T. (Ed.), *The Perception of the Environment: Essays in Livelihood, Dwelling and Skill*. London, UK: Routlege. pp. 132–151.

Jayne, M. 2005. *Cities and Consumption*. London, UK: Routledge.

Kealiikanakaoleohaililani, K., N. Kurashima, K. S. Francisco, C.P. Giardina, R. P. Louis, H. McMillen, C. K. Asing, K. Asing, T.A. Block, M. Browning, K. Camara, L. Camara, M. L. Dudley, M. Frazier, N. Gomes, A. E. Gordon, M. Gordon, L. Heu, A. Irvine, N. Kaawa, S. Kirkpatrick, E. Leucht, C. H. Perry, J. Replogle, L.-L. Salbosa, A. Sato, L. Schubert, A. Sterling, A. L.Uowolo, J. Uowolo, B. Walker, A. N. Whitehead & Yogi, D. 2018. Ritual + sustainability science? A portal into the science of Aloha. *Sustainability* 10, (10): 3478. https://doi.org/10.3390/su10103478.

Kepe, T., McGregor, G. & Irvine, P. 2015. Rights of 'passage' and contested land use: Gendered conflict over urban space during ritual performance in South Africa. *Applied Geography*, 57: 91–99.

Laird, S.A., Awung, G.L., Lysinge, R.J. & Ndive, L.E. 2011. The interweave of people and place: Biocultural diversity in migrant and indigenous livelihoods around Mount Cameroon. *International Forestry Review* 13(3): 275–293.

Loures, L. Santos, R. & Panagopolos, T. 2007. Urban parks and sustainable city planning – The case of Portimâo, Portugal. *WSEAS Transactions on Environment and Development*, 10(3): 171–180.

Maffi, L. 2005. Linguistic, cultural, and biological diversity. *Annual Review of Anthropology*, 34: 599–617.

Maffi, L. & Woodley, E. (Eds.). 2010. *Biocultural Diversity Conservation: A Global Sourcebook*. Washington, DC: Earthscan.

Marschall, S. 2010. *Landscape of Memory. Commemorative Monuments, Memorials and Public Statuary in Post-Apartheid South Africa*. Leiden, The Netherlands: Brill.

Martin, S.K. 2004. Monuments in the garden: The garden cemetery in Australia. *Postcolonial Studies*, 7(3): 333–352.

Milton, K. 2002. *Environmentalism and Cultural Theory: Exploring the Role of Anthropology in Environmental Discourse*. London, UK: Routledge.

Merçon, J., Vetter, S., Tengö, M., Cocks, M., Balvanera, P., Rosell, J.A. & Ayala-Orozco, B. 2019. From local landscapes to international policy: Contributions of the biocultural paradigm to global sustainability. *Global Sustainability* 2, e7, 1–11. doi: 10.1017/sus.2019.4.

Meskell, L.M., & Scheermeyer, C. 2008. Heritage as therapy: Set pieces from the new South Africa. *Journal of Material Culture*, 13: 153–173.

Peter, K., Elands, B. & Buijs, A. 2010. Social interactions in urban parks: Stimulating social cohesion? *Urban Forestry & Urban Greening*, 9(2): 93–100.

Peires, J. 1990. Suicide or genocide: Xhosa perceptions of the Nongqawuse controversy. *Radical History Review*, 46(7): 47–57.

Posey, D. 1999. *Cultural and Spiritual Values of Biodiversity*. London, UK: United Nations Environmental Programme, UNEP and Intermediate technology Publications.

Posey D.A., Frechione, J. Eddins, J. & da Silva, L.F. 1984. Ethnoecology as applied anthropology in Amazonian development. *Human Organization*, 43(2): 95–107.

Pretty, J., Adama, B., Berkes, F., de Althayde, S., Dudley, N., Hunn, E., Maffi, L., et al. 2009. The intersection of biological diversity and cultural diversity: Towards integration. *Conservation and Society*, 7(2): 100–112.

Pungetti, G. 2013. Biocultural diversity for sustainable ecological, cultural and sacred landscapes: The biocultural landscape approach. In Fu B. & Jones K. (Eds.), *Landscape Ecology for Sustainable Environment and Culture*. Dordrecht, The Netherland: Springer. pp. 55–76.

Puttick, J.R., Hoffman, M.T. & Gambiza, J. 2011. Historical and recent land-use impacts on the vegetation of Bathurst, a municipal commonage in the Eastern Cape, South Africa. *African Journal of Range & Forage Science*, 28: 9–20.

Radebe, D. 2018. An assessment of amount, distribution and use of urban green spaces in small towns of the Eastern Cape. M.Sc. thesis, Rhodes University, Grahamstown. 130 pp.

Rössler, M. 2012. World heritage cultural landscapes: A UNESCO flagship programme 1992–2006. *Landscape Research*, 31(4): 333–353.

Sampson, C. & Goodrich, K. 2009. A place for community: West Coast, New Zealand. *Society & Natural Resources*, 22(10): 901–915.

Schama, S. 1995. *Landscape and Memory*. London, UK: HarperCollins.

Swanwick, C., Dunnett, N. & Woolley, H. 2003. Nature, role and value of green space in towns and cities: An overview. *Built Environment* 29(2): 94–106.

Skinner, R. 2017. Every bite buys a bullet: Sanctions, boycotts and solidarity in transnational anti-apartheid activism. *Moving the Social*, 57: 97–114.

Smith, L. & Waterton, E. 2011. Constrained by common sense: The authorized heritage discourse in contemporary debates. In Camran, J., Skeats, R. & McDavid, C. (Eds.), *Oxford Handbook of Public Archaeology*. Oxford, UK: Oxford University Press. pp. 153–171.

Taylor, K. & Lennon, J. 2011. Cultural landscapes: A bridge between culture and nature? *International Journal of Heritage Studies*, 17(6): 537–554.

Victor, S. 2019. *Curator of History*, Amathole Museum, King William's Town.

Voss, M. 2012. Urbanizing the North-Eastern frontier: The Frontier Intelligentsia and the Making of Colonial Queenstown, c. 1859-1857. Doctoral dissertation, University of Cape Town, Cape Town.

Walker, S.E. & Duffield, B.S. 1983. Urban parks and open spaces – An overview. *Landscape Research*, 8(2): 2–12.

Wentzel, M. & Tlabela, K. 2006. Historical background to South African migration. In Kok, P. Gelderblom, D., Oucho, J.O. & van Zyl, J. (Eds.), *Migration in South and Southern Africa: Dynamics and Determinants*. Pretoria, South Africa: HSRC. pp. 71–96.

Williams, D.R., Patterson, M.E., Roggenbuck, J.W. & Watson, A.E. 1992. Beyond the commodity metaphor: Examining emotional and symbolic attachment to place. *Leisure Sciences*, 14: 29–46.

Williams-Bruinders, L. 2013. Making spaces or building places? A look at social sustainability in low cost housing, Port Elizabeth, South Africa. *Environmental Economics*, 4(3): 51–58.

Woodley, E. 2010. Lessons learned from the projects. In Maffi, L. & Woodley, E. (Eds.), *Biocultural Diversity Conservation: A Global Source Book*. Washington, DC: Earthscan. pp. 155–173.

Wotshela, L. 2009. Land redistribution politics in the Eastern Cape midlands: The case of the Lukhanji municipality, 1995–2006. *Kronos*, 35(1): 142–174.

Zhang, H., Chen, B., Sun, Z. & Bao, Z. 2013. Landscape perception and recreation needs in urban green space in Fuyang. *Urban Forestry & Urban Greening*, 12: 44–52.

Zituta, H.M. 1997. The spatial planning of racial residential segregation in King William's Town: 1826–1991. PhD Dissertation, Rhodes University, Grahamstown.

7 Nature on the border

The reconfiguration of biocultural experiences by migrant women on the southern border of Mexico

Ana Lucía Lagunes Gasca and Juliana Merçon

Introduction

> It does not matter where you are or where are you are going because the planet is still the same one. You move from one city to another but you are still on the same planet; there is no other one. It would be nice to be able to tell your experience in the place where you are. It is nice if you bring something good from another place that could help care for that new region. Even if you do not stay in that place, you are cooperating to take care of the planet. If you live in the United States, in Mexico, or anywhere else, the impact of your help is global, because the problem is that we, human beings, we have ruined our home. This has nothing to do with being a migrant, from one country or the other.

Valentina Fuentes' words invite us to rethink both the physical borders of our actions and their repercussions. Valentina is of Maya Man descent, originally from Las Majadas, Guatemala, and she now lives in Tapachula, a city within the southern border of Mexico. Her experience, like those of many other women in situations of migration, allows us to understand human mobility in relation to people's agency to bring about change. Inherent in the experience of geographical migration is the potential to contribute to collective life in the new location. This positive potential is frequently overlooked in places where migrant women arrive. As Valentina said: "It would be nice to be able to tell your experience in the place where you are". So, we have made a place here, in this chapter, to listen to her experiences and that of many other women who bring good things from other places to help us look after our common home, that is, the planet. Their stories might guide us towards knowing and imagining other relationships between culture and nature. Their particular biocultural experiences may somehow be linked to those of thousands of people who migrate between cultures and between rural and urban areas, experiencing high vulnerability as they traverse and in many cases defy political borders.

The International Organisation for Migration reports that 3.4 per cent of the world population lives outside their country of origin – this corresponds to 258 million people in a situation of migration. Out of these, 48 per cent are women and 14 per cent are children. The fact that the migration process costs some people their lives attests to the starkly unequal conditions under which people from different parts of the world experience migration (IOM 2017).

Women have made up the majority of internal migrants in Latin America over the past 50 years. While some have migrated from city to city, the primary movement is from rural to urban locations for domestic labour and work in manufacturing (De los Reyes 2001, Pessar 2005). In spite of the discrimination and harsh conditions that a great number of migrant women face when travelling without official documents, their decision to migrate has been identified as a form of resistance to hegemonic politico-economic models and gender inequality (Monzón 2006, VMAPM et al. 2016).

Decolonial feminists have analysed the feminisation of migration and the complexity of power relations that constitute the experience of migrant women through the lens of intersectionality (Espinosa et al. 2014, Magliano 2015). This approach responds to the need to understand social positionality. It does so by using a combination of identification processes and has contributed to the study of migration as a phenomenon in which a variety of forms of inequality converge. A number of different studies show how social classifications of gender, class, nationality, race, ethnicity, age, migration status and religion directly affect the daily life of migrants and influence their access to rights and opportunities, as well as their acts of resistance and strategies towards change (Crenshaw 2012, Magliano 2015).

Social classifications play a crucial role in border cities, considering that borders are not simply geographic or geopolitical delimitations, but also have symbolic significance, constituting communities of belonging and interchange, as well as exclusion (Rodríguez 2014). Some of the material and symbolic ways through which indigenous migrant women inhabit border cities correspond to culturally inherited, non-instrumental connections with the land or territory, which are often suppressed by the main urban discourses and practices. Guatemala Maya K'iche human rights defender, Chávez Ixcaquic, (2014, p. 10) claims, for example, that:

> Our territories are for us, indigenous women, constructions of vital spaces for the communities. In our territories, historicity, culture, identity, education, politics, economy coexist with beings such as animals and plants, including also minerals and the cosmos. Our conception of the territory goes beyond the physical space: it includes existential elements that nourish life, such as the bonds that unite people who are still active in processes to defend life in that place, in spite of not living there any longer.

In this sense, far from being an inert space, the land or territory is constantly being produced through the interactions between geo-biophysical and socio-symbolic components, with certain biocultural conceptions and practices prevailing over others. In many cases, people who are no longer living in their territories of origin maintain strong connections with their communities and contribute to shaping both physical and symbolic dynamics.

The construction of a hospitable place where indigenous and non-indigenous migrant women can feel safe and fully respected in their ties with the new territory is one of the objectives of the civil society organisation Fray Matías de Córdova A.C. Human Rights Centre (http://cdhfraymatias.org/web), working in the city of Tapachula in the Mexican southern state of Chiapas. The creation by this organisation of T'ja Xuj (Maya Mam for "Women's House") has made possible positive connections between the migrant women and the city of Tapachula. Discriminatory treatment against people in situations of migration abounds, despite human mobility having characterised the region for centuries (Flamtermesky 2014, Moreno 2017).

The voices of 18 migrant women who participate in T'ja Xuj are shared in this chapter. Their biocultural experiences both in their places of origin and in Tapachula allow us to understand how connections with care, food, sacredness, healing practices and green spaces are maintained, reconfigured or constrained in the new territory, showing significant differences and hybridisations between rural and urban natures. Before we hear their voices, we explore some ideas around the notions of biocultural experience and territory. In the following sections, we also describe the city of Tapachula and T'ja Xuj as a place of hospitality.

Biocultural experiences and the territorialisation of nature

The notion of "experience" encompasses knowledge and skills that are developed from doing and seeing as well as that which happens and affects one's life. Biocultural experiences are those lived processes that reveal how the intricate relationships between cultural practices and nature are embodied in a person's life and identity, and diversely manifested through knowledge, practices, values and feelings within a certain territory. By focusing on the crucial but not commonly addressed connections between experience and territory in order to approach biocultural phenomena, we intend to contribute to more effective ethico-political forms of understanding and action. Feminist perspectives assist us in weaving these theoretical and practical threads.

In general, feminism can be defined as a movement of resistance against gender domination (Bartra 2012, Blazquez 2012). This movement has allowed for a broader understanding of multidimensional violence against women and for the transformation of power relations, which are conditioned by patriarchy in its close relation with capitalism and colonialism, through the social construction of gender, class and cultural groups' asymmetries (Espinosa et al. 2013). As ecofeminist thinkers propose, the same heteropatriarchal system that sustains the oppression of women and other non-hegemonic genders operates over nature through relations of exploitation and domination (Plumwood 1991, Shiva 1998).

In line with feminist thinkers (Blazquez 2012, Gargallo 2014), we highlight the epistemological, ethical and political potential of the notion of experience in order to contribute to the recognition, understanding and transformation of power relations. What does the focus on particular experiences allow us to know and do?

- The concrete experiences of women in a situation of migration (and other people who are positioned at the margin of the dominant social order) allow us to understand certain aspects of reality, which are not commonly perceived by scholars, policy-makers and other social actors who occupy more privileged positions in society.
- Discourses based on personal experiences configure knowledge production as a situated process. Knowledge created from particular experiences does not have the pretension to universality, even though singularity is necessarily connected with many other experiences.
- Personal experience is always political. In other words, no point of view or locus of enunciation is neutral or beyond some particular power construction. We can only think, feel and live in the world the way we do because we are positioned in a certain cultural group, nationality, class, gender, age, occupation, migratory situation etc. The focus on people's experiences highlights this political positionality.
- Beyond what can be known through dialogues around personal experiences, the dialogic process itself is a significant personal experience, especially for people who are not usually listened to or socially valued.
- The content of experiences and the act of narrating them allow us to acknowledge the agency of the narrators, and can provide significant elements for inter-personal caring and for transformative collective action.

As it is not possible to speak from a neutral position, decolonial and community feminist thinkers have highlighted the importance of acknowledging "the geopolitical and corporeal-political location of the subject who speaks from certain coordinates of global power" (Magliano 2015, p. 696). From this perspective, nature becomes an abstract concept that needs to be territorialised to allow for more effective comprehension of the biocultural experiences and the power relations that enable or hinder such experiences.

To territorialise nature means to link its biophysical aspects to the geopolitical processes that constitute it. In this sense, territorialised nature would encompass the broad relationship between landscape and political exercise, corresponding to how the social and biological components of the place are organised to respond to specific interests. In other words, the territorialisation of nature refers to the efforts of individuals and groups to appropriate a place for themselves; that is, to affect, influence or control people, phenomena and relations through the delimitation and control of a particular biophysical and geographic area (Sack 1986). Biocultural experiences allow us to know how rural, urban and border natures are territorialised. They inform us, for example, about diverse forms of relationship between cultures and the territory, emphasising their dynamic, heterogenic and disputable nature.

Tapachula: a border city

The nature-territory to which all biocultural experiences in this chapter refer is a Mayan cultural zone. In 1882, a political border of 139 kilometres was established separating the countries that we now know as Mexico, Guatemala and Belize (Hernández 2012, De la Torre 2015). For the Jakalteko, Kaqchikel, Mam, Mochó, Qánjoba´l and Chuj people, this division meant the beginning of a long history of violent campaigns of acculturation.

Tapachula is located between the Pacific Ocean and the Sierra Madre Occidental, a volcanic mountain range. It is part of a landscape with high biocultural diversity, expressed in the relationships that cultural groups have for millennia maintained with forests, rivers, beaches and other landscapes. Tapachula is considered the most important city of this transborder region, with significant commercial activities around coffee, cacao, banana, papaya, mango and soy (De la Torre 2015). It has 348,156 inhabitants, with 59 per cent of the population being younger than 35 years old (INEGI 2015).

While the presence of migrants is essential in the zone, not all foreigners are equally received, registered or incorporated into the city's dynamics or its official statistics (Anguiano 2008). In the 1990s, Tapachula became the main migratory corridor for Latin American people whose destination was the United States. In 2017, the Mexican Commission for Refugee Aid reported 14,596 asylum requests from adult people and families, and 259 requests for unaccompanied children and adolescents (UNHCR 2018). The United Nations High Commissioner for Refugees (2018) reported that there were 68.5 million forcibly displaced people in the world during 2017. Out of them, 25.4 million are recognised as refugees and 33.1 million are requesting international protection. The epicentres are the Mediterranean, the Andaman Sea and Central America.

Over half (55 per cent) of the total foreign population settled in the state of Chiapas is made up of women (Instituto Nacional de Estadística y Geografía 2015), with the majority being from Guatemala (66 per cent), Honduras (11 per cent), El Salvador (6 per cent), and the remaining 17 per cent from diverse nationalities. Most of them are young, of productive and reproductive ages, and with high levels of illiteracy. They mostly work in the agricultural sector, the sexual and entertainment industry, and in domestic jobs (UN Women 2015). The statistics do not show, however, the multiple situations of gender inequality that women in situations of migration face or the strategies they create to cope with discrimination. Their biocultural experiences and contributions are also mostly invisible to the eyes of official institutions and local populations. Cocks et al. similarly provide insight into how previously marginalised communities biocultural experiences remain unacknowledged in town and city landscapes as Eurocentric constructs continue to define the types of nature and ways of being in nature within towns and cities landscapes (Chapter 6).

T'ja Xuj: house of women

T'ja Xuj opens every Sunday, at the Miguel Hidalgo main square, in the centre of Tapachula (Figure 7.1). It is a gathering place for women in any migratory situation; a place where relationships of care and dignity are fostered by and for all visitors. The house is supported by the Fray Matías de Córdova Human Rights Centre, a non-governmental organisation founded in 1997.

T'ja Xuj was initially created to receive women coming mainly from Guatemala who had been hired as domestic workers in the city. These women constitute a migrant group, which, despite the conditions of labour exploitation and social vulnerability they bear, is crucial to the local economy. Currently, the two main groups who gather at the house every Sunday are adult women with children, mainly from El Salvador and Honduras, and young and adult women from Guatemala, who are mostly domestic or commerce workers, with a circular or repeat migration dynamic. Some of them have just arrived and have no support network while others have lived in the city for many years.

T'ja Xuj's work encompasses four main areas: gender, human rights, interculturality and autonomy. Considering these dimensions, some collective

Figure 7.1 Tapachula's main square. Source: Lagunes Gasca, Ana Lucía.

processes have focused on the exchange of knowledge between those who visit the house. In these exchanges, knowledge has been approached as types of understanding (content) and cultural practices (know how), which are related to previous experiences in the territories of origin and their recreation through migratory processes. By sharing their knowledge, the women who gather at T'ja Xuj have been empowered and have fostered new relationships with each other and the city (Figure 7.2).

To learn from the women's biocultural experiences, we conducted four interviews, one focus group and one workshop in which maps, photographs and various texts were used to elicit the women's perceptions of the territory. During these exchanges, we were guided by principles from decolonial feminist pedagogies (Espinosa et al. 2013) and feminist action-research (Biglia 2014, Flamtermesky 2014), such as personal relevance, care, deconstructing hierarchy and deconstructing dichotomic thinking.

Fourteen women of ages ranging from 20 to 70 participated in the focal group, some agreed to have their surnames used and others did not. They were Floriberta Roblero, Rosmery Paz, Hilda Pineda, Ruth Escobar, Masiel Labarca, Aidé, Griselda, Lidia Henríquez, Hilma, Digna, Karla Alfaro,

Figure 7.2 "All different, all with rights". Women in T'ja Xuj. Source: Lagunes Gasca, Ana Lucía.

Elizabeth, Maribel and Rachel. Their interventions are identified in the text as "focal group". Only one woman is originally from Tapachula, two are from Guatemala (Maya Mam communities), and the other women are asylum seekers or have already been recognised as refugees, originally from El Salvador, Honduras and Venezuela. Three of them grew up in urban zones, and eleven come from rural areas. Long interviews were held with Andrea Lorenzo (Mayan Kakchiquel), María Nolasco (Mayan Mam), Valentina Fuentes (Mayan Mam) and Carmen López (Mayan Kakchiquel), who chose to have their names shown in this writing.

Biocultural experiences: rural, urban and transborder natures

From the interviews and focal group dialogues carried out with the 18 women who regularly visit T'ja Xuj, we highlight some biocultural experiences that show differences and also hybridisations between rural and urban natures. The flux between territories allows us to think about the construction of a transborder version of nature that combines aspects from the original rural culture and local urban conditions, the latter often characterised by restrictions, discrimination and exclusions. To show these differences and overlaps, the sequence in which the quotes are presented reflects a transition in the experiences from rural to urban contexts. The biocultural experiences narrated by the women of T'ja Xuj reveal (1) manifestations of integral care and reciprocal relations between nature, self and others; (2) differences in the relationship with food in rural and urban contexts; (3) rituality, healing practices and connections with sacred places and (4) how these biocultural experiences are geopolitically positioned and shaped by social conditions.

Integral care: nature, self and others

The stories told by women of T'ja Xuj show that life in the countryside is characterised by multiple expressions of care, many of which can be understood as a form of reciprocity between humans and nature. Trees, rivers and the land are looked after with the understanding that these caring practices also maintain human life and constitute one's identity. Andrea is a 20-year-old woman of Mayan Kaqchiquel descent, who explains that nature and selfhood are closely intertwined: "Nature is a fundamental part of your life. Nature is part of who I am. You learn to love it; you are part of it too". Cocks et al. found similar sentiments being shared by isiXhosa residents who felt that that "nature is a part of who I am" and that "nature is an important aspect of my culture/traditions" (Chapter 6).

People commonly invoke the ever-present interdependence of natural elements to teach about the close connection between one's life and surrounding manifestations. Carmen is 52 years old and migrated from Mayan Kaqchiquel territory in Guatemala at the age of 16 to work as a domestic labourer in Tapachula. She said her father taught her to respect the trees, water and wind, and remembers that he used to say:

> [W]e shouldn't complain about the wind. We shouldn't say 'it is too windy' or 'what terrible wind' because it is thanks to the air that we breathe and complaining about it is as if we were rejecting it or not honouring it.

These relations of care, respect and reciprocity were also mentioned in the focal group in which one participant commented:

> Nature teaches you something personal. Every day it teaches us that if we tend to it, we will have a better life, our children and us. Water comes from nature, and water is life. If we look after it, we will continue to have it for a long time.

Growing and buying food

A remarkable difference between life in rural areas and the city lies in the relationship with food. In rural nature, food is planted in direct relation with the earth; and sources of nutrition are supplemented in exchanges with other community members, while in the city food is bought and there are fewer interactions around food with the neighbouring community, as discussed in the focal group:

> In the countryside, we grow many things. They teach us the perfect time to plant, according to the moon, so that we can have a good harvest. People in the city buy everything while in the countryside we harvest. If you want a 'chayote' (*Sechium edule*), you go and get it. If you want some malanga (*Xantyosoma sagittifolium*), you go and pull it out. You sell your coffee and keep your money.

Life depends on direct activities with the earth and surroundings, and also on bonds that are locally nurtured: "In the rural villages, if someone does not have something and the other person has it, that's it; we share and support each other. There is more sharing and interaction. We know each person of the community" (focal group participant). This socio-natural network of support makes life possible beyond one's dependency on the monetary economies that characterise urban sociality. Carmen illustrates this difference: "Here, in the city, if we have money, we eat. If we don't have money, we don't eat".

Valentina Fuentes is 47 years old and comes originally from the county of San Cristóbal Cucho, Mayan Mam territory in Guatemala. She explains that the new relations with food in the city sometimes encompass feelings that are not as present in rural areas. Pity, compassion and discomfort towards the death of animals are some of these emotions as she further describes:

> When we cooked maize in my house, we had to be careful not to put too much lime to nixtamalise it.[1] We also used to take the corn to the mill or

> grind it at home. Here in the city you buy already-made tortillas.[2] Meat comes already cut into pieces, and everything is very clean. I was the oldest of my siblings, and I had to do everything! I learnt by watching … my mum taught us how to kill a hen, and it never crossed my mind to feel pity or '"oh, poor thing", I just did it and poom! Now I try and avoid it, because it feels strange … in the city, the story is a whole different thing … the chicken comes ready to cook; you don't need to cut its head. But this was normal for me before. They taught me that first we have to feed the animals and eat them later. But now in my head I have this feeling of pity for them. When I first arrived in Tapachula I saw that all you have to do is to take the chicken out of the tray, season it and put it in the oven. Well, of course, I liked it! You save a lot of work. However, I also miss the natural way; it has another flavour, another quality also.

Despite some significant changes, for many T'ja Xuj women food growing persists in the urban context. While space is reduced and work routines limit the time dedicated to gardening, the women continue to find natural ways to cultivate and prepare food, from biological control of plagues to using firewood for cooking. Andrea is 20 years old and has been living in Tapachula for the last three years. Every month she travels back to her home in the Mayan Kaqchiquel region of San Pablo, Guatemala. Despite the lack of space in her workplace in Tapachula, she continues to grow some fruit:

> I go looking for plants, and I arrange them in the garden. I planted papaya, and it is already big, producing much fruit. As there is a little garden there, I took advantage of it. I also planted bananas; they are still growing.

Carmen cultivates a similar relationship with eatable plants in the city. When her plants are affected by some infestation, she does not apply synthetic products, but tries her own natural solutions:

> While I work, I also try and see my plants there at home. I like growing my own plants. The bad thing is that sometimes there is zompopo (*Atta laevigata*)[3] – the other day it ate all my chilli plant! A friend told me to put a plastic bag under the plant so I did it and it worked. Then they came back, but I took the ashes and put it under the chilli plant and it finally worked. Nobody told me to do that; I invented it myself. The ashes are bitter, so I thought they would not like it.

The enjoyment of food in the countryside remains and co-exists with the adaptation to the city. Women with circular migratory experiences, reconfigure their ways of living, valuing the particularities of each place. Valentina provides a good example of how biocultural experiences are reconfigured or re-signified through migration:

> Every time we go back home, my mum receives us with a cooked meal, with some yerbamora (*Solanum nigrum*) soup, beans and tortillas. Cooking with firewood makes it taste so different! So you go back and enjoy all those moments. Handmade tortillas are so good; there is so much care in the whole process. I appreciate the meat tamales[4] more now! And it is nice to change from one place to the other. You've already adapted to the new place, but when you go back you can enjoy it too. Before I left, it all looked normal and I didn't appreciate it. Now that I live far away, I appreciate it and enjoy it when I am there.

Rituality, healing practices and sacred places

The T'ja Xuj women narrate how sacred relations with nature, which are so vividly experienced in rural areas, are modified in the city. Nature's sacred character remains in urban zones; however, the effectiveness of natural healing practices and the experience of certain rituals are altered. In contrast with urban biocultural experiences, María explains that in her Mayan Mam community of origin, some desired natural phenomena and key agricultural events are linked to the sacred through rituals performed by the shaman:

> When it doesn't rain for a long time and the crops start drying, we call the shaman who will then do his job from the top of the hill: he kills a chicken, asks for turkey eggs, asks for a drink, and a lot of people go there with those things to ask God for a good harvest. The shaman kills those animals, burns them and does his thing. He does it in the afternoon and at night it begins to rain. I've seen that happen!.

Traditional midwives in indigenous communities also hold significant biocultural knowledge and perform a key social role. Valentina describes the special attention she received after the delivery of her daughter. Her bathing practices in the city were perceived here as part of the problem:

> When my daughter was born in the village, I didn't feel one of my legs and my midwife said to me that this was happening "because you had your pregnancy in Tapachula and you used to shower with cold water. I am going to bathe you". I then felt that she was going to burn me … she bathed me with nearly boiling water prepared with different types of herbs, starting by my feet. At that time I didn't have any milk for breastfeeding. Well, she finished bathing me, wrapped me in a blanket to make me sweat out all the cold I had inside my body, and laid me down. My mother heated a beer and cooked me some ranch chicken broth. They fed me and gave me something to drink. My body relaxed in an impressive way, helping me heal after giving birth. It was wonderful to have all that attention.

Traditional healers are also present in the city of Tapachula, where they perform similar rituals to those held in rural natural environments. However, despite the similarities in form, María tells us that the healing work done in the city is not as effective as that of her Mayan Mam territory of origin:

> In spring break I went to the beach with a friend of mine. We went swimming and there was a big hole in the sea and we started drowning! A woman pulled us out and helped us. I felt very scared then. After that, I started having strange dreams, I felt that someone was coming to get me; a black woman was coming to get me. I felt so scared one morning … the lady for whom I work asked me "what happened, what happened?" I told her that every time I fell asleep I dreamt that I was in the water. And she told me that she knew someone who could cure me of fright. She took me there and I was cured. I then went back home and told my brother about what had happened. He told me to go to another man. I went there and I was cured again. The one in Tapachula cured me only with plants. He sprinkled basil on me and blew into my face after having some alcoholic drink. The one from my village killed a rooster and told me to give him some turkey eggs. He also asked me for a litre of alcoholic drink with ruda (*Ruta graveolens*) and black pepper and blew it onto my face after drinking it. I felt good with the one from Tapachula but the one from my village cured me really well. He also gave me some of the mix to drink.

The T'ja Xuj women report that traditional healing practices are less common in the city, mainly because it is easy to purchase and use industrialised medication. In the focal group, a woman explained that she usually gives medicine from the pharmacies to her children because "it is easier to give them some pill than to find the plant and prepare it". She admits that "this is how we start losing our customs and ways of healing". Healing practices are often associated with nature's sacredness, whose expression in the city occurs mainly in green areas and rivers. This provides Andrea with reasons to ask us to respect these places: "Nature is sacred and we have to take care of everything. Rivers and the little parks are also sacred. These plants are sacred, so we need to respect them. We should not harm them or throw litter on them".

Despite the acknowledgement of the sacred in urban spaces, pre-Hispanic rituals of great cultural importance are not performed in these contexts due to the fear of people's prejudice. Carmen explains that in her Mayan Kaqchiquel village every 24th of June water springs are beautifully decorated:

> People take their coffee and bread to the forests and share their food by the springs. It is really beautiful. And at the beginning of the year, during the fiesta patronal [annual patron saint festival], we present seeds, machetes and hoes to thank God for all he gives us to eat and to be alive, for all those things that are given by the earth. Here in Tapachula we don't celebrate

> the 24th of June. We don't decorate the springs, because we don't know the places well and what would people say! People would ask for an explanation and we wouldn't know what to say. This is a custom that comes from our parents, our grandparents ... I don't know where it comes from.

Liberation theology (a religious movement based on a synthesis of Christian theology and Marxist theory) has played a significant historic role in the organisation of indigenous and peasant peoples' struggles for a dignified life in the region. The work of T'ja Xuj and Fray Matías Human Rights Centre finds its origin in this movement. However, in the conversations held with the women's group, they highlight their lack of trust in Catholic priests: the prejudice and negligence the church has shown towards people in situations of migration, and the prevailing disconnection between Catholicism and the sacredness of nature. Carmen tells us about her own experience:

> here in the city, I don't trust the church priests and I don't talk with them or ask them if I could take seeds or the harvest to be blessed. They don't pay attention to simple people like me; they only notice the rich people.

Similar examples were provided by women in the focal group one of whom remembered that "when the migrant caravan arrived at the church, the priests closed it". Another woman said: "I thought it was strange because that was where people wanted to go to thank God for having arrived". Another woman in the focal group claimed that the church does not revere nature: "The church tells us what is sacred, but nature is also sacred! They worship God, but they don't love what God created. Without nature, we wouldn't be here, we wouldn't breath or eat".

Biocultural geopolitics: urban nature and territorial injustice

The types of space one can access and the kind of use one makes of urban areas are linked to the socio-political organisation of territories, allowing or inhibiting certain biocultural experiences. The manifestation of this biocultural geopolitics in the city of Tapachula reveals inequities in the access and use of urban nature, as we show further below. Firstly, we explore some of the socio-spatial dynamics that occur in the city's main square, a hardscape with some greenery, a fountain and concrete benches (see Figure 7.1). This is an important gathering place, especially for workers and people in a situation of migration in the region, as Valentina describes:

> When we just arrived in Tapachula, the park was really nice. We would get together there, tell our stories, how was your week, everything. We enjoyed the place and its water fountain. We had some good moments and could forget about everything; about work .... It was a way of relaxing,

> meeting friends, talking with someone who is not your employer. It was a safe place. Back then, we were many women, you would always find the same people there.

de Mendonça (Chapter 9) also provides examples of how the socio, political and economic factors influence the types of biocultural experiences that are and are not permissible on the beaches in Rio.

Tapachula is a city with high temperatures throughout the year and very few green areas. The shade and freshness of trees are thus disputed and not equally distributed. This unequal use of space can be analysed through the perspective of intersectionality to show how, even in contexts where similar migratory conditions are present, different genders and places of origin count as criteria for the reproduction of socio-political hierarchies. María illustrates how these territorial power dynamics are driven by social classifications based on places of origin:

> There are many trees in the park, where one can go for a walk, but since our brothers and sisters from other countries have arrived, sometimes they are using these places. So one has to go to the front area of the county hall or come to T'ja Xuj.

In the focal group, the social-political use of space was described through an unclear combination of categories of gender and place of origin:

> We women have always met in front of the church, where there are those trees in a half moon shape. But now that place is full of them [other people in situation of migration], so we have to look for another place because some of them are really rude, they start saying things and telling us off. I feel people from Honduras are rude … so I look for shade somewhere else.

The experience of nature in Tapachula is also contingent on the place where you live, where you work and your mobility possibilities. These conditions are influenced in turn by race, gender and class. In other words, the kind of relations that you can build with places, such as the possibility to live in particular places and move around the city, depends on your appearance, income, social status, where you are from and the migratory documents you hold. In this sense, Andrea states: "I know that there are beautiful places outside Tapachula. But because I work so much and it is difficult to go there, I never enjoy all of that". The working conditions that many women in a situation of migration endure are often an obstacle for enjoyment and rest in urban nature. Labour exploitation has direct implications in the availability of time, energy and economic resources, which favour broader relations with the territory, as Carmen shows us: "Here we only work and sleep. One is kept away and worries about doing all the things that we have to do: to wash, clean, iron, prepare food. When we are done, we can't even see the sun". When there is some time for leisure, it is often difficult to reach desired places, as Carmen narrates:

> From time to time I asked the Mrs for permission to go with my friends to the river, by the fields. That's where we enjoy going. But we had to walk more than an hour to get there and it was very tiring because we had to carry our food and other things. Sometimes the Mrs told the driver to pick us up at the highway and allowed us to cook at her house in the fields.

In addition to conditions related to social class (high workload and low income), these particular cases of territorial injustice or unequal use of space are further driven by the migratory status of the women of T'ja Xuj:

> People like us, who come from Central America without documents, can't go everywhere. We are not free. We walk around afraid of being arrested, of being removed from the public transport. This is one of our main problems: not having documents. It is a problem for us and for people from other parts of the world.

Actions towards hospitable cities: territorialising nature and justice – some final remarks

> Our culture is missing here
>
> Valentina Fuentes

The conversations held with the women who regularly visit T'ja Xuj allow us to understand that many of the relations that they established with nature in their rural communities contrast significantly with those that they maintain in the cities. We noticed, for instance, that many biocultural experiences in rural territories are characterised by caring relations that express interdependence and reciprocity between humans and non-human entities. In similar ways, food growing, the direct experience of animal death and the mutual support to supplement nutrition are practices that are considerably different to those taking place in the city. Differences around rituality and the effectiveness of healing practices in the countryside and the city are also remarkable. In a broad sense, we could maintain that rural and urban natures involve distinct types of biocultural experiences.

However, there is a series of transborder or hybrid biocultural manifestations that combine rural and indigenous practices with non-indigenous and urban ones. In the experiences of the women from T'ja Xuj this reconfiguration of biocultural experiences becomes more evident in relation to food growing in small spaces, traditional healing practices and the sacredness that continues to be subjectively attributed to urban nature. Other biocultural experiences are suppressed in the urban context, such as the rituals that celebrate the sacredness of water bodies.

The social prohibition of certain biocultural manifestations, as well as the limitations in access and use of particular spaces on the basis of place of origin, gender and class, offer examples of the unequal distribution of territorial justice in Tapachula. Intersectionality plays an important role in the socio-political analysis of territorialised nature. The biocultural experiences of women in situations of migration in Tapachula, most of whom are employed as domestic workers, provide a particular perspective on the lives of this vulnerable group and their views. The knowledge generated in the dialogues with the women of T'ja Xuj opens new areas of enquiry and possibilities for action. The thorough attention to how both vulnerability and agency are embodied by these women can thus lead to forms of understanding and action that more adequately respond to their experiences, needs and dreams.

It is also important to note that the neglect by public institutions of the biocultural contributions from people in situations of migration implies not only a breach of the right to cultural diversity (cultural reproduction, non-discrimination, among others) but also a significant inability to act upon enriching opportunities for the construction of more hospitable, just and sustainable cities for the whole population. Biocultural expressions of reciprocity, respect and sacredness towards nature can contribute significantly to infusing meaning into urban nature and instilling practices of greater care by all.

In our dialogues with the women of T'ja Xuj, we asked them about the necessary actions to be considered for better care of urban nature. They promptly mentioned two main strategic areas: waste management and food growing in the city. They expressed concern about the problematic waste management and non-existent recycling policies in Tapachula. Despite the city's production of huge volumes of rubbish and industrial waste, the disposal systems are inadequate with negative effects on the soil, the rivers and the health of the population. Given that these women have a strong connection with water, they expressed particular concern for the polluted state of the rivers in the region.

As a result of these reflections, the members of T'ja Xuj decided to organise a campaign to collect rubbish and lead a series of actions focused on public servants and other social groups in Tapachula. The women of T'ja Xuj also recognised that the soil in Tapachula is really fertile and their enthusiasm resulted in a decision to create a collective vegetable garden, based on the knowledge they bring to the city. The potential of community gardens for expressing and strengthening biocultural relations with the territory is immense. Besides allowing the sharing of traditional knowledge around plants, cultivation practices and gastronomy, vegetable gardens constitute a key space for collaboration, sociality and amusement, also contributing to the greening of cities.

The coordinating team of T'ja Xuj reported that this reflexive process has reinforced the importance of biocultural experiences and territorial connections in the southern border area as a means to promote dignified living conditions for women in situations of migration. Their existence and resistance in these places is important for the construction of a more diverse world: a world in which many life worlds can flourish. We hope that these biocultural experiences contribute to the construction of more diverse and hospitable cities, greater justice in relation to the use of nature and sustainable transborder versions of natures in other regions of Mexico and beyond.

Acknowledgements

This chapter integrates knowledge that was co-produced with participants of T'ja Xuj Women's House. We express our deep gratitude to Floriberta Roblero, Rosmery Paz, Hilda Pineda, Ruth Escobar, Masiel Labarca, Aidé, Griselda, Lidia Henríquez, Yelma, Digna, Karla Alfaro, Elizabeth, Maribel, Rachel, Andrea Lorenzo, María Nolasco, Valentina Fuentes, Carmen López, Janny Cruz, Concepción Kanayama and Margarita Morales. We also wish to thank Sara Sotto for the translation of this chapter and Griela García Borges for her invaluable collaboration in the dialogues.

Notes

1 Nixtamalisation (hominy) is a traditional process for the preparation of maize (corn), in which the corn is soaked and cooked in an alkaline solution, usually limewater (but sometimes wood ash lye), washed, and then hulled.
2 Tortilla is a traditional staple in Mexico and Guatemala. It is unleavened, thin and round bread prepared from cornmeal and baked on flat plates of earthware or iron.
3 Leafcutter ant with a smooth shining head. It has been eaten for hundreds of years by different cultural groups in the Americas.
4 Tamales are a traditional Mesoamerican dish made of corn dough and steamed in a corn husk or banana leaf. The leaf wrapping can be used as a plate. Tamales can be filled with meat, vegetables, cheese or fruit.

References

Anguiano, T.M. 2008. Chiapas: territorio de inmigración, emigración y tránsito migratorio. *Papeles de población*, 14(56): 215–232.

Bartra, E. 2012. Acerca de la investigación y metodología feminista. In Blazquez, N., Flores, F. & Ríos, M. (Eds.), *Investigación feminista: epistemología, metodología y representaciones sociales*. México City, México: UNAM. pp. 67–78.

Biglia, B. 2014. Avances, dilemas y retos de las epistemologías feministas en la investigación social. In Azkue, I.M., Luxán, M., Legarreta, M., Guzmán, G., Zirion, I. & Carballo, J.A. (Eds.), *Otras formas de (re)conocer. Reflexiones, herramientas y aplicaciones desde la investigación feminista*. Don Ostia San Sebastian, Spain: SIMRF.

Blazquez, N. 2012. Epistemología feminista: temas centrales. In Blazquez, N., Flores, F. & Ríos, M. (Eds.), *Investigación feminista. Epistemología, metogología y representaciones sociales*. México City, México: UNAM. pp. 21–38.

Chávez, A.L. 2014. Mujeres indígenas: cuerpos, territorios y vida en común. *Feminista Siempre*, 19: 1–33.

Crenshaw, K. 2012. Cartografiando márgenes: interseccionalidad, políticas identitarias y violencia contra las mujeres de color. In Platero, R. (Ed.), *Intersecciones: cuerpos y sexualidades en la encrucijada*. Barcelona, Sapin: Bellaterra. pp. 87–122.

De la Torre, A. 2015. *La negociación del tratado de límites entre México y Guatemala –1853 a 1902*-Por Nuestra Tierra Chiapas A.C., Tapachula.

De los Reyes, P. 2001. Women and migrants: Continuity and change in patterns of female migration in Latin America. In Sharpe, E.P. (Ed.), *Women, Gender and Labour Migration*. London, UK: Routledge. pp. 275–289.

Espinosa, Y., Gómez, D. & Lugones, M.K. 2013. Reflexiones pedagógicas en torno al feminismo descolonial. Una conversa en cuatro voces. In Walsh, C. (Ed.), *Pedagogías decoloniales: practicas insurgentes de resisitir, (re)existir, y (re) vivir. Tomo I.* Quito, Ecuador: Abya-Yala.

Espinosa, Y., Gómez, D. & Ochoa, K. 2014. *Tejiendo de otro modo: Feminismo, epistemología y apuestas descoloniales en Abya Yala*. Popayán, Colombia: Editorial Universidad del Cauca.

Flamtermesky, H. 2014. Mujer Frontera. Experiencia de Investigación de Acción Participativa Feminista (IAPF) Con Mujeres Víctimas de la Trata de Personas. *Athenea Digital*, 14(4): 389–400.

Gargallo, F. 2014. *Feminismos desde Abya Yala. Ideas y proposiciones de las mujeres de 607 pueblos en Nuestra América*. Ciudad de México, México: Editorial Corte y Confección.

Hernández, R. 2012. *Sur profundo. Identidades indígenas en la frontera Chiapas-Guatemala*. México City, México: Publicaciones de la Casa Chata, CIESAS, CDI.

Instituto Nacional de Estadística y Geografía (INEGI). 2015. *Encuesta Intercensal*. Gobierno Nacional, México: Instituto Nacional de Estadística y Geografía de México.

International Organization for Migration (IOM). 2017. Migration and migrants: A global overview. In Migration, I.O. (Ed.), *World Migration Report 2018*. Geneva, Switzerland: IOM.

Magliano, M.J. 2015. Interseccionalidad y migraciones: Potencialidades y desafíos. *Revista Estudios Feministas*, 23(3): 691–712.

Monzón, A.S. 2006. *Las viajeras invisibles: mujeres migrantes en la región Centroamericana y el sur de México*. Guatemala: Consejería en Proyectos de Centroamerica y México.

Moreno, R. 2017. Pedagogía hermenéutica del lugar: estudio narrativo sobre la relación entre lugaridad, alteridad, ciudad y escuela. *Actualidades Investigativas en educación*, 17(2): 1–34.

Pessar, P. 2005. *Women, Gender, and International Migration Across and Beyond the Americas: Inequalities and Limited Empowerment*. Mexico City, Mexico: UN Population Division.

Plumwood, V. 1991. Nature, self, and gender: Feminism, environmental philosophy and the critique of rationalism. *Hypatia*, 6(11): 3–37.

Rodríguez, R. 2014. *Epistemología de la frontera. Modelos de sociedad y políticas públicas*. Mexico City, México: EON.

Sack, R.D. 1986. Defining territoriality. In Sack, R.D. (Ed.), *Human Territoriality: It's theory and History*. Cambridge, UK: Cambridge University Press, (pp. 18-20).

Shiva, V. 1988. *Staying Alive: Women, Ecology and Development*. London, UK: Zed Books.

United Nations Women. 2015. *Derechos Humanos de las Trabajadoras Migrantes*. Retrieved from: http://mexico.unwomen.org/es/digiteca/publicaciones/2016/01/derechos-humanos-de-las-rabajadoras-migrante.

UNHCR. 2018. *Global Trends Forced Displacement in 2017*. Geneva, Switzerland: United Nations High Commissioner for Refugees.

Voces Mesoamericanas Acción con Pueblos Migrantes A.C. (VMAPM); Centro de Derechos Humanos Fray Matías de Córdova A.C. & Equipo de Estudios Comunitarios de Acción Psicosocial. 2016. *Estado del arte. Investigación: Violencia contra las mujeres en contextos de migración*. International Development Research Centre. Retrieved from: https://idl-bnc.idrc.ca/dspace/bitstream/10625/55953/1/IDL-55953.pdf.

8 Translating spiritual experience into environmental stewardship at Jamaica Bay, New York City

Zachary Garcia, Sonya Sachdeva, Lindsay Campbell and Erika Svendsen

Introduction

> There is a certain emotional draw to praying at the Bay, even if it is just standing there, closing my eyes and chanting a mantra. The world could not survive without our bodies of water; through them life is sustained.
>
> (Aminta Kilawan, co-founder of Sadhana)

Globally, populations are increasingly migrating to urban environments (United Nations 2017), requiring a rethinking of the management of and access to shared natural resources (Atkinson 2003, Roy 2009). Furthermore, shifts in migration patterns and the introduction of new communities in urban contexts also require an investment in cultural resources (Peters et al. 2010). As communities settle in new environments, cultural factors such as language, traditions and religious practices are critical in laying a foundation and developing a sense of place (i.e. the meanings attached to a spatial setting) but may need to be adapted and transformed to fit new contexts (Jorgensen & Stedman 2001, Antonscich 2010). In some cases, these transformations draw upon new landscapes and natural resources (Berkes 2012, Anderson 2014) to create cultural keystone places – defined as "places of high cultural salience for a particular group of people at a particular time and critical to their identity and wellbeing" (Janowski & Ingold 2012, Cuerrier et al. 2015). In one such case, Indo-Caribbean Hindu communities have adopted the waterfronts of Jamaica Bay in Queens, New York as a cultural keystone place, and specifically as a proxy for the sacred Ganga River in India. The present chapter draws upon fieldwork, including participant observation and semi-structured interviews, conducted during the summer of 2018 with members of the Indo-Caribbean Hindu community in New York City. We describe how Jamaica Bay has been imbued with spiritual value for this community over the past 30 years or more, and perhaps more significantly, how the adoption of Jamaica Bay as a symbol for the sacred Ganga River (also called The Ganges in anglicised form) has resulted in an emergent grassroots movement towards environmental stewardship (Semple 2015).

Background

The Indo-Caribbean community can be traced back to the early 19th century when the British established a system of indentured servitude where Indians were, at times, forcibly removed from India to work in the Caribbean territories to fund the growing sugarcane plantations (Younger 2010). Typically, these working agreements were comparable to slavery (Tinker 1974). These forms of "indentured labour" continued until 1917, during which time the influx of Indian communities to countries such as Guyana, Trinidad and Suriname brought with them their sense of home in the form of traditions, cultures, languages and religious practices (Manuel 1998, Verma 2008). Indo-Caribbean Hindu communities exemplified their cultural resilience by adapting and transforming within these Caribbean environments, re-articulating their Hindu traditions to correspond and reflect within these new environments (Haraksingh 1984, Ramnarine 1998). Surrounded by the *Kala Pani* (black water; i.e., the ocean), the ancestral home was distant, yet the power of beliefs sustained until one incident in Guyana changed the Hindu religious landscape. A prominent priest from Guyana, Jamsie Nadioo, orated this story to Stephanides in the 1980s. The story was frequently mentioned by interviewees. It tells of the emergence of Mother Ganga (also referred to as Maa Ganga or Ganga Ma), a key Hindu goddess, coming to her devotees (Stephanides & Singh 2000). Here is an excerpt of the translation:

> The people who came from India brought the book with them. They came to Albion Estate as "bound coolies" and worked six days a week. One or maybe two months after they arrived, one of the children became sick and no doctor could find the remedy. When they were not working on Sunday, they all went to the river bank to chant to the Mother Goddess. They prayed to Mother Ganges:
>
> "Divine Mother Ganges, we left India and came so far. We came to a different country and we have problems. Will you please listen to our prayers and shower us with your blessings. Oh Mother Ganges, please come to us. We have problems".
>
> Then Mother Ganges came, and one of the gurus began to read the mantra from the Mariamma Talatu and one of them became ecstatic with her presence. Mother entered his body and everybody prostrated themselves before her and asked her:
>
> "Divine Mother, can you tell us what is wrong?'"
>
> "You have all forgotten me and you have neglected my worship, so I made this child sick. Now, go to a clean place where there are plenty of cattle, and make my temple there and pray to Mother Kali. Get a bucket of water with turmeric, neem leaves, and flowers, bathe the child with it, and the sickness will pass away".
>
> They did so and the child got well. They then promised to worship the Mother from that time on. That is how puja came to this country.

This ancestral anecdote is one that is retold today in many Indo-Caribbean Hindu *mandirs* (Hindu temples) and *kovils* (Shakti temples). The memory of worshipping at the waterfronts of these powerful and sacred rivers in India, such as the Ganga, was realised and replicated by these Hindu communities. The healing effects of Mother Ganga, as a deity residing in the water, was a critical moment in the way water worship became an intimate and continued process for the Indo-Caribbean Hindu community.

After immigration reforms in the late 1960s and early 1970s, it became somewhat easier for immigrants to settle in the United States. People from the Caribbean began to migrate north (Roopnarine 2003). At this point, Indo-Caribbean communities started settling in the Borough of Queens in New York, predominantly in Richmond Hill, Ozone Park, and Jamaica. For these twice-migrants, the divine Hindu reality of worshipping the natural elements of air, water, space, fire and earth accompanied them to the Caribbean and to New York City (Min 2013). In New York City, certain areas of the Gateway National Recreation Area within Jamaica Bay became cultural keystone places for worshipping, and more importantly, a proxy for the sacred Ganga River (Figure 8.1).

This ancestral memory has persisted and culminated in distinct expressions and direct experiences that are now present at Jamaica Bay. Hindu devotees have been coming to these beaches to pray, perform *puja* (worship), seek guidance and truth, and offer sacred items for any number of Hindu deities (e.g., Mother Ganga, Ganesha, Hanuman, Sarswati). Scattered along these beaches, sacred relics vary from fruits, *diyas* (small clay pots), *saris* (fabric), plastic and aluminium items, *jhandi* flags (a flag symbolising a deity) and *murthis* (statues) of Hindu gods and goddesses (Bisram 2011) (see Figures 8.2 and 8.3). By leaving these offerings at the beaches, Jamaica Bay has become a central sacred space for Indo-Caribbean Hindus to interact with the divine. Once these offerings are placed in the waters, they are purified, and it is generally understood that the sacred relics should remain in the pure sacred waters. See de Mendonça (Chapter 8) for a description of how devotees of Afro-Brazilian religions make use of beaches along Copacabana beaches in Rio, Brazil to worship and pray to the goddess of the sea.

The adoption of Jamaica Bay as a sacred space by the Indo-Caribbean Hindu community has not been without tension, however. Over the past decade or so, conflict within the Hindu community, between other residents, and with government officials arose due to sacred items being left as offerings at the beaches (Van Hooreweghe 2012). Many within the Indo-Caribbean Hindu community believe that touching or throwing these sacred offerings away is sacrilegious. For the surrounding non-Hindu community, there is a disproportionate amount of blame placed on Hindu communities for the pollution of the beaches. Though these sacred items are not the primary cause of the pollution at Jamaica Bay (Hartig et al. 2002), nor are they the only types of items left behind (i.e., there are many secular examples as well, such as wrappers, soda bottles, etc.), they are visible reminders of the presence of the

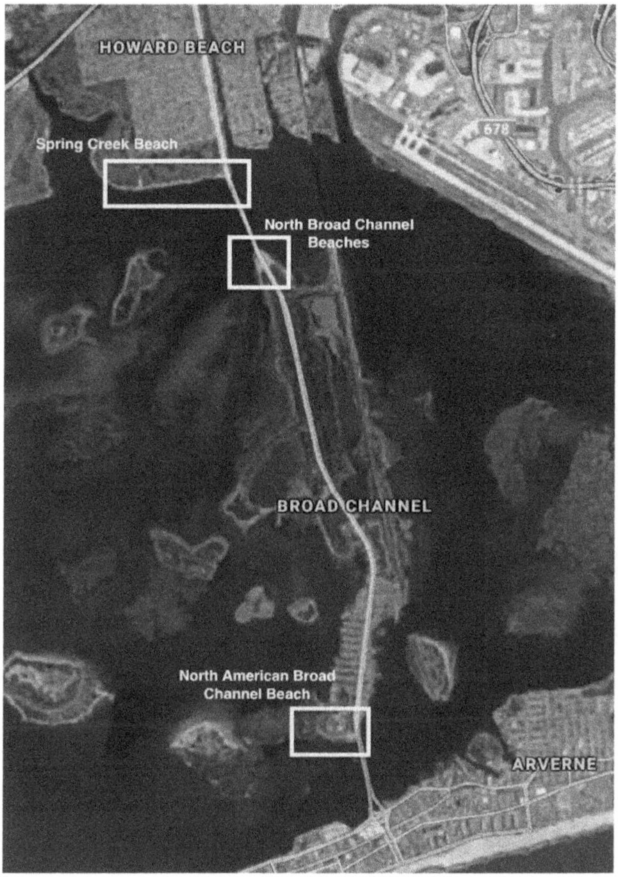

Figure 8.1 Map of Gateway National Recreation Area with the three areas of worshipping demarcated in white.

Indo-Caribbean Hindu community (Stabroek News 2011). The authoritative manager of Gateway National Recreation Area, the National Park Service (NPS), is often the first to encounter these visible reminders at several worshipping sites at Jamaica Bay. For these NPS rangers, they are left in a precarious position, because the Indo-Caribbean Hindu community have the right to worship freely at these beaches. As one former chief of Gateway National Recreation Area stated:

> My feeling has always been you can handle this [leaving offerings at the beaches] very easily – you talk to people. Every time I've spoken with people they're very polite and very nice, you know look you cannot leave this trash here – the tinfoil, the saris, the ... you gotta take it out, it's a National Park. And most people do it.

Figure 8.2 *Murthis* are left on the beaches after a *puja* at the Gate National Recreation Area. Source: Garcia, Zach.

NPS has made attempts to productively bridge barriers between resident groups and the Indo-Caribbean Hindu community, but our work and others suggest that natural resource managers in urban ecosystems may need to develop new approaches for engaging with multiple users who construe these areas in diverse ways (McMillen et al., Chapter 12). With the reinterpretation of sacred spaces in cities, communities and managers may have to continue to revisit their role and adapt their approaches to natural resource management, by drawing on cultural resources and community engagement.

Within the Indo-Caribbean Hindu community, a group of community organisers, community experts and organisations, such as Sadhana (https://www.sadhana.org) and the United Madrassi Association (UMA; https://www.madrassi.org), are pushing the boundaries of their religious practices at the beaches to explore the interface between spiritual and environmental accountability. Sadhana is a non-profit coalition of Hindus worldwide wanting to

Figure 8.3 The beach at Gate National Recreation Area has sacred items scattered along the beach front. Source: Garcia, Zach.

assert a progressive Hindu voice into public discourse. Sadhana is a platform for Hindus to advocate for progressive social values and social justice. Since 2012, Sadhana has been conducting monthly cleanups at Jamaica Bay which were inspired by the annual Earth Day cleanups conducted by the Federation of Hindu Mandirs composed of Indo-Caribbean Hindu elders. Similarly, the UMA is a progressive non-profit organisation committed to bringing about unity in the community, within the Madrassi diaspora, the greater Hindu community, and other NGOs, NPOs and grassroots organisations. They have hosted clean-up events, known as *Dharti Amma* Day (Mother Earth Day), several times a year where volunteers join and clean the oceans and beaches of all trash – religious and non-religious items. UMA volunteers are taught to do the cleanings with zero judgement towards those who have left offerings or littered the ocean or beaches. UMA's goal is to inspire change and bring an end to the practice of throwing unwanted and broken *murthis* in the ocean. One youth group, Shakti Samaj, has also played a significant role in teaming up with both Sadhana and UMA at their various community events. Both of these organisations, Sadhana and UMA, organise regular clean-up efforts (Sadhana 2018) in partnership with NPS and other community groups. These organisations reach out to various temples and other Indo-Caribbean organisations (Shakti Samaj) to promote environmental awareness at the beaches while collecting sacred items and disposing of trash.

In the current study, we ask what values drive these individuals (interviewees) and organisations (Sadhana, UMA and Shakti Samaj) to invoke environmental stewardship at Jamaica Bay as a form of their spiritual experiences. One important component of the Indo-Caribbean Hindu environmental groups is promoting an understanding of Hindu spiritual epistemologies of Jamaica Bay. We also investigate what this means for natural resource managers who supervise the land and residents who use the same space for recreation.

Gateway National Recreation Area, Jamaica Bay

This work took place in the Borough of Queens, New York, within the Gateway National Recreation Area (managed by NPS) as well as other open spaces under different jurisdictions in this area of Jamaica Bay. Various spots within Jamaica Bay and other waterways in New York City are used for worship, of which we looked at four (refer to Figure 8.1). These sacred areas were chosen due to their relatively high level of engagement by devotees.

Spring Creek Beach

This area sits at the north end of the Joseph P. Addabbo Memorial Bridge (Cross Bay Boulevard). The jurisdiction of this land belongs to the Metropolitan Transportation Authority (MTA). The beachfront on the southern and south-western fronts are the responsibility of NPS. Accessibility to the beaches can be complicated – there is limited access on either side of the bridge. The bridge covers a strip of beach, providing privacy for worshippers to commune with the divine waters (Figure 8.4). On the eastern side of the bridge, individuals can park on the highway or access street parking. Both sides of the bridge have small human-made paths which guide the user to the beachfront. The stretch of beachfront at Spring Creek is 0.9 km. Apart from worshippers, recreational users such as fishermen, couples and water-sport users occasionally use this beach. As of August 2018, trash receptacles were not present at this beach.

North Broad Channel Beaches

This area, which is at the end of the Joseph P. Addabbo Memorial Bridge (Cross Bay Boulevard) is split into two. The jurisdiction for these areas falls under NPS. One beach resides on the western side, and another is on the eastern side. Both beaches possess a parking lot that can accommodate roughly

Figure 8.4 Spring Creek Beach. Source: Garcia, Zach.

Figure 8.5 Western area of North Broad Channel. Source: Garcia, Zach.

30–40 vehicles. With these parking lots, it is easily accessible to worship in these areas with a large group. The two beaches are connected by a small strip of narrow land underneath the Joseph P. Addabbo Memorial Bridge – this area is low-hanging and not welcoming for individuals hoping to pass underneath easily. The main open beach areas are for various gatherings: *pujas*, picnics and parties.

Western Area of North Broad Channel

This area has a small concrete gazebo structure to the northern end. Worshippers occasionally have ceremonies at the gazebo. In front of this structure, there are two post boards that contain a variety of information (Figure 8.5). This beach area stretches for 0.4 km. There were eight trash receptacles as of August 2018 around the parking lot.

Eastern Area of North Broad Channel

This beach has a similar layout to its western counterpart, but without a gazebo structure. This stretch of beach faces John F. Kennedy Airport with planes taking off every few minutes. This beach stretches for 0.3 km. The opposite side (southern end) of the beachfront is forested area (Figure 8.6). There were six trash receptacles as of August 2018 around the parking lot.

Figure 8.6 Eastern area of North Broad Channel. Source: Zach Garcia.

Figure 8.7 North American Broad Channel Beach. Source: Garcia, Zach.

North American Broad Channel Beach

This area is about 6.1 km south from the other worshipping areas. North American Broad Channel Park is at the beginning of Cross Bay Veteran's Memorial Bridge. There is a parking lot at the northern edge of the park. From the parking lot, there is a concrete pathway that leads to three baseball fields, after which narrow footpaths develop and they lead straight to the water. There is a small walk through grasses that eventually turn to a sandy cove; a private and ideal place for worship (Figure 8.7). This park is managed by different entities. The baseball fields are managed by the New York City Department of Parks and Recreation (NYC DPR), while the beach fronts are the responsibility of the NPS. Like the other beaches in Jamaica Bay, the Cross Bay Veteran's Memorial Bridge provides privacy for worshippers. The stretch of beach is fairly short – 0.3 km.

Vegetation provides a safe barrier from the curious or invasive gaze of onlookers. Worshipping is generally done in the morning, because of morning *surya* (sun) oblations and because it is less populated with people who could potentially disrupt the *puja*. Ceremonies vary depending on the time of year and circumstance. Permit prices differ based on gathering size and whether an NPS staff member is needed on the premises during the event.

Devotees worship at Jamaica Bay on a weekly or monthly basis. Apart from these worships, the sacred relics being bequeathed to the water vary as well. Items left for the divine include: fruits (bananas, limes, coconuts), milk, leaves (neem), *kheer* (rice pudding), incense, *ghee*, cane juice, *saris*, *murthis*, *jhandi* flags, *jhandas* (bamboo sticks) and flowers. There are also other items left with spiritual or religious intent in the vicinity. However, one cannot always definitively distinguish between Hindu and non-Hindu offerings left at the beaches. Due to the multiuser interface at these beaches, various spiritual and religious individuals and groups (Spiritual Baptist, Shango Baptist, Santería practitioners, and Voodoo and Hoodoo practitioners and followers) practice ceremonies and also offer items to the ocean at Jamaica Bay (Vertovec 1998).

Examining worship practices

For this study, we interviewed nine individuals from the Indo-Caribbean Hindu community who regularly worship at Jamaica Bay (three women and six men). The two distinct denominations in Indo-Caribbean Hindu culture were represented, *Sanatan* and *Shakti* (see Kloss 2016 for descriptions). All interviewees have experienced worshipping, praying and conducting rituals at the beaches for the majority of their lives. Indo-Caribbean Hindu organisations gave us the names and contact information of 14 Indo-Caribbean Hindu temples in the Borough of Queens. Interviewees were selected by contacting these *Sanatan* and *Shakti* temples to see if anyone was available to speak about worshipping practices at the beaches. After contacting these temples, a list of individuals who frequently pray at the beaches and a list of community leaders were given. This study had an uneven gender distribution of interviewees because of cultural norms within the community, which made it simpler for the first author (who identifies as male) to contact and meet with men than to conduct interviews with women. Additionally, one pandit (*Sanatan* priest) from Chennai, India, was interviewed (man), as were one worker from NPS and one community organiser from Jamaica Bay (both men). All quotes mentioned in this study are gathered from Indo-Caribbean Hindu interviewees: Aminta Kilawan, Vijah Ramjattan, Devin Udairam, Varuna Sahabir and Malini Chunnilal. Interviews were conducted either in person or over the phone. These confidential interviews were semi-structured with the intention of exploring engagement at Jamaica Bay, connections to the ecosystem and personal anecdotes at the beaches. Observational data was based on visits to each of the beaches more than 25 times from 22 May 2018 until 19 August 2018. Photos of the beaches, surrounding areas and contents left by worshippers of various spiritual and religious sects, and recreational users were taken. Detailed ethnographic notes were taken after trips to the beaches and attendance at community events (including religious events and visits to temples).

After observable data was collected, weekly debriefs with United States Forest Service (USFS) research partners were conducted to discuss these notes.

Temples and other religious and spiritual events were documented and then synthesised based on our observations. These observations provided a reliable context of cultural elements within the community and grounded our interpretations. They allowed us to communicate competently with Hindu temples, leaders and community members in a well-intentioned manner.

Spiritual experiences and environmental values in Jamaica Bay

Our qualitative analysis identified key elements involved in transforming environmental values among these community organisers and organisations (Sadhana, UMA and Shakti Samaj). Our results are grouped into two larger themes of *spiritual experiences* and how these experiences inform the *transformation of environmental values*. Within the first theme of spiritual experiences, interviewees placed heavy emphasis on the cultural, religious and historical background of worshipping at the waters, and the relationships being developed by worshipping, communing and healing through various acts to the spiritual resource of the Mother. In the second section of our results, we consider how environmental values are being transformed based on these spiritual experiences.

Spiritual experiences at Jamaica Bay

Community members were drawn to interact and convene at Jamaica Bay for many reasons. As described,

> There are forms of creation after we pray. Yeah … there is creation. There, on a spiritual level, it uplifts you, in terms of creation, for whatever reason, when you're done praying at the beach when you leave and you go home, you see a change in a lot of things, you see like good stuff happening, you get good omens all around you. So, it's like a creation of good – and the destruction of bad. So, everything you went there for is abolished. The negativity is abolished.
>
> (Devin Udairam, a devotee)

The experiences at the beaches culminate as a call, reaction and then response from the divine. As one worships in this natural environment, a call is given to the divine. Interviewees then describe the unique reactions occurring at the beaches within the context of their specific Hindu denominations, *Sanatan* or *Shakti* worship. These unique reactions can be in the form of simply praying to a deity or a ritual practitioner manifesting a deity. Interviewees then elaborated on the powerful responses that come from the deities that reside in the waters – healing effects, abolishment of negativity, etc. It is the interconnectivity of all flowing water that directly connects interviewees to the divine power of the Mother. The relationship of calling, reacting and responding with the divine Mother lays the foundation to the transformative environmental changes (clean-up efforts) that are occurring at the beaches.

Worshipping, communing and healing

> The Ganga Maa is in my heart, the purity of the divine Mother is in my heart, and all of the Devatas and Devis, Christ, and Allah, they're all in my heart. I could just close my eyes and be anywhere in the world.
>
> (Vijah Ramjattan, president of United Madrassi Association)

When asked how long have they been coming to the beaches to worship, most stated that they have been coming as long as they can remember. This pattern is one that runs deep within the devoted community – the water at Jamaica Bay flows with the ritual memory of ancestral traditions. Ways of worshipping at the beaches vary for each devotee. Arguably, each *puja* is unique. Larger groups gather at the water to perform *pujas* for special ceremonial occasions such as *Visarjan*, *Durga Puja* or *Karagam Puja*. Hindu ritual celebrations of life called *sanskaars*, performed from birth to death, also take place here. These events can hold special significance in one's life – for example, the *mundan sanskaar*, the ritual cutting of a child's hair. In Indo-Caribbean Hindu culture, interviewees mentioned the necessity to either have this event performed at the waters of Jamaica Bay or place the hair in the water after the ceremony. The generational reoccurrence of these ceremonies at the waterways of Jamaica Bay illustrate how these practices have grounded and oriented generations of devotees (Chapple & Tucker 2000).

One unique feature of the region is the diversity of beliefs, even within the Indo-Caribbean Hindu community. Two forms of worship dominate this community – *Sanatan* and *Shakti* worship. Both of these forms of worship follow the Vedic texts (Hindu scriptures). It is not to say that there is no overlap between these groups – many people subscribe to certain rituals from both forms of worship; however, there are worshippers who strictly adhere only to *Sanatan* or *Shakti* worship. Strict *Sanatan* worship conforms to the Vedanta scriptures and the deities within those texts. *Sanatan* worship could be simplified as being more of a symbolic semblance of the limited pantheon of Hindu deities within the Vedic scriptures. For example, when *Sanatan* worshippers commune at the beaches, mantras will be read and there is a symbolic offering of items to the water via fruits, flowers and other small sacred articles. *Shaktiism* includes the Vedic Hindu pantheon, but it also incorporates South Indian folk deities (Mariamma, Kateri and Sangili Karuppan); and for some *Shakti* temples in Queens, they include a wide range of spiritual figures such as the prophet Mohammed, the prophet Jesus and Buddha.

What distinguishes *Shaktiism* from *Sanatan* worship is the centrality of the Mother – the female energy, particularly Mother Kali. In addition, manifestating a deity in the body of a ritual practitioner is a significant feature in *Shakti* worship. In a *Shakti* worship at the beach, *tappu* drums (South Indian snare drums) will beat fiercely while rituals are performed to guide the deity into the human vessel – blessed coconut milk, turmeric and ocean water are mixed to provide a concoction of energetic power which is poured over the deity's head to stay connected to the vessel. At this point, the devotees approach the divine and call upon the deity for guidance and healing. This interaction is known as "looking after". The spiritual conversation that is held for a *Sanatan* or *Shakti* worshipper with the Mother is an integral part of one's connection to the actual place, Jamaica Bay. The response given by the Mother in whichever form brings forth a sense of personal wellbeing.

Natural resource: the Mother

> When I was younger … I was born with asthma, severe asthma. And, the doctors gave me one month to live. My mother and father went to a priest, a *pujari*. And the *pujari* told them take me to the waters, bathe me in the water. And pray to the Mother, to purify the child to live. So said, so done … one month they gave me to live – they [the doctors] took it away. They [the doctors] said, they don't know how. They [the doctors] don't know why, but the child's healthy again. I opened my eyes, I wasn't breathing properly. I was incubated. And then all of a sudden, I came back. So that's my significance to Mother, she is my mother. And I put everything in my life. I owe it to her, the Mother of Ganga.
>
> (Devin Udairam, founder of Shakti Samaj)

The image of the Mother takes various personal forms for devotees, just as the Mother, herself takes various spiritual forms. One may hear her name as Mariamma – which is the South Indian folk goddess. It is believed that the characterisation of Mariamma gave rise to Mother Kali (McDermott & Kripal 2003). Therefore, there is the tradition of conceptualising both of these goddesses as one; and then, the other forms (Ganga and Kateri) branch from the centrality of Mariamma (Kali).

A popular form of the Mother that takes shape in *Sanatan* and *Shakti* worship is Mother Ganga. Mother Ganga is the presiding deity over the Ganga River in India (Chapple & Tucker 2000) – her symbolic presence is encapsulated in the waters at Jamaica Bay. She is known as the healer of blockages, sustainer of life and a force of absolute purity. Water is an aspect of Mother Ganga's absolute purity – in this way, the water is a purifying force for the devotee as well. This purification for the devotee takes shape in cleansing oneself with the water. In *Sanatan* worship, common offerings to Mother Ganga include items from *pujas* or by-products from prior *pujas* – these items (e.g. *havan* ritual can include any of the following: seeds, incense, *ghee*, milk or curd) cannot be disposed of in the trash; therefore, these *puja* items are taken to the waterfronts at Jamaica Bay.

When invoking deities in the physical or symbolic sense, there is an understanding that the direct link to the supreme, in any form, takes place at the waters. The devotional pilgrimages (either daily, monthly or yearly) are based on an intimate cause and effect relationship with the deity. Devotees share the causes of pain in their lives, either in the symbolic or manifested (physical) form, with the deity. The action of communicating with the deity through *pujas*, prayers or bathing elicit a spiritual call that releases directly into the urban environment (Svendsen et al. 2016). Interviewees mentioned a variety of responses from the divine, which include blessings, inspiration, protection, peace, success, happiness and enlightenment. For example, Mother Ganga is represented in the Indo-Caribbean Hindu community as a goddess of fertility. For devotees wishing to conceive, prayers to Mother Ganga bring the hope that conception will occur. In a broader sense of this relationship, there is a positive connection being continually strengthened with the Mother and the devotee at the beaches. The healing properties stimulated by these prayers cannot be understated. The Mother in all senses is an all-powerful entity that provides strength, guidance and healing to her devotees when called upon. This relationship takes form in a filial bond with the deity, her residence in the natural environment, and for some devotees, this process of Mother and child translates into a filial reciprocity with the surrounding ecosystem.

Environmental values informed by the spiritual experience

> So it's not only a Hindu issue, it's a human issue, it's a society issue, it's a Jamaica Bay issue, Jamaica Bay is made up of everybody, not only Hindus.
>
> (Vijah Ramjattan, president of United Madrassi Association)

Based on these spiritual experiences encountered at the waters at Jamaica Bay, the individuals and organisation leaders within the Indo-Caribbean Hindu community are actively attempting to articulate a new understanding within their religious ecology. They are attempting to build upon their deep-seated Hindu connections to the natural world to evoke environmental change, by taking steps to initiate discussions on environmental accountability in terms of their spiritual beliefs and experiences. The environmentally driven spiritual values evoked by these individuals and organisations allow other devotees to become more ecologically grounded and aware. In other words, they enrich environmental stewardship by employing rhetoric around one's obligation to the Earth and to the universe through eco-spiritual worship (Hitzhusen & Tucker 2013). The values of responsibility, loyalty and volunteerism are becoming entrenched within Sadhana, UMA and Shakti Samaj. However, in terms of their own Hindu beliefs, key tenets of *seva* and *bhakti* are being revisited. *Seva*, defined as a "loving service", is a service performed unto many (Haberman 2016). The sense of *bhakti* is a "devotion to a god in the Hindu tradition. This devotion takes various forms, emotional outpouring in song or ecstatic dance, literary production, or exchange with images of a deity"

(Chapple & Tucker 2000). *Bhakti* and *seva* in a worshipping sense are rearticulated to encompass an eco-obligation. That is to say, *bhakti* and *seva* are actively performed for not just the sake of the deity, but it is reciprocally performed to ensure the purity of the environment, the Mother and the self. Our interviews revealed four distinct elements that harness this new sense of eco-obligation to directly transform environmental values within the community: people-place relationship, heightened sense of environmental hazards/perceptions, spiritually driven environmental participation, and knowledge reconfiguration and sharing, which are each highlighted below.

People-place relationship

> I want them to know a lot of us West Indians are doing this damage; there are West Indians who are trying to clean it up, I don't want us to be viewed in any which way that we are bad people.
>
> (Devin Udairam, founder of Shakti Samaj)

The importance of the waters is spiritually valuable, yet, the religious ecology – Jamaica Bay itself – is important for these worshippers as well. A sense of ownership is formed by devotees at Jamaica Bay, as they emphasise the historical legacy and resiliency of worshipping at the waters in the Caribbean and in Queens (McMillen et al. 2016). There is a strong sense of love and respect for this area, as one individual expressed, "I want mainly, I want to see the beach cleaned because of the religious and spiritual and emotional feelings I have towards it". This strong sense of people-place relationship informs how devotees shape their experiences and understand the cultural significance within these spaces.

Ownership over one's community, traditions, and consequently, sacred offerings are also a part of this religious ecology. When asked about the most serious environmental issues at the beach, interviewees mentioned that the sacred items left at the beach are a major concern. For those individuals and organisations initiating discussions at community events and at temples, there is a larger topic around communal transgressions at the beaches. There is a sense of ownership and guardianship over the sacred offerings because the items are a part of the community. The offerings are transfixed within the forms of *bhakti* and *seva* to the deity, and in turn, are an ecological imprint within the spiritual landscape. For example, at UMA cleanups, Shakti worshipping volunteers are encouraged to be dressed in their traditional *Kovil* attire – yellow and white garments. Cleaning the beaches dressed in the same outfits worn while conducting rituals at the beaches and in *Kovil* is aimed at placating community tensions. The local community will see that those, who they think are responsible for the pollution, are also engaged in cleaning up the environment. UMA beach cleanings also involve ritual *tappu* drumming, singing, music and traditional food. For these individuals and organisations attempting to promote environmental discussions, the spiritual landscape is not separate from the ecological landscape, it is one. Therefore, there is a push for *bhakti* and *seva* to uphold both the spiritual and ecological vitality of the environment. As one *pujari* mentioned, the ownership he feels over his temple congregation overflows into the environmental integrity of the beaches, "But now, you know it's my people, so I make sure that they pick up after themselves, I'm very strict on it now". This sense of guardianship over one's congregation connects to the environmental attitudes this *pujari* holds to ensure that this place of worship remains free of items. His sense of *bhakti* and *seva* extends to incorporate an eco-obligation for the environment. This form of accountability for the beaches by Indo-Caribbean Hindus works to the advantage of NPS, because as one NPS employee mentioned, "I mean we just don't have staff to do anything else, we have too many sites to take care of, we just, we need the community help to do it". The individuals and organisations interviewed focus on the sheer importance of their worshipping space, and the protection of it – a religious ecology where their grandparents and parents communed.

Heightened sense of environmental hazards/perceptions

> People don't know these things and I think with communication and awareness, then they'll realise the effect that they're having on the environment. But they are thinking, oh, my little bit of, my one yard of cloth or whatever, what damage can it do?
>
> (Varuna Sahabir, pandita at Shri Lakshmi Narayan Mandir)

Interviewees shared their perceptions of the environmental effects both sacred items and general pollution had on flora and fauna at the beaches. We observed a spectrum of perspectives of what should and should not be left at the beaches by NPS, community groups and Indo-Caribbean Hindu individuals and organisations. For instance, one prevailing notion is that *prasad*, a religious offering, which can be in the form of rice, fruits or other blessed foods given to the water eventually become food for wildlife at Jamaica Bay. For these devotees, the act of making these offerings is one that nourishes the Earth and its creatures – a loving form of *seva*.

However, others believe that leaving any food item at the waters disrupts the ecological cycle for flora and fauna due to the pesticides and other residual chemicals on these foods. Sadhana and UMA strive to educate worshippers that the good intention (*seva*) placed at the waters is sufficient – the physical act of feeding wildlife is not necessary. Another individual expressed that he came to worship one day, and he saw that the same garbage bags from the following week was still there. It was a moment for this devotee to begin thinking about the impacts of slow accumulation. Another devotee expressed similar thoughts that if everyone brought a sacred offering one weekend – the cumulative impact of those offerings in one weekend would compromise the ecosystem.

Spiritually-driven environmental participation

> Because you're at the water, the Mother is right next to you. Pray to her while you're doing these things, certain things that you're picking up – just pray to her while you're doing it.
>
> (Devin Udairam, founder of Shakti Samaj)

For Sadhana and UMA, a large part of their work has been focused on spiritually driven environmental participation with the belief that everyone is endowed with a sacred presence within. When communing at the beach, a devotee acknowledging a deity is a greeting of one's inner ecology with that of the outer religious ecology. If environmental efforts are undertaken at the beaches, the devotee is performing *seva* on behalf of the deity and its divine Earthly creations. Consequently, the devotee is also benefitting, because their spirit is gaining fulfilment by providing *bhakti* to the deity. As stated before, there is a call, reaction and response process that takes place at the beaches. In spiritually driven environmental participation, one more step is added – devotee reaction. This process becomes a call to the divine, a reaction from the divine, a response from the divine and then a reaction by the devotee which becomes the eco-obligation of *seva* or *bhakti*.

This reactionary process was beautifully captured by one worshipper, "You see the Mother, you see the ocean, you're looking for this truth, you have to find it – but without realising it, it's in you". This process is seen as an innate sense of finding truth through one's own actions. This reactionary step can be expanded as a way of accessing new pathways to guide other devotees to partake in caring for the environment as a part of spiritual devotion. *Bhakti* evokes the filial piety of the Mother-child relationship where the Mother looks after the child and the child, in turn, cares for the Mother. In the push for environmental stewardship, *bhakti* and *seva* become a force of caring for the Mother as an environmental figure. When environmental outreach programmes are completed at the beaches, such as cleaning the beaches, these organisations incite the symbolic acknowledgement and appreciation from the Mother for having done a service unto her, the Earth, and oneself. These individuals and organisations urge devotees to trust based on the notion that the divine wants the worshipping area clean.

A more encompassing environmental take on *bhakti* and *seva* is situated within these discussions. As one individual described, "I say devotion in motion. I let that be my way of praying – active worship. I feel like that purpose is higher". Eco-obligations of *seva* and *bhakti* are intertwined with the active forms of environmental cleanup. Rather than offering a sacred item, the devotee is offering themselves as a vessel to clean the beaches. UMA beach clean-ups are unique in this way, because the participating *Kovils* and devotees are instructed to view the clean-ups as an act of *puja* where they serve the divine by cleaning and serving what the divine has created.

Knowledge reconfiguration and sharing

> It's said that one should lead by example, so why not start now? Every small change we make today, creates a bigger and better change for tomorrow. With this, we can ensure that we create a better world for not only now, but for generations to come. The example we set today, will lay the foundation and instill the values for tomorrow's generations. If we all go a little, we can make a big difference. God gave us this world and it's our duty to protect and take care of it. That's how I look at it.
>
> (Malini Chunnilal, Shakti Samaj clean-up participant)

We observed two forms of knowledge reconfiguration and sharing. On one hand, existing beliefs are reconfigured and retold to take a more eco-friendly form by bridging science and spiritual beliefs. On the other, organisations, individuals and temples challenge the status quo by sharing eco-centric information across various platforms. There have been discussions in these communities of reworking how deities are viewed and perceived. For example, as one *Shakti* worshipper explained the connection between Mother Ganga taking various divine forms at the beach and the various molecular forms she can also take – water as a liquid, solid and gas. With this reinterpretation, there is a re-emergence of the Mother and her direct being as water. This perspective could reformulate how water, in general, as a healing power, must also be respected and pure in all forms. There is a new relationship created whereby the spiritual entity of Mother Ganga, not only resides in the water but she, herself, is the water.

This reworking can take shape in how devotees perceive offering items. Rather than placing objects in the water and harming it, offering a "pure mind" could be sufficient devotion. In the above example, there is a blending of the scientific and Hindu cosmology to promote environmental awareness. Another act of knowledge reconfiguration comes in the form of restoring the *murthis* and *diyas* left at the beaches. Items that were left at the beaches due to their "brokenness", are collected during clean-up events and restored (painted, glued, decorated etc.) to have a new sense of "sanctity" or sacred life. Sadhana has hosted several art workshops in the community and at the Jamaica Bay Visitor Centre to restore these offerings. These workshops fill a gap in the community in terms of access to art opportunities, but they also help to develop personal connections to these items. In 2019, a Canadian Indo-Caribbean artist, Dr. Andil Gosine, developed an exhibit to explore the deep connections of the offerings left at the beaches and to understand the connections of nature and faith (Vasquez 2019).

Knowledge sharing comes in the form of pamphlets and social media. Sadhana developed two comprehensive handouts to share with both worshippers and NPS staff. The first pamphlet incorporates the reconfiguration of what should and should not be left at the beach. The second pamphlet is unique in that is was co-developed with NPS – this pamphlet has a tone that not only speaks to Hindu devotees, but it conveys the sanctity of Jamaica Bay to a non-Hindu visitor. These pamphlets are given to devotees that come to the beach and to temple congregations.

UMA also offers educational material through literature to educate the community on eco-friendly worship. UMA began the *Murthi* Rescue Mission: Our *Murthis* Are Not Garbage campaign where unwanted *murthis* are picked up free of charge and confidentially from anyone in the community (https://www.madrassi.org/). UMA has also posted videos on social media to show how sacred items left at the beach can negatively impact the ecosystem and beach users. Thousands of viewers have watched these videos, sparking hundreds of comments. Indo-Caribbean Hindu communities in Ontario, Canada and Florida also have commented on these social media posts, mentioning the same situation at lake and ocean fronts. These approaches enlist the eco-obligation to the Mother and the environment, in the form of *seva* and *bhakti*, but also enlighten people on the ecological impacts of the offerings.

Conclusion

On the shores of Jamaica Bay exist rich and varied practices of spirituality. The Indo-Caribbean Hindu community of the greater New York area view this site as a manifestation of Mother Ganga, a key goddess within Hindu theology. Jamaica Bay has become enmeshed within the cultural, traditional and religious practices of this community, not, however, to the detriment of its ecological integrity, as some have suggested (Dolnick 2011). Rather, our work suggests that this community's religious practices can lead to ecological connections and deepen community members' sense of place. These environmentally inspired values are being kindled amidst temples, personal discussions and over social media. Many of these values are outlined in this study, yet personal and familial values for these water-worshipping practices transcend this study. Reshaping and re-articulating environmental values within a spiritual context, where nature is already worshipped, is one route for environmental promotion, like in the case of Indo-Caribbean Hindu devotees who also practice environmental stewardship. There are a number of paths forwards. Some of our non-governmental research participant-informants indicated a need to conduct a quantitative assessment of Hindu artefacts left behind in these areas compared to general litter left at the beach. The comparison of the number and volume of sacred items versus general litter may help (a) understand the impact that clean-up efforts sponsored by Hindu organisations in the area are having; (b) identify pre-existing ecological issues at the Bay, distinct from the spiritual practices conducted by Hindu communities and (c) promote discussions between Indo-Caribbean community members and other local communities (fishers, boaters, bird watchers etc.) who can sometimes be at odds with the Hindu practices observed at the beach (Landor 2008, Van Hooreweghe 2015). McMillen et al. (Chapter 12) similarly explores how creating platforms to share sacred relationships provides opportunities to engage with a diversity of environmental issues.

Future research might also involve a larger investigation of other spiritual systems engaged at Jamaica Bay and their environmental attitudes towards the Bay – some of these spiritual systems include Voodooism, Hoodooism, Shango Baptist and Spiritual Baptist. Additionally, a wider exploration with Indo-Caribbean individuals and organisations in Canada, Florida and California who also engage in worshipping practices at waterbodies might broadly demonstrate the role of community-based structures in shaping religious ecologies. Finally, future work should aim to understand and bridge the needs, priorities and key concerns of the diverse set of land managers, stewards and stakeholders involved in the Bay. In addition to the need to research and understand managers' perspectives that were not covered in this chapter, there is an opportunity for co-learning and knowledge exchange that could be facilitated in a collaborative workshop, discussion or "competency group" setting. Such an engagement would build upon the existing collaborative relationships that have been developed, for example, between Sadhana and NPS in co-hosting stewardship events and co-creating brochures for the public. A forum for open dialogue could be used to build further trust and a shared vision for Jamaica Bay.

References

Anderson, E.N. 2014. *Caring for Place: Ecology, Ideology, and Emotion in Traditional Landscape Management*. Walnut Creek, CA: Left Coast Press.

Antonsich, M. 2010. Searching for belonging – An analytical framework. *Geography Compass*, 4(6): 644–659.

Atkinson, R. 2003. Domestication by cappuccino or a revenge on urban space? Control and empowerment in the management of public spaces. *Urban Studies*, 40(9): 1829–1843.

Berkes, F. 2012. *Sacred Ecology* (3rd ed.). New York, NY: Routledge.

Bisram, V. 2011. Hindu leaders in New York must come together and lobby for space to conduct shore rituals. *Stabroek News*. https://www.stabroeknews.com/2011/opinion/letters/04/24/hindu-leaders-in-new-york-must-come-together-and-lobby-for-space-to-conduct-shore-rituals/ (Accessed: 19 April 2019).

Chapple, C.K. & Tucker, M.E. 2000. *Hinduism and Ecology: The Intersection of Earth, Sky, and Water*. Boston, MA: Center for the Study of World Religions.

Cuerrier, A., Turner, N.J., Gomes, T.C., Garibaldi, A. & Downing, A. 2015. Cultural keystone places: Conservation and restoration in cultural landscapes. *Journal of Ethnobiology*, 35(3): 427–448. doi: 10.2993/0278-0771-35.3.427.

Dolnick, S. 2011. Hindus find a Ganges in Queens, to park rangers' dismay. *The New York Times*.

Haberman, D.L. 2016. Hindusim – devotional love of the world. In Jenkins, W.J., Tucker, M.E & Grim, J. (eds). *Routledge Handbook of Religion and Ecology*. Routledge, London. pp. 35–42.

Hartig, E.K., Gornitz, V., Kolker, A., Mushacke, F. & Fallon, D. 2002. Anthropogenic and climate-change impacts on salt marshes of Jamaica Bay, New York City. *Wetlands*, 22(1): 71–89.

Haraksingh, K. 1984. The Hindu experience in Trinidad. In *Proceedings of Conference on 'East Indians in the Caribbean-beyond survival', August 28th – September 5th*, University of the West Indies, St. Augustine.

Hitzhusen, G.E. & Tucker, M.E. 2013. The potential of religion for Earth stewardship. *Frontiers in Ecology & the Environment*, 11(7): 368–376.

Janowski, M. & Ingold, T. 2012. *Imagining Landscapes: Past, Present and Future*. New York, NY: Routledge.

Jorgensen, B.S. & Stedman, R.C. 2001. Sense of place as an attitude: Lakeshore owners' attitudes toward their properties. *Journal of Environmental Psychology*, 21(3): 233–248.

Kloss, S.T. 2016. *Fabrics of Indianness: The Exchange and Consumption of Clothing in Transnational Guyanese Hindu Communities*. Palgrave MacMillan, New York, NY.

Landor, L. 2008. Gateway, volunteers forced to clean litter left on shores after rituals. Queen Chron. https://www.qchron.com/editions/south/gateway-volunteers-forced-to-clean-litter-left-on-shores-after/%20article_70957027-43d0-5149-b638-c5ea3a76c118.html (Accessed: 19 April, 2019).

Manuel, P. 1998. Music, identity, and images of India in the Indo-Caribbean Diaspora. *Asian Music*, 29(1): 17–35.

McDermott, R.F. & Kripal, J.J. 2003. *Encountering Kali in the Margins, at the Center, in the West*. Los Angeles, CA: University of California Press.

McMillen, H., Campbell, L., Svendsen, E. & Reynolds, R. 2016. Recognizing stewardship practices as indicators of social resilience: In living memorials and in a community garden. *Sustainability*, 8(8). doi: 10.3390/su8080775.

Min, P.G. 2013. The attachments of New York City Caribbean Indian immigrants to Indian culture, Indian immigrants and India. *Journal of Ethnic & Migration Studies*, 39(10): 1601–1616. doi: 10.1080/1369183x.2013.833688.

Peters, K., Elands, B. & Buijs, A. 2010. Social interactions in urban parks: Stimulating social cohesion? *Urban Forestry & Urban Greening*, 9(2): 93–100. doi: 10.1016/j.ufug.2009.11.003.

Ramnarine, T.K. 1998. Historical representations, performance spaces, and kinship themes in Indian-Caribbean popular songs texts. *Asian Music*, 30(1): 1–33.

Roopnarine, L. 2003. Indo-Caribbean migration: From periphery to core. *Caribbean Quarterly*, 49(3): 30–60.

Roy, A. 2009. Why India cannot plan its cities: Informality, insurgence and the idiom of urbanization. *Planning Theory*, 8(1): 76–87. doi: 10.1177/1473095208099299.

Sadhana. 2018. Project Privithi. from http://www.sadhana.org/project-prithvi/

Semple, K. 2015. Promoting environmental awareness among practicing Hindus. *The New York Times*. https://www.nytimes.com/2015/11/18/nyregion/promoting-environmental-awareness-among-practicing-hindus.html (Accessed: 19 April, 2019).

Stabroek News. 2011. Guyanese Hindu rituals in NY lead to environment clash. *Stabroek News*. https://www.stabroeknews.com/2011/news/guyana/04/22/guyanese-hindu-rituals-in-ny-lead-to-environment-clash/ (Accessed: 19 April, 2019).

Stephanides, S. & Singh, K. 2000. *Translating Kali's Feast: The Goddess in Indo-Caribbean Ritual and Fiction*. Amsterdam: Brill.

Svendsen, E.S., Campbell, L.K. & McMillen, H.L. 2016. Stories, shrines, and symbols: Recognizing psycho-social-spiritual benefits of urban parks and natural areas. *Journal of Ethnobiology*, 36(4): 881–907. doi: 10.2993/0278-0771-36.4.881.

Tinker, H. 1974. *A New System of Slavery: The Export of Indian Labour Overseas, 1830 – 1920*. London, UK: Oxford University Press.

United Nations. 2017. *World Population Prospects: The 2017 Revision, Key Findings and Advance Tables* (p. 248). New York, NY: Department of Economic and Social Affairs.

Van Hooreweghe, K. 2012. *The Creeks, Beaches, and Bay of the Jamaica Bay Estuary: The Importance of Place in Cultivating Relationships to Nature*. New York, NY: City University of New York.

Van Hooreweghe, K. 2015. Place and exclusion in New York City's Jamaica Bay. *Journal of Environmental Studies Science*, 6: 753–758. doi: 10.1007/s13412-015-0256-6.

Vasquez, V. 2019. Deities (Part 1) opens – discovering sacred offerings at Jamaica Bay. *The Wave*. https://www.rockawave.com/articles/deities-part-1-opens/ (Accessed: 19 May, 2019).

Verma, M.M. 2008. Indo-Caribbean Hindu practice in Queens: ethnomethods of constituting place, practice and subjects. PhD thesis, Columbia University, Ann Arbor.

Vertovec, S. 1998. Ethnic distance and religious convergence: Shango, Spiritual Baptist, and Kali Mai traditions in Trinidad. *Social Compass*, 2(45): 247–263.

Younger, P. 2010. *New Homelands: Hindu Communities in Mauritius, Guyana, Trinidad, South Africa, Fiji, and East Africa*. New York, NY: Oxford University Press.

9 Biocultural diversity at Rio de Janeiro's urban beaches

Wellbeing, belonging and conflict

Guilherme Cruz de Mendonça

Introduction

"No mar estava escrita uma cidade" (In the sea a city was written). These words are inscribed on a bench located on the sidewalk of Copacabana beach. The author is Carlos Drummond de Andrade, perhaps the most famous Brazilian poet. He was born in Minas Gerais State, but moved to Rio de Janeiro and lived for many decades along Copacabana beach. In the mornings, he used to cross Avenida Atlântica and sit on a public bench, admiring the land and seascape, watching people passing by, sunbathing and probably finding inspiration for his work. Years after his death, the local government built a bronze statue seated on the same bench in his honour. The Carlos Drummond de Andrade statue now pays tribute to human-nature relationships in beach environments located in Rio de Janeiro.

Beaches and nature play a central role in Rio de Janeiro's identity. The city is located along the Southeast Brazilian coast within the Atlantic Forest Biome, one of the most diverse Brazilian biomes, which follows the country's coastline from the south up to the Amazon border. The city boasts 77 beaches categorised into two main types: bay and oceanfront. The most famous bay beaches are Copacabana, Ipanema and Arpoador. The oceanfront beaches are located in the South Zone which is the most economically developed area of the city. The districts that have beaches are the most valued areas in the city. Living near the beaches is highly appreciated by the citizens and is reflected in the local identity of *cariocas* (a *carioca* is person who was born or has lived for a long time in Rio) whose lifestyle and culture has been moulded around these beaches.

Today, Rio's beaches are considered one of the most democratic spaces in the city, with the free access allowing for a variety of cultural groups of different social and economic standing to mingle. Beaches have been chosen as the point of entry for this study because of their centrality for the city and its inhabitants.

Conceptual perspective and study information

Rio de Janeiro's rich biological and cultural diversity, which offers a laboratory for the study of biocultural diversity, is recognised by the Brazilian state and UNESCO. In 2012, the World Heritage Committee included Rio de Janeiro in the list of World Heritage Sites, recognising it as both a natural and cultural landscape, which was the very first time that the concept of 'cultural landscapes' had been applied to a big city.

Inherent to the contemporary concepts of both 'complexity' and biocultural diversity is a recognition that associations between nature and culture need to be restored. Complexity theory requires thinking and practices that overcome the dichotomy that society has created between nature and culture. Within this perspective, the recognition, understanding and protection of biocultural diversity is seen as essential for the resilience of social-ecological systems and for the sustainability of life.

Biocultural diversity linkages are now recognised as being present and reflected within cityscapes (Mendonça 2014), as the many chapters of this book illustrate. Such linkages and relationships have been brought into the cityscape through processes of migration. Traditional people migrating from rural areas to cities often bring their knowledge, perceptions and ways of interacting with nature into the city. Descendants of traditional people may rescue older traditions or recreate traditions from rural communities in cities. Additionally, as cities expand, they may incorporate more traditional rural areas, with the rural-urban divide thus becoming less pronounced and biocultural diversity being enriched. However, is it not only indigenous and traditional cultures that contribute to the network of linkages that make up biocultural diversity; it is all communities, including those associated with highly complex modernised societies (Cocks 2006, Elands et al. 2015, Mendonça 2014). These may be relevant factors for building a process of wellbeing and belonging, as well as for producing conflict.

This chapter applies a biocultural diversity lens to the urban beach environments of Rio de Janeiro, considering them as a nature-culture-city hybrid. It will attempt to answer the following questions: how differently do urban residents view, engage with and give meaning to beach environments in Rio de Janeiro? How do the inextricable links between nature and culture within beach environments strengthen the sense of wellbeing and belonging among different social groups within the city? And through the lens of political ecology, considering the beach environments as resources, what are the conflicts over them? In considering these questions, the chapter will examine the diversity of knowledge, practices, meanings and lifestyles (i.e., cultures) associated with beach environments in Rio de Janeiro's urban context.

The two beaches chosen for the study

Of Rio de Janeiro's 77 beaches, the case study focussed on Copacabana and Leme, which are among the most frequented beaches by locals and tourists. Both beaches have historical roles and contain elements that have contributed to local and Brazilian identity and culture. Together, Copacabana and Leme form a continuous natural stretch of beach 4.7 km long. The Copacabana section, which is 3.6 km in length, is separated from the 1.1 km Leme section by a busy street. The division between the two beaches is therefore a social construction and not a natural characteristic of the coastline. The Copacabana district is located in the South Zone of Rio de Janeiro and is the second most densely populated district in Brazil and most densely populated in Rio de Janeiro with 161,000 people living in 4.2 km^2. Copacabana, known by locals and nationals as "Princesinha do Mar" (Little Sea Princess) is the most famous beach in the country. The beach and the adjacent avenue, Avenida Atlântica, is the best-known area in the district.

The chapter includes geographical and historical information which gives context to the city's beach environment and case study information about biocultural diversity at the city-beach interface. For the case study, participant observation and survey methodologies were used to gather data on the diversity of ways in which urban residents interact, engage with and give meaning to Rio's beaches. Observations on the of uses and relationships with and on the beaches was done over various weekends between November 2018 and June 2019. In addition, the results and data from three academic works conducted by the author (Mendonça 2014), or supervised by the author in 2018–2019 have been drawn from.[1] The author was born and raised in Rio de Janeiro and has been residing in a neighbourhood close to the study area. The observation and analysis also reflect the author's experiences, engagements and interactions with beach environments as a "local" or "native".

Historical background

When the first European settlers (Portuguese and French) arrived in Rio de Janeiro in the 16th century, native indigenous communities occupied both sides of Guanabara Bay and its entrance. Conflicts to settle and establish the foundation of the city occurred within the bay. In the past, the beach that is today called Copacabana could only be reached via the ocean or by a long horse ride. The first building along the beach front was a small church in honour of Nossa Senhora de Copacabana, a Bolivian saint. The church overlooks Lake Titicaca. In the local Aymara language "Copacabana" means blue viewpoint. There are references from the 19th century to explorations that took place in the area. The first settlers to occupy the area included a small group of fishermen that used the beach as a point of support. Temporary huts and lodges were built.

The Copacabana district was officially created in 1892 and a tunnel called "Old Tunnel" was built to connect Copacabana to the rest of the city. In 1910, a second tunnel called "New Tunnel" was built. The opening of the two tunnels contributed to the occupation and development of the Copacabana district and the movement of residents and commercial outlets to the coast. As the perception grew of beaches as desirable places for recreation and swimming, rather than threats to humans, so new areas were conquered for social expansion. By the end of the 19th century, bathing and swimming in the sea had been attributed therapeutic and recreational value and social and economic elites appreciated the possibility of living near the ocean and having beach houses in Copacabana.

Rio de Janeiro was the country's capital until 1960. Copacabana was considered the most influential district during the first half of the 20th century, as it hosted the economic, social, intellectual and cultural elites. Copacabana beach has played a significant role in shaping the Brazilian concept of urban beaches as representing glamour, with luxury hotels and expensive buildings created for and occupied by the elite. It has also shaped the Brazilian practice of bathing and swimming in the sea. In 1960, when Brasilia replaced Rio de Janeiro as the country's capital, Copacabana retained its cultural status, but partially lost its centrality as a political power, signalling the decay of Copacabana. During the 1960s and 1970s local elites began to move to the neighbouring beaches of Ipanema and Leblon, and lower- to middle-class residents began to move into Copacabana.

Today Copacabana is one of most diverse districts of the city in terms of class, race, religion and gender: rich people live alongside poor people, Whites alongside Blacks, Catholics alongside Jews and Protestants, heterosexuals alongside members of the LGBTI+ community and tourists alongside locals. Local expressions to reflect this diversity include the "asphalt" (paved areas for the middle classes) and the "slum" (favela). Copacabana is surrounded by hills that are occupied by three "favelas", while Leme has two. In the 2000s, a new subway was built which increased access to Copacabana from the suburbs and increased social diversity in the area. While Brazil has one of the highest levels of inequality in the world (OXFAM 2017), Copacabana has come to represent the inequality and the cultural diversity of the entire region; this diversity is also represented on its beaches.

The four primary elements of the urban beach area

The beach area is composed of four elements: (1) the frontline buildings and two traffic lanes which run parallel to the beach; (2) the boardwalk which consists of a bicycle path, kiosks and bars; (3) the beach – sandy strip and (4) the ocean.

The frontline buildings are predominantly residential. In the area there are also many hotels, a commercial building, a consulate, a public high school, and several restaurants serving Brazilian and international food for locals and tourists. Avenida Atlântica (Atlantic Avenue) is considered by the locals to be a sophisticated area to live in, particularly if the apartment has an ocean or beach view.

The *boardwalk*, built in 1906 by Portuguese engineers inspired by Rocio Square in Lisbon, is made from white calcite and black basalt imported from Portugal. The contrasting white and black represents the place where the dark waters of Tejo River meet the clear waves of the Atlantic Ocean. Today, the boardwalk design is a relevant part of the city's identity, serving as an icon to identify not only the city, but also its inhabitants. The Copacabana boardwalk design can be seen in many products such as entrance mats, bags, towels and souvenir items in touristic shops.

Along the boardwalk many kiosks are located offering different kinds of food and drinks. During the holiday season, bands perform live, from Brazilian bossa nova and samba to international pop hits. Several kiosks receive financial support from businesses such as TV broadcasters and beer breweries, to host events such as parties or yoga classes. In 2018, one of the local kiosks hosted "philosophy at the beach" workshops where teachers and researchers gave free lectures on philosophy and literature to the public. Under the kiosks, there are public paid toilets and shower facilities.

From Leme to Pontal beaches,[2] there are 27 life and rescue stations, called "Postos de Salvamento" in Portuguese. In Copacabana, there are six such stations. At each station there are also public toilets and showers, which can be used for a small fee. The Postos, not only in Copacabana, but along all Rio de Janeiro beaches, also have a social function as place references and markers of identity. The most famous Posto in Copacabana is the sixth one. Some factors which explain the buzz around this beach area are the quieter sea, the landscape with Sugar Loaf Mountain in the background, the fisherman community and the market.

The *beach* was originally a thin strip, but in the 1970s sand was brought in from other areas to widen it. It now offers bathing and sports facilities as well as infrastructure, including tents from which entrepreneurs can rent out chairs and beach umbrellas for bathers. The *ocean* is considered hazardous as it is subject to strong currents sometimes inundating the boardwalk and roads. The sea temperature is cold, due to the cold ocean currents that run off the coast of Brazil. The curvature of the bay means that waves are calmer at the end of the beach.

Use of the urban-beach interface by different groups

Users of these four elements include residents who live near the beach as well as those who come from the North and poor zones of the city. Copacabana has one of the highest concentrations of hotels in the city, attracting tourists from within Brazil and abroad. Professional athletes and amateurs practising different kinds of sports also use the beach. Members of the LGBTI+ community concentrate around one of the kiosks which has a rainbow flag. At the end of Copacabana beach, a fishing community has obtained the legal right to occupy the end of the boardwalk and retain their traditional boats and run a small fish market.

The boardwalk of Copacabana beach is a promenade where people can see and be seen. Locals and tourists alike often stroll along the boardwalk without a specific destination in mind: the local term for someone strolling in this manner is *flaneur*. Street artists such as musicians, performers and sculptors are also present. Sand sculptures are built on the beach by artists. In the evenings, the part of the boardwalk that extends into the road lanes is occupied by a street market where food, souvenirs and art are sold.

Two sculptures of Brazilian writers have been erected on the boardwalk. The one pays tribute to the famous Brazilian poet, Carlos Drummond de Andrade. He lived most of his life in Copacabana and would be seen regularly walking on the boardwalk and seated on the beach in front of his apartment building, reading or writing poems. The other sculpture is in honour of the well-known Brazilian author, Clarice Lispector, who used to live in Leme and enjoyed walking along the boardwalk to view the sunset over the beach. Both sculptures represent their subjects seated on benches and it has become a ritual for locals and tourists to wait in long lines to take photographs and selfie images of themselves seated next to the sculptures.

Bathers represent a diversity of social classes and come from different parts of the city. They interact with the beach environment in different ways. There are those who swim for leisure, while others engage in sporting activities, such as swimming, surfing or paddling. Others prefer not to enter the ocean at all as they fear swimming, particularly when the currents are strong or when the sea is polluted. Some prefer to simply socialise and sunbathe. During the summer months the beaches are crowded with women of different ages lying on their beach towels enjoying the sun. Public policies and campaigns against skin cancer have not dampened people's desire for a tanned skin. Many young women will use commercial or homemade products to speed up the tanning process.

Beaches are also places to meet up with friends and spend time with family and relatives. In the same way that visits to parks provided places for families and friends to get together in cities in Europe (Elands et al., Chapter 11) and in Helsinki (Vierikko & Jaatsi, Chapter 5). Cocks et al. (Chapter 6) describes similar activities taking place at dams which are situated within the municipal boundaries of two towns. In Eastern Cape, South Africa, it is common to observe groups of families and friends demarcating circular spaces with beach umbrellas and beach chairs with their cooler boxes and sand towels in the middle. Some even bring blow-up pools for their young children to play in. Another frequent practice is arranging social gatherings, which include meeting up with friends at the beach, particularly in the evenings for a date or for drinking soirees.

The space provided on the beach is free and many social groups occupy the beach. While certain groups tend to congregate in particular places, people's relationship to the beach environment has a democratic atmosphere and relationships to beach environments are dynamic. Copacabana beach has a high concentration of LGBTI+ in front of Copacabana Hotel. A popular crowd frequents the beach in front of the streets that lead to subway stations or some bus stops. Recently, middle-class youth have also started occupying the beginning of the beach (Leme).

The beaches also provide spaces for commercial outlets to operate. Those offering food and drink and other beach-related products are the most dominant. It is very common to observe fried shrimps, pastries and famous "Biscoito Globo"[3] being sold. "Mate Leão"[3] and beers are the most popular drinks available. Some of these commercial practices and products, such as, "Biscoito Globo" and "Mate Leão" were recognised by the City Hall as local cultural heritage and have legal protection.

Sporting activities are a key dimension of people's interaction with the Copacabana beach environment. On the boardwalk, there is a bike lane used by cyclists and runners. On Sundays, the road lanes are closed to motor vehicles so that they can be used for running and cycling. On the beach, public facilities to play sports such as soccer and volleyball have been erected. Private businesses have set up schools to teach people how to play frescoball (paddle ball); footvolley and "altinha". Altinha is a game played near the water with a soccer ball. The game's goal is to keep the ball high ("alta") and not to let it fall on the sand. All players contribute to the goal, so there are no opposing teams playing against each other. There are no winners or losers. It is a collaborative practice, connected to the democratic spirit that is present in Rio's beaches.

Parties, rituals and protests

It is possible to observe biocultural diversity practices taking place on a daily basis at Copacabana beach. It is also important to draw attention to specific events that highlight the diversity of social group interactions that take place. These include New Year's Eve celebrations, Lemanjá processions and political demonstrations.

The night of December 31 is very special for Brazilians and cariocas alike. Each year, Brazil's most famous New Year's Eve Party, known by the French word "Réveillon", is hosted at Copacabana beach. After the Rio de Janeiro Carnival (just before Lent), this is the second biggest party that attracts people to the city. The preparations for this evening start well in advance. Usually the hotels in Copacabana are fully booked and tourist buses from different parts of Brazil and South America form long lines in the streets of nearby districts. Groups and caravans arrive early in order to be guaranteed a good space to watch the fireworks. Usually three big temporary concert stages are built on the beach. The concerts officially start at 6 pm however music is played the whole day, since artists check their instruments and practice the songs.

At 6 pm, all traffic is closed and no private cars are allowed to enter the district. The flow of people is intense; official statistics state that two million people attend the Réveillon in Copacabana beach every year. They are mostly concentrated at the beach, but also in the frontline buildings, and on transatlantic ships and private boats. A number of ritual practices are practised by Brazilians during the Réveillon. Many Brazilians attending will wear white clothes as the colour white is associated with peace. Different colour underwear is worn to symbolise personal wishes for the New Year. For example, if a person wants to find love, he or she will wear red or pink. Blue symbolises health. Green, hope. Yellow and gold, money. Silver prosperity. Black or dark colours are not worn as they are thought to bring negativity and even death.

People gather in groups with family or friends to see in the New Year. Food and drinks are brought to the beach. The types of food vary between the social classes. Seafood, such as shrimps, cod fish and tuna, is usually favoured by the rich and middle classes, while chicken and ham is brought by the poorer classes. Grapes are also popular, as it is believed they attract good luck. Sparkling white wine is the commonly drunk alcoholic beverage. Others prefer to have their dinner at home and simply come down to the beach for festivities.

While waiting for the midnight fireworks to begin, offerings to Lemanjá, the goddess of the sea, are made. This includes putting flowers and plants on the sand or throwing them into the sea. Some believe that jumping over seven little waves will attract good luck for the next year. All these rituals are strongly connected to Afro-Brazilian religions such as Candomblé and Umbanda. The presence of people of Afro-Brazilian descent along with their rituals and practices has steadily decreased. While historically, members of these traditional religions occupied the beach for the Réveillon, since the financing of the first fireworks display by two well-known hotels in 1970 their presence has decreased.

Nowadays, the climax of the festival is the fireworks displays, which illuminate the skies of Copacabana beach and nearby districts for 20 minutes each year. The displays are very colourful and in the 2019 Réveillon, gold, red, green and silver were the dominant colours. As the fireworks exploded in the air, they produced different shapes in the sky with hearts, smiley faces, flowers, stars and arches being most common. Many of those attending the fireworks comment on the sensual experience as the sound and vibration of each firework's explosion is so intense that people can feel it in their bodies. Media also publicise the impressions of specialists, celebrities and journalists during and after the party.

The occasion generates a strong sense of belonging and identity across the different cultural and social groups and it is common to observe people who do not know each other greeting and wishing each other collective peace and good wishes for the New Year. After the fireworks, concerts and private parties keep revellers celebrating until early morning.

Another expression of urban biocultural diversity at Copacabana beach is the Lemanjá procession, which takes place on the 29th of December. Lemanjá in Afro-Brazilian religions is the goddess of the sea. She is a very popular *orisha* (a sprit sent by higher divinities) in Brazil. In 2003, a fire destroyed Madureira Municipal Market, located in Rio's North Zone. The district was famous for samba and Black music street parties and was well patronised by people of Afro-Brazilian decent. The market was well-known for its size and diversity. It is possible to find cheap Chinese products and local herbs and artefacts used for religious practices at the market. After the fire, some market sellers carried out a ceremony to ask Lemanjá for protection. The December celebrations start inside the market with a 2.2-metre tall Lemanjá statue being placed at the entrance of the market at around 10 am. The procession of worshippers then makes their way across the city until they reach Copacabana beach. Along the way musicians play religious songs. See Garcia et al. (Chapter 8) for a description of how Indo-Caribbean Hindu devotees come to beaches along Jamaica Bay to pray, perform puja (worship), seek guidance and truth, and offer sacred items to Hindu deities.

The worshippers, dressed in blue and white, sing in Portuguese and Yorubá languages. African music is played and danced to. Rituals for protection are performed, which includes giving offerings to Lemanjá in the form of flowers, mirrors and cider bottles. Some worshippers take to the sea with boats and flowers to make offerings to the goddess. The ceremony lasts the whole day, finishing around 9 pm.

The third event to be discussed is the LGBTI Pride Parade also held at Copacabana beach. In October 2018, the beach hosted the 23rd Parade, organised by Rainbow Group, an NGO dedicated to sexual diversity activism. The parade starts at the end of the beach (Posto 5) and moves to the front of Lido Square based at Posto 2. Approximately one million people participated in the 2018 Parade, following the five "Trio Elétricos",[4] which was slightly fewer people than the previous year, when 1.3 million people attended with 10 "trios". As in different parts of the world, the LGBTI Pride Parade is a big party to advocate for LGBTI rights and equality. Copacabana beach was chosen because of the diversity of cultural and social groups it attracts and historically the Copacabana has offered many members of LGBTI community bars and nightclubs to frequent. While similar parades are held in other districts, by far the biggest and most traditional is the one held at Copacabana beach. The area occupied by the Rainbow kiosk in front of Copacabana Palace Hotel has been a meeting point for the LGBTI community since 1950 (Green 2000).

Considering Rio de Janeiro beaches through a biocultural diversity lens

Beaches play important roles in the daily life of cariocas. As already described, these spaces along the seashore are used and occupied by different social groups for a diverse range of social practices and activities, including leisure, social activities, sport and spiritual practices. The results show that nature itself is an important element of culture in that nature, namely the beach, is the basis for the diversity of cultural expressions that occur there and alongside it. One does not find the same activities and expressions in locations away from the beach. The elements of nature at Copacabana beach are part of the culture of urban residents and the city. Such an example stands in stark contrast to how the cultural values, held by former marginalised communities to natural elements are not a part of town's reflected culture and heritage (Cocks et al. Chapter 6).

The sand and the sea are spaces for cultural diversity, since different social groups are connected to them and express their cultures in these spaces. These expressions, which illustrate the links between nature and culture, range from the activities of the fishing community, who use the sand and the sea for their fishing practices, to those who use the space to celebrate, socialise, relax, exercise, enjoy nature and express themselves artistically and spiritually.

The experience of going to the beach, the perceptions of the "beach" as a place, and the practice of going to the beach make it possible to read codes, to build knowledge and actions related to the sea and the beach. There are social codes that are read and interpreted by locals, who are considered "traditional people", but also as modern urban citizens. The Copacabana and Ipanema beaches are symbols that identify Rio de Janeiro and Brazil. Even some of the elements of this stretch of beach have been declared part of the city's cultural heritage and have legal protection. As mentioned, they are represented in several products such as works of art, design objects, or beach towels, cups, souvenirs and t-shirts. Thus, this case study also supports the assertion of Cocks (2006) that the concept of biocultural diversity "is also applicable to non-indigenous traditional communities". Therefore, experiences, knowledge and practices (Ingold and Kurttila 2000) enable the social appropriation of the sea and the beach (Leff 2016) in different ways, by diverse groups, which vary in space and time. Therefore, interactions between urban residents and beach environments – both tangible (e.g., swimming) and intangible (knowledge about the currents) and at the beach itself (relaxing on the sand) or away from the beach (symbolic representations) – are relevant to the building of views about nature, culture and their relationships, in such urban beach environments. Interactions, experiences and worldviews play important roles in the process of wellbeing, building identity and a sense of belonging. Natural and cultural heritage are important factors for this process. See Cocks et al. (Chapter 6) and Nesbitt et al. (Chapter 10) for additional examples of this interplay. Consequently, the inextricable links between nature and culture in urban beach environments strengthen the process of wellbeing and belonging for social groups and for the city, though recreation, workshop and socialisation. This confirms the idea of the city as a hybrid natural and cultural entity (Swyngedouw 2009).

Diversity underpinning conflict

The diversity of relationships between nature and culture may also be a focus of conflict. Social groups that interact in different ways with urban beaches may dispute these environments and their constituent elements. Although the beaches of Rio de Janeiro have democratic access and a positive image of open public spaces, there have been social, racial and gender disputes. These disputes between groups or "tribes" (as referred to by locals) have been over space and access to nature. Such disputes have occurred because of the centrality of the beaches and the different interests they attract. People come who want to relax while others come to work both formally and informally. For example, the fisherman with their traditional boats are competing with bathers who are swimming and sportsmen and women who are using their stand-up paddle and yachts in the sea. On the sand, those engaged in sports or kiosk activities may be a disturbance to those seeking to relax or sunbathe. These conflicts are deeper than simple conflicts over space at the beach and sea. They reflect the structural class, race and gender conflicts that exist in Brazilian society. The New Year and Festa de Lemanjá events reveal these tensions. For example, the majority of Afro-Brazilian rituals performed by religious worshippers have lost the right to sufficient space as firework displays are dominating the use of the beach. All these changes have been supported by local government and the tourism industry which is interested in the creation of new "traditions" and other ways to occupy a central and symbolic space in the city, revealing the power and political dimension of biocultural diversity in Rio de Janeiro.

Before firework displays were introduced into the Réveillon festival, Afro-Brazilian religious worshippers wearing white would occupy the beach to make offerings on New Year's Eve. Such practices were criticised for polluting the ocean and sand along the beach. The pollution argument reveals a structural racism and social fears regarding Afro-Brazilian religions. In any case many efforts have been made by worshippers and organisers to reduce the impact of these rituals.

In recent years local government has adopted various policies in an attempt to alleviate the tensions. These have ranged from the recognition to the prohibition of traditional religious rituals, or even designating far away beaches for poorer classes. The beaches located within the Guanabara Bay are frequented by the poor, of whom the majority are Afro-Brazilians. The designation of the Afro-Brazilian religious rituals to beaches that are "not good for swimming" reinforces social exclusion, effectively prohibiting the worshippers from practising their religion in Copacabana, and requiring them to remain at the beaches already occupied by poorer residents. A complex point is that the majority of Afro-Brazilian religious worshippers are of African origin; however, Brazil is a multicultural society and religious syncretism is relatively common, which means that there are also White Afro-Brazilian worshippers. Since the religion has African roots, the White worshippers are socially considered to be part of the Afro-Brazilian culture and are in turn also oppressed. However, the colour of their skin partially reduces the prejudices they experience, which allows them to access some political spaces better than the black worshippers. Nonetheless, the relationships between sectors of civil society and the State are culturally oriented.

This conflictual relationship was reflected in the way the 2018 Lemanjá celebration was controlled. The local major, an evangelical bishop, prohibited the procession from taking place from the Madureira district to Copacabana beach and initially he also prohibited the use of the beach for rituals. Then, a couple of days before the festival, the local government gave permits for rituals to take place at the beach; however, the procession remained prohibited. The worshippers then conducted the celebratory rituals at both the Madureira market and Copacabana beach, but without as much infrastructure as in the previous years. Furthermore, the organisers were put under pressure to change the date of the celebration from New Year's Eve to the 29th of December. In the previous ten years, the Lemanjá celebration party has always happened on the 31st of December. As one religious leader declared at the 1918 party, "we were taken away from 31st and they do not respect our space" (Carvalho & Mendonça 2019). Another dimension of conflicts over beaches is connected to urban violence. Rio has high rates of crime and gang attacks on Copacabana and Ipanema beaches are a frequent occurrence, particularly in summer. Locals and tourists are victims of such attacks. The majority of Rio's residents are poor and of African descent. Race and class have come to dominate discourses on urban violence. Within such a framing some people, particularly those from elite economic and social sectors, have advocated limiting access to certain beaches of people from poor communities and peripheral suburbs.

One expression of this conflict is related to urban mobility. Copacabana and Ipanema can be accessed by subway and bus systems, from different parts of the city including the suburbs and peripheries. When the crime rate at these beaches increases or media attention is given to crimes committed, measures to combat crime usually include changes in public transport access to the beaches, such as rescheduling times, or limiting the number of passengers. The subway is managed by state authorities whereas the bus system, which is usually more vulnerable to manipulation, is managed by local authorities. However, the design of the most recent subway system was an indication of the race and class dimension. One day before the 2016 Olympics Opening Ceremony, a new and long-awaited subway line was inaugurated which has had the effect of making access from poor parts of the city to the coastal areas of the town more difficult.

Conclusion

To guide the research on biocultural diversity in urban beach environments, the case study attempted to answer three questions, which are restated here: (1) how do urban residents view, interact and engage with, and give meaning in different ways to beach environments in Rio de Janeiro? (2) how do the inextricable links between nature and culture in the beach environments strengthen the process of wellbeing and belonging to social groups and to the city? (3) through the lens of political ecology, considering the beach environments as resources, what are the conflicts over them?

In order to answer these questions, a case study was carried out in Copacabana beach. Urban biocultural diversity at Copacabana beach was identified mostly at a cultural landscape level. Further research is needed to identify it at species and genetic levels. The results show that urban biocultural diversity is composed of interactions and engagements with the beach, such as leisure, sports, spirituality or sociability. These activities are important to build views about natural landscapes, culture and their relationships, and to give meanings to them in the urban and beach environments. Therefore, interactions, experiences, worldviews and symbolic representations have roles in the process of building identity. Biocultural diversity in the urban beach environments strengthens the process of wellbeing and belonging to social groups and to the city. Although the study reflects a specific case, it highlights the richness of beach environments to biocultural diversity studies. Considering the increase of beach culture in different parts of the world, more research is needed to build knowledge about a diversity of ways in which social groups interact with, engage with and view urban beach environments, and in order to facilitate conservation and management strategies for such important environments.

Notes

1 Vivarini, Vivian. *Biocultural Diversity in Leme district in Rio de Janeiro* (2018–2019, work in progress) and Reis, Alexia. *Spirituality and Environment: The Iemanjá party of Madureira Market in Copacabana beach*, (2018–2019, work in progress).
2 "Do Leme ao Pontal" is a very popular song by Tim Maia.
3 Biscioto Globo is a salty, tapioca baked biscuit/cracker that can be either sweet or sour; usually shaped in a ring. Matte Leao is a traditional flavoured tea.
4 Big trucks that have been transformed into mobile stages for big concerts.

References

Carvalho, A.R. & Mendonça, G.C. 2019. *Fé e espiritualidade como fator de preservação: estudo da Festa de Iemanjá do Mercadão de Madureira*. Rio de Janeiro, Brazil: Curso Superior de Tecnologia em Gestão Ambiental, Instituto Federal de Educação, Ciência e Tecnologia do Rio de Janeiro (IFRJ), Campus Rio de Janeiro. 54 p.

Cocks, M.L. 2006. Biocultural diversity: Moving beyond the realm of 'indigenous' and 'local' people. *Human Ecology*, 34(2): 185–200.

Elands, B.H., Wiersum, K.F., Buijs, A.E. & Vierikko, K. 2015. Policy interpretations and manifestation of biocultural diversity in urbanized Europe: Conservation of lived biodiversity. *Biodiversity & Conservation*, 24(13): 3347–3366.

Green, J.N. 2000. *Além do carnaval: a homossexualidade masculina no Brasil do século XX*. São Paulo, Brazil: Ed. Unesp.

Ingold, T. & Kurttila, T. 2000. Perceiving the environment in Finnish Lapland. *Body & Society*, 6: 183–196.

Leff, E. 2016. *A aposta pela vida: imaginação sociológica e imaginários sociais nos territórios ambientais do Sul*. Petrópolis, Brazil: Vozes. 510 p.

Mendonca, G.C. de. 2014. *Diversidade biocultural, direito e cidades: implementação do marco jurídico sobre diversidade biocultural na cidade do Rio de Janeiro*. Rio de Janeiro, Brazil: Tese Doutorado em Meio Ambiente – Universidade do Estado do Rio de Janeiro. 254 p.

Oxfam Brazil. 2017. *A distância que nos une: um retrato das desigualdades brasileiras*. São Paulo: Oxfam Brasil.

Swyngedouw, E. 2009. A cidade como um híbrido: natureza, sociedade e "urbanização-ciborgue". In Acserald, Henri (org). *A duração das cidades: sustentabilidade e risco nas políticas urbanas*. Rio de Janeiro, Brazil: Lamparina. 256 p.

10 Intercultural learning in contested space

The biocultural realities of global cities through the lens of Vancouver, Canada

Lorien Nesbitt, Cecil Konijnendijk, Nathanael Lauster and Hyeone Park

Prelude

Jim Leyden (Míx̌ał Kanim, Black Bear Warrior) walks in the shade beside a row of cherry trees in Oppenheimer Park in Vancouver, British Columbia (BC), Canada. The park is busy with people sitting, walking and chatting, and the sound of traffic buzzes nearby. There is coffee available just outside the community centre building and two young volunteers offer food to people in the park.

> The traditional name for this area, translated, means "grove of many maples" or "grove of many trees". This whole area was covered in cedar trees, and the like, for generations and generations, and in the development of the city, that's all been wiped out,

Jim explains. This scene is in many ways emblematic of the history of the Vancouver metro area and the intertwined biocultural discourses and diversities of the region. Oppenheimer Park sits on the traditional, unceded territory of the Squamish (Sḵwx̱wú7mesh), Tsleil-Waututh (Səl̓ílwətaʔł) and Musqueam (xʷməθkʷəy̓əm) First Nations. It is also surrounded by the Downtown Eastside, Canada's poorest neighbourhood. The people in the park represent some of Canada's most marginalised citizens and are disproportionately indigenous. The area is also the site of the former Japantown, a hub of Japanese culture before the Second World War, and it sits alongside Vancouver's historic Chinatown and the rapidly gentrifying Gastown neighbourhood. Oppenheimer Park is a meeting place of various histories, cultures and biocultural discourses, reflecting the synergies and tensions of this multicultural city.

"Within the West Coast nations, my elder talks about the cedar tree as being the Tree of Life", says Jim.

> So, we brought in a cedar tree. We brought it in, we brought an elder in from the Squamish Nation and did a ceremony around the planting of the tree and made sure it was planted in a good way and a healthy way. Now

> it's like, a small tree in the Downtown Eastside is going to be part of our process of bringing this stuff back.

Jim is an elder in the Squamish First Nation, whose traditional territory covers much of Vancouver. Through his work with his nation and organisations in Vancouver's Downtown Eastside, Jim is also a leader in the movement to reclaim and rediscover urban indigenous biocultural relationships and stewardship practices in a city that has eroded them over time. Jim and his colleagues teach marginalised urban populations how to cultivate medicinal plants and use them in traditional medicines, and they lead groups of downtown residents on trips in the North Shore mountains and Sto:lo territory to help people get out into the more natural forests surrounding the city. They are also leaders of the indigenous-led resistance to the Trans-Mountain Pipeline that is proposed to transport diluted bitumen from Alberta, Canada, to Burrard Inlet in the Vancouver metro area.

As a growing, multicultural region, the Vancouver metro area is a microcosm of the opportunities and challenges faced by cities and urban societies around the world. It is the site of rich indigenous history and current culture, colonisation by European settlers and rapid immigration. Recently, it has also developed strong green city discourses while experiencing population growth and densification. Drawing on media and document analysis, and supplemented by interviews with key Vancouver residents and urban forestry actors, this chapter explores the biocultural diversity of the region. The names Vancouver and Vancouver metro area are used interchangeably in this chapter, to refer to the region as a whole and acknowledge that municipal boundaries are constructs of the dominant form of government in the region but don't reflect, for example, indigenous claims to land or natural processes.

In the beginning: indigenous cultures and colonisation in Vancouver

A gentle breeze moves through the canopy and leaves rustle as Robert Nahanee stands at the base of a large cedar tree. Robert is an elder of the Squamish Nation and a leader in his community.

> I'm from the Squamish Nation", he explains. "I live in the village of Utslawn, what we know as North Vancouver today. This is where my grandfather was born, he was born in Stanley Park, and I love it here. It's relaxing because we use the trees, this area, the green, as a place of time out, for people to come and take quality time and sit with a tree, to sit here and breathe in this way, where it's clear, your thoughts are clear, your feelings are clear, you're able to cry if you want to cry, or you're able to laugh if you want to laugh. These trees create balance in our life, and they create balance on this Earth. Right behind me here is the cedar tree. It gave our old people everything that they needed. It gave them

> transportation, canoes, it gave them housing, our long houses, it gave them clothes through the bark here, and it gave us heat, it gave us tools, it played a major role in our spirituality, and so we really put faith into this tree that we know, the Tree of Life, the cedar tree.

The Coast Salish peoples of Vancouver have existed within, interacted with, and been stewards of this landscape since at least 16,000 BC (Kheraj 2013, Musqueam First Nation 2019a). The Pacific Northwest coastal ecosystems are uniquely biologically diverse and have thus supported large indigenous populations for millennia. The Vancouver metro area sits on the unceded traditional territories of 11 First Nations, including the Hwlitsum, Katzie, Kwantlen, Kwikwetlem, Matsqui, Musqueam (xʷməθkʷəy̓əm), Qayqayt (qiqéyt), Semiahmoo, Squamish (Sḵwx̱wú7mesh), Tsawwassen (sc̓əwaθenaʔɬ təməxʷ), and Tsleil-Waututh (səl̓ílwətaʔɬ) Nations. The City of Vancouver comprises three unceded territories, those of the Musqueam, Squamish and Tsleil-Waututh Nations (Figure 10.1). These territories are referred to as "unceded" because treaties were not negotiated between settler governments and these indigenous nations during colonisation. As Robert's great great grandfather, Chief Capilano, stated when he led a delegation to England in 1906 to assert that Aboriginal title to the lands had never been extinguished, "I'm still alive and I'm still on the land". The unceded nature of these lands, and the continued existence of Aboriginal title, has been acknowledged by various decisions of the Supreme Court of Canada (UBC First Nations Studies Programme 2009).

These nations were, and continue to be, culturally complex, artistic societies with economies and practices closely tied to the lands with which they live (Kheraj 2013, Tseil-Waututh Nation 2019, Musqueam First Nation 2019b). Prior to contact with Europeans, the indigenous communities in the Pacific Northwest region were some of the more complex, populous societies in North America. Tsleil-Waututh oral history establishes the original population of that nation at 10,000, the Musqueam Nation tells of large pre-contact

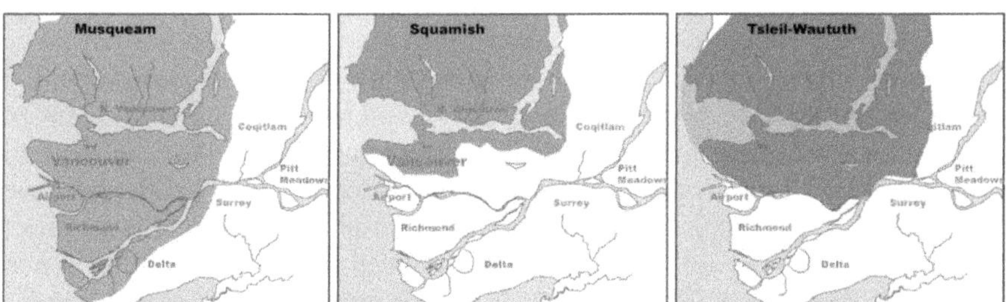

Figure 10.1 Maps of the traditional, unceded territories of the Musqueam, Squamish and Tsleil-Waututh Nations, British Columbia, Canada. Source: johomaps.com.

populations that were decimated by small-pox, and Robert Boyd provides a conservative estimate of 183,000 people living along the coast from northern California to Alaska before European contact (Boyd 1999, Musqueam First Nation 2019a, Tseil-Waututh Nation 2019). These nations describe their biocultural discourses as revolving around responsible stewardship and reciprocal relationships with the natural environments in which they live (Musqueam First Nation 2019b, Squamish Nation 2019, Tseil-Waututh Nation 2019). In particular, indigenous biocultural discourses in the region have developed with the local (urban) forest. The cedar tree (*Thuja plicata*) and a range of other native plants, such as Salmonberry (*Rubus spectabilis*), red huckleberry (*Vaccinium parvifolium*), deer fern (*Blechnum spicant*), stinging nettle (*Urtica dioica*) and many others, have played a central role in Coast Salish culture and are intertwined with indigenous biocultural relationships, with cedar trees providing material for clothing, tools and shelter, plants providing sacred medicines and healthy sources of food, and indigenous communities playing a stewardship role in relationship with the land and water that supported the sustainability of local ecosystems. See also Cocks et al. (Chapter 6) for similar attachments to specific species.

Indigenous biocultural discourses, broadly defined by responsible stewardship, integration with natural systems and an understanding of human populations as part of rather than separate from the "natural" world, are in contrast to the biocultural discourses of European and other settlers that subsequently came to dominate the Vancouver landscape. Where indigenous discourses were broadly focused on relationships of balance and care, settler cultures were more focused on dominating, "taming" and extracting consumptive value from the lands and waters within which they lived. These ecosystems were seen as separate from human populations, and the influence of this biocultural discourse can still be seen in Vancouver today.

A case that illustrates the conflicting biocultural discourses of indigenous and settler cultures in Vancouver is Stanley Park, one of the most well-known parks in North America and the "green jewel" of Vancouver. The modern history of Stanley Park has been one of simultaneous subjugation and preservation of a "natural" ideal. Located next to the downtown core, the area was originally set aside as a government reserve in 1859 and its use as a public park was granted to the City of Vancouver by the federal government in 1887 (Kheraj 2013). The management of Stanley Park focused on maintaining a naturalistic landscape that would still allow for recreation (Kheraj 2013). Interestingly, a central focus of this movement to maintain the "natural" character of the park was the creation of the seawall to prevent the erosion, by natural forces, of the park shoreline (Birnie 1964). The process of constructing the seawall began in 1914 as a project to control shoreline erosion and gradually produced the seawall more or less as we know it today by 1971 (Kheraj 2013). This remarkable move to preserve the natural state of the park by fundamentally changing its natural shoreline to prevent environmental change represents an important part of settler biocultural discourses in relation to urban green spaces: modification for static preservation.

What is particularly striking in this case is that this "natural", "wild" park was previously inhabited by indigenous communities for millennia (Kheraj 2013). The oral traditions of local First Nations and archaeological records confirm that Stanley Park was home to multiple indigenous settlements, the most prominent of which was Whoi Whoi. Whoi Whoi was one of the largest settlements of the Squamish people and current members of the nation remember recent generations living in Stanley Park. Stanley Park was not a "wilderness" location but a place of active stewardship where humans lived in relationship with local ecosystems. This contrasting view of human relationships with the non-human environment, integrated and interconnected on the one hand, and separated on the other, is illustrative of indigenous and settler biocultural discourses and their relationship with each other.

As European colonisation progressed, the colonisers brought with them a conception of the relationship between the land and the people that saw the land as a natural resource to be extracted or preserved. In particular, the development of industrial mining, forestry and fisheries in the region redefined the governance and composition of local ecosystems through the lens of settler biocultural discourses. The British Crown created the colony of British Columbia in 1858. As local industries developed, the governor at the time, James Douglas, opened the region to settlement under his authority as governor of the colony (Kheraj 2013). During the 1860s, Europeans developed settlements alongside the indigenous communities of the region. Aside from the removal of vegetation to make way for urban development and commercial forestry, the settlers also intentionally brought plants and animals from Europe to produce familiar landscapes. This development followed the model already established in earlier colonies in eastern and southern America and was focused on creating a "Neo-Europe" in the new colony (Crosby 1986). These imports were initially mostly livestock and garden plants, including both ornamental and edible plants, but subsequently included trees commonly found in European and eastern North American cities, such as London plane trees (*Platanus x acerifolia*), hawthorns (*Crataegus spp.*), tulip trees (*Liriodendron tulipifera*), American elm (*Ulmus americana*), horse-chestnut (*Aesculus hippocastanum*) and black locusts (*Robinia pseudoacacia*) (Heritage Vancouver 2019). This replacement of the urban canopy with exotic tree species was part of the City Beautiful movement in North American cities that was focused on beautifying cities through street tree planting and park establishment that represented European and settler aesthetic ideals. However, native species were also an important part of the street tree population in this period. In particular, bigleaf maple (*Acer macrophyllum*) was a prominent part of Vancouver's pre-clearance canopy and its early street tree canopy (City of Vancouver and Vancouver Board of Parks and Recreation 2018, Heritage Vancouver 2019). The urban trees introduced to the city during this period continue to form an important part of the tree canopy in the Vancouver area (Heritage Vancouver 2019).

What came next: social and environmental change from the 19th century to the present

The cherry trees beside which Jim was walking, a species not native to the area, symbolise part of the more recent cultural heritage of the neighbourhood and the metro region. The neighbourhood of Oppenheimer Park was once called "Japantown" or "Little Tokyo" and was one of the centres of Japanese settlement in the area. Despite the important role of Japanese immigrants in the development of the Vancouver region, Japanese Canadians' property was confiscated and these populations were moved to internment camps in Interior BC during the Second World War. Although the role of Japanese culture in Vancouver has eroded since the internments, Japanese settlers were an important part of the immigrant population in Vancouver in the late 19th and early 20th centuries, and evidence of this role remains in Japanese buildings, businesses and traditional gardens, as well as the 17,000 flowering cherry trees (*Prunus serrulata*) and over 13,000 flowering plum trees (*Prunus spp.*) that are visually similar to cherry trees and virtually indistinguishable to the general public (Whysall 2013). Cherry tree planting began in the 1930s when the Japanese cities of Kobe and Yokohama gave 500 trees to the Park Board to plant at the Japanese cenotaph in Stanley Park to honour and commemorate Japanese Canadians who served in the First World War. Another 300 trees were subsequently donated by the Japanese consul and have been a popular planting choice ever since (Whysall 2013).

Vancouver is a city shaped by immigration. In the 19th century, Asian settlers were an important part of the early colonial immigrant population arriving in the 1860s and were involved in the extensive land clearing and industrial development that accompanied these arrivals (City of Vancouver and Vancouver Board of Parks and Recreation 2018). However, by the late 19th and early 20th centuries, immigration policies sought to restrict immigration from non-European countries and focused on securing workers from Britain, northwestern Europe and the United States (Green & Green 1999). Immigration from Asian, African and Southern European countries was increasingly restricted in the early and mid-20th century up until 1962, when Canada abandoned its racist immigration policies and people from non-European countries were allowed to settle permanently in Vancouver in larger numbers (Green & Green 1999), including from Afghanistan, China, Colombia, India, Indonesia, Iran, Iraq, Myanmar, Philippines, South Korea, Sudan, Uganda and Vietnam. Immigrants from Asia account for nearly four-fifths of recent immigrants to the metro area. The vast majority of residents speak English. However, other linguistic communities are also sizeable, and Mandarin, Cantonese and Punjabi speakers all surpass French, Canada's second official language. Tagalog, Hindi, Spanish, Korean and Farsi speakers also constitute large communities and there are many more small communities speaking diverse languages.

Recent immigration patterns are reflected in the composition of Vancouver's urban forest, as well as its ornamental gardens and parks. As cherry trees mark the importance of Japanese culture in the city's biocultural diversity, traditional and botanical gardens such as the Dr. Sun Yat-Sen Classical Chinese Garden and the Nitobe Memorial Garden also offer representations of classical Asian biocultural discourses in the city. Ginkgo trees (*Ginkgo biloba*) are increasingly planted as culturally relevant urban trees with small growth forms that are more and more appropriate as the city densifies and the potential for infrastructure conflict increases.

Biocultural discourses, diversity and conflict

Expressions of biocultural diversity and sites of intercultural exchange

Sitting at a picnic table in Town Centre Park, Coquitlam, BC, Lanny Englund describes his professional approach to managing urban forests for local benefits.

> Probably, a lot of people don't know about all the benefits that an urban forest provides. So, there's a whole range of benefits, starting, you know, simply with aesthetics. It makes a place more beautiful, and if a place is more beautiful, it's more enjoyable.

Lanny is the Pacific Region Representative of the Canadian Urban Forest Network, and a practising urban forester. In the shade of an Autumn Blaze maple tree (*Acer freemanii*), a commonly planted cultivar chosen for its brilliant red autumn foliage, he articulates an approach to understanding urban forests based on the ecosystem services framework.

> There are the sort of green infrastructure benefits, terminology that you're hearing more and more, storm water capture, reducing the urban heat island effect. Of course, what's popular now is climate change and carbon sequestration. Another issue is ground-level pollution or air pollution in urban areas and trees really provide a big benefit there as well by taking up those pollutants rather than us breathing them in.

Lanny's explanations reflect the dominant institutional and professional focus of urban forestry in Vancouver, centring around urban liveability, sustainability, ecosystem services and the benefits that urban vegetation provides to residents. Although this is the dominant discourse in Vancouver's biocultural landscape, there are multiple examples of biodiversity expression, creation and exchange across the city that together inform Vancouver's diverse biocultural discourses.

Vancouverism and the Greenest City discourse

Vancouver's environmentally conscious, "living first" approach to urban and especially downtown development and planning has a long history. In the early 1970s, decision makers and planners started to focus on making the city more liveable. This resulted, for example, in abandoning initial plans to build a freeway through the city (Kluckner 2011). A new planning approach that became coined as "Vancouverism" was developed. Liveability and later also sustainability were turned into core discourses driving planning and architecture, and Vancouver's unique situation within stunning natural surroundings obtained greater appreciation. Protected sightlines or "view cones" were set up, and tall, slender residential towers became the norm to maintain views of the nearby mountains (Figure 10.2). These towers became interspersed with low-rise buildings, public spaces and small parks, as well as pedestrian- and (later) bicycle-friendly streetscapes (McCann 2013; Beasley 2019). According to former city planner Larry Beasley (2019), the activities and international attention resulting from Expo 86 played an important role in further advancing the Vancouverism planning philosophy, which was conceptualised as being collaborative and inclusive, working together with, for example, private developers and local communities.

Vancouver's "green" and sustainability discourse was strengthened in the period leading up to the hosting of the 2010 Olympic Winter Games. Then-mayor Gregor Robertson launched his plan for Vancouver becoming the world's Greenest City while attending the 2009 Copenhagen climate summit (McCann 2013). A strategy and action plan were developed to drive forward

Figure 10.2 False Creek, downtown Vancouver and the north shore mountains. Source: Nesbitt, Lorien.

the city's green development, focusing on planning and design of dense, residential downtown neighbourhoods, intended to encourage walking, cycling and transit use instead of automobile dependency.

According to McCann (2013), Vancouverism and the Greenest City policy are examples of a discourse that represents a new "extrospection", where Vancouver brands and profiles itself in a global arena. McCann talks about "policy boosterism" as a subset of traditional branding and marketing activities that involves the active promotion of local policies, programmes or practices across wider geographical fields as well as to broader communities of interested peers. However, a consequence of the new global focus may be a disconnect with the city's residents and communities, as well as some of its local biocultural discourses. Moreover, with the growing global reputation and attractiveness of the city and resulting higher property prices, lower income residents face increasing challenges and a division can emerge between the "haves" and "have nots", as well as between private property development and public interests (Kataoka 2009).

Institutional policies and plans

An examination of the institutional and professional focus of urban forestry in the region also provides some insight into the dominant biocultural discourses. Vancouver's urban forest strategy, finalised in 2018, aims to "protect, plant, and manage trees to create a diverse, resilient, and beautiful urban forest on public and private lands across the city" (City of Vancouver and Vancouver Board of Parks and Recreation 2018, p. iv). The strategy has five goals: (1) to protect the urban forest during development, (2) to plant trees to grow the urban forest, (3) to manage trees for health and safety, (4) to engage citizens in the urban forest and (5) to monitor the status and condition of the urban forest. These goals articulate a focus on the growth and maintenance of the urban forest, with a secondary focus on citizen engagement. Similarly, Vancouver's Greenest City 2020 Action Plan focuses on spatial access to parks for urban residents and planting 150,000 new trees by 2020 (City of Vancouver 2012). The regional Connecting the Dots strategy aims to increase urban green connectivity in the region, and the Vancouver Biodiversity Strategy seeks to support biodiversity and enhance access to nature for residents (City of Vancouver and Vancouver Board of Parks and Recreation 2018, Metro Vancouver 2015). All four documents illustrate an institutional biocultural discourse focused on managing an urban forest resource for the broad benefit of urban residents. Implicit in the goals, and discussed in both documents, is a focus on urban nature as a source of benefits and value for urban residents. Ecosystem services or green infrastructure are the dominant discourses in these institutional documents and provide the conceptual framework that underpins the goals and actions prioritised by the city.

The focus on ecosystem services is also apparent in the academic literature, and perhaps the institutional focus reflects the common academic discourses in the field. Ecosystem services research has been dominant in the field of urban forestry throughout the 21st century, and has produced a large body of literature, particularly focused on regulating urban ecosystem services, such as improved microclimates, reduced air pollution and flooding reduction (Hotte et al. 2015, McPherson et al. 1997, Nowak et al. 2006). While the ecosystem services terminology has led to increased prominence for urban forests in municipal decision-making and budgeting, it also highlights a relationship to and conception of urban forests that is transactional in nature. The ecosystem services concept commodifies urban forests and assigns value to them based on their utility for human populations. This somewhat extractive view is similar to traditional colonial conceptions of nature and humans' relationships to it, and appears to be an important part of the biocultural discourse of Vancouver's dominant settler culture and the profession of urban forestry (Konijnendijk et al. 2006).

The ecosystem services framework and concept has become so dominant in recent years, that it is unlikely that municipal governments, such as the City of Vancouver, are aware of their adherence to this concept, to the potential exclusion of others, or of its potentially exclusionary nature. And yet, alternative conceptions of and relationships with urban nature operate within a mosaic of places and projects across the city.

Iconic parks and gardens

Stanley Park, mentioned earlier as a key area of indigenous settlement and culture, is Vancouver's most iconic park. Although it has been a site of settlement and human disturbance, it contains some of the last remaining old growth stands in the city and is celebrated for its natural and "wilderness" character (Kheraj 2013). It is considered representative of Vancouver's natural identity and it is primarily composed of native vegetation (Kheraj 2013). Stanley Park's iconic place within the urban forest makes it an important site for tourism and environmental education. It is a favourite destination for local and foreign tourists, attracting more than eight million tourists per year (Pemberton 2013). Beyond its wilderness values, the park also offers nine kilometres of paved seawall, waterfront restaurants, an aquarium, a small golf course, an outdoor summer theatre, an outdoor swimming pool, totem poles, rose and rhododendron gardens and a miniature railway. And yet, the focus of park management and education is the natural forest ecosystem represented by the park. Restoring natural ecosystems in the park is a goal of Vancouver's Urban Forest Strategy (City of Vancouver and Vancouver Board of Parks and Recreation 2018), and the Stanley Park Ecology Society offers environmental education programming that highlights the importance of the park for maintaining local ecosystems and native biodiversity (Stanley Park Ecology Society 2019b). A major activity of the Society is the removal of invasive, non-native vegetation, such as Himalayan blackberry, and reestablishment of native species.

In reflection of the indigenous settlement history and the continued existence of Aboriginal title described above, local indigenous heritage is also celebrated in Stanley Park. The park is home to nine totem poles at Brockton

Point. The poles, the last of which was added to the site in 2009, are intended to welcome visitors to the traditional lands of the Coast Salish people (City of Vancouver 2018). The Stanley Park Ecology Society also offers programming focused on indigenous perspectives and ways of knowing. This includes guided walks describing indigenous plant use, workshops on traditional medicine making and a native plant demonstration garden (Stanley Park Ecology Society 2019a). These opportunities to engage with and learn about indigenous biocultural relationships suggest that Stanley Park informs local biocultural discourses not just as a natural park with representative native vegetation, but also as a site where indigenous heritage is acknowledged and celebrated.

The Dr. Sun Yat-Sen Traditional Chinese Garden, discussed earlier as an important symbol of the history of Asian immigration and culture in the region, is a semi-private garden in downtown Vancouver that celebrates and provides education about traditional Chinese garden styles and biocultural discourses. Built with joint support from the governments of Canada and the People's Republic of China, the garden symbolises the meeting of cultures in the historic Chinatown neighbourhood (Dr. Sun Yat-Sen Classical Chinese Garden 2019). The garden is modelled after the Ming Dynasty scholars' gardens in the city of Suzhou and is built from traditional materials sourced from China while the garden plants were sourced locally. It provides a celebration of aspects of traditional Chinese biocultural relationships and a space for contemplation in a natural space.

Vincent Kwan, Executive Director of the garden, explains the biocultural philosophy underlying the design and management of the garden:

> In a lot of gardens, the way that the landscape is designed or managed is through a set of principles and guidelines that are meant to provide the shape that people believe a garden should be. For a classical Chinese garden, that is somewhat different in that the way that the plants are managed is to reflect not the control, but how humans or people are meant to be in harmony with nature. So, that principle is something that informs on a daily basis how the gardeners would choose how the plants are maintained or where plants are put and how it compliments and not compliments the different aspects of the garden, including the rocks, including the architecture. So, everything is meant to be working with and serving nature instead of people controlling the shape and form of all the horticulture as a way to define what a garden should be.

Biocultural exploration and intercultural exchange

The University of British Columbia (UBC) Farm, located beside the Vancouver campus of UBC, is the site of projects and programmes aimed at exploring and expanding residents' relationships with urban nature and offering opportunities for intercultural learning (University of British Columbia 2019).

One unique space for biocultural exploration and learning at UBC Farm is the Tal A'xin: Maya in Exile Garden. Started in the mid-1990s by five Maya families who came to Canada as refugees, the Maya Garden allows its founders to practice their Mayan farming traditions and share those traditions with others (UBC Communications and Marketing 2016a). The Maya Garden grows ancestral crops known as the "Three Sisters", corn, beans and squash. The garden also produces a range of colourful flowers and medicinal plants that are sold along with the food. Beyond food and flower production, the goal of the garden is to celebrate Mayan knowledge and traditions and share these with local people. Members of the garden offer cooking and farming workshops to the public and volunteers are welcome to help out in the field. The garden offers a relaxed and friendly atmosphere for knowledge exchange and the celebration and sharing of a unique biocultural discourse. Similar opportunities of relaxation and sharing exist among migrant gardeners living in European cities (Elands et al. Chapter 11).

The families responsible for the Maya Garden continue to practice and teach their traditional farming culture, based on principles of familial relationship with the natural world. Maximo Morales, one of the leaders of the garden, sits on the grass, looking out over the field as he takes a brief break from tending the crops:

> Before beginning to work, we do a Mayan ceremony. We do this ceremony because in our culture, this is how we ask the Creator, nature, if we can begin to work the earth. We know that for us, the earth means mother. It is like us, it is like a mother. That is why we say "mother earth" because the earth feeds us. If there is no earth, we cannot not live, there is no food. So, for us, we really respect mother earth.

The UBC Farm is also home to indigenous gardens focused on helping indigenous populations, and particularly indigenous youth, reconnect with and explore their natural heritage and develop biocultural relationships based on traditional farming and herbal practices. The Tu'Wusht Garden is an initiative of the Vancouver Native Health Society to support the health and wellbeing of marginalised urban indigenous peoples. The name Tu'Wusht means "We Belong" in the Tla'amin Coast Salish dialect and reflects the focus of work in this garden (UBC Communications and Marketing 2016b). Urban indigenous youth and elders work together to grow, prepare and eat food, and through this process share traditional knowledge and skills. While the focus of the garden is on intergenerational learning between indigenous elders and youth, farm volunteers are welcome to participate in work in the garden, giving it an intercultural education function.

The Cottonwood Community Garden is situated in one of the lowest-income neighbourhoods in the region and started as guerrilla gardening by residents who wished to improve food security and transform an illegal dumpsite into a biologically diverse green space that reflects their cultures and needs (Park et al. 2018). The Cottonwood Community Garden houses not only individual plots but also a Native garden, an Asian garden and a Hispanic/Latin garden, which are collectively managed by members and volunteers. Similarly, the Riley Park Community Garden was planted to meet the community's desire for an alternative community-gathering space in the neighbourhood. Members and volunteers collectively maintain the garden and plants, and share produce. The garden has designated plots for certain cultural groups and hosts events such as a native plant mapping workshop, Syrian multicultural lunch, Latino Dance and watering night, Chinese Lind Dancing night and a Daikon and Japanese Culture workshop. Elands et al. also provides examples of different types of community garden projects across cities in Europe (Chapter 11).

The common thread that unites these sites of biocultural exploration and exchange is the opportunity to co-create a relationship with the natural world that celebrates diverse knowledge forms and practices. The examples described above centre around food and medicine production, drawing on local and international indigenous and settler knowledge sources. Teaching, learning and sharing are central tenets of the programming on these sites, and offer an inclusive exploration of what it means to be in relationship with the urban natural world.

Biocultural diversity as a source of conflict

Despite the many examples of positive biocultural synergies and intercultural exchange that can be found in Vancouver today, biocultural diversity can also be a source of conflict in a multicultural city. The diversity described above does not always work in synergy. The conflict among divergent biocultural discourses in Vancouver is showcased by two current cases: the Trans-Mountain pipeline and the otter versus koi saga.

Trans-Mountain Pipeline

The Trans-Mountain Pipeline is a pipeline project planned to transport diluted bitumen from the Alberta oil sands to Burrard Inlet in the Vancouver metro area. It proposes to twin an existing pipeline and more than double the tanker traffic in the Burrard Inlet. Although many Canadians and the federal government believe that the pipeline is in the national interest, there is strong resistance to the pipeline project from some BC First Nations, local Vancouver residents and multiple environmental organisations (Jubas 2018). The source of this conflict is a fundamental disagreement about the nature of the project's environmental risks and the importance of protecting the natural environment from contamination. The project has been a flashpoint for conflict between those who support energy resource development and extractive biocultural relationships, and those who oppose further oil and gas development, want to see strong action on climate change and who support more reciprocal biocultural relationships.

The Trans-Mountain Pipeline project is emblematic of the potential for conflict among fundamentally divergent biocultural relationships. The support for the pipeline can be seen as a continuation of the extractive biocultural discourses that arrived in Vancouver with European colonisation. According to this worldview, the natural world contains "natural resources" that should be extracted for human benefit and the financial gain to be made from this extraction justifies the potential environmental harm that may accompany it. The resistance to the pipeline is unified by a more relational or reciprocal biocultural discourse, in which humans are a part of nature and are in a respectful, stewardship relationship with it. According to this worldview, humans are responsible for protecting the natural world of which we are a part, and harmful natural resource extraction is not worth the potential economic gain.

Jim Leyden has been a leader in the resistance to the Trans-Mountain Pipeline project and is the Coast Salish Watch House (Kwekwecnewtxw) elder (Protect the Inlet 2019). He sits outside the Watch House on a sunny Saturday afternoon, surrounded by birdsong as local cyclists and walkers pass by along a nearby trail.

> My question is: can Canada afford the risk that the environmentalists will be right in this one and that the oil spill will come in? When the salmon aren't running up the Fraser anymore, are your minimum wage security officers going to be able to feed your people? And that's a legitimate question. We deal with the two different ways of looking at things. The traditional way, in which we're here to save this world. Everything around us, we have a natural law, we are in relationship with them, with all of this. This is a tree. I actually come to pray to this tree all the time. I introduced myself when I came here and that tree introduced me to everything else around, and they know I'm good. So that's what we do .... Kinder Morgan isn't a person. Do you know why it isn't a person? It doesn't have a heart, doesn't have a soul, doesn't have a spirit. It is responsible for making as much money as it can and reporting to people who are there to get money. It's not responsible for anything else. And the people who are feeding their families over there, and I'm meaning the Becks and Kinder and Morgan and all that, they're doing it the best they know how. And they live in their heads and they haven't experienced the heart part. They live in their accounting and they live in their "how can we make more money out of this and that". And so, anyone in Kinder Morgan is going to have to follow that line of thought, that heartless line of thought. And that's what allows them to put fences up in the salmon streams. That's what allows them to order their workers to climb up a tree and take down an eagle's nest. That's what allows them to look us bold-faced in the eye and say "There's no threat here, these tankers will never sink".

Otter vs. koi

The otter vs. koi saga is an interesting showcase of the biocultural tensions that can arise in a multicultural region. In the fall of 2018, a wild river otter (*Lontra canadensis*) entered the city and discovered the Dr. Sun Yat-Sen Traditional Chinese Garden and the pool of ornamental koi fish (*Cyprinus carpio*) maintained by the garden. Over the subsequent days and weeks, the otter began to hunt and eat the koi in the garden, eating ten koi in total. Although parks staff and professional wildlife biologists were brought in to attempt to catch the otter, it continued to elude capture and eat the fish (Chan 2018). Ultimately, the garden was forced to move the fish to save the remaining population (Takeuchi 2018).

When the story made the news, many Vancouver residents were excited to hear that an otter had seemingly taken up residence in the city and saw it as an example of urban rewilding. The story gained increasing coverage as the otter continued to catch and eat the koi, avoiding all traps, sometimes stealing bait, and becoming a symbol of the resilience of the natural world. Public support for the otter particularly grew when an anonymous Twitter user set up a humorous account for the otter (@ChinatownOtter, #otterwatch2018) to mock the officials attempting to catch it. Posts such as "More traps means more tasty baits" and "I watched the news conference from my secret hiding place" delighted members of the public and the account quickly gained 2,395 followers. Unfortunately, the gradual disappearance of the koi was a negative and painful experience for some residents, in particular for garden staff who had cared for some of these fish for many years. Eventually, Team Otter (#TeamOtter) and Team Koi (#TeamKoi) were established on social media to unite residents supporting either side. Although the social media accounts and coverage were generally humorous, they were also seen by some as representative of tensions between indigenous and settler populations on one hand and more recent immigrants from Asia on the other.

The fact that this conflict was between a wild otter and a pool of foreign koi was particularly emblematic in this case. The otter is a native species in the Pacific Northwest ecosystems that has been hunted for its valuable fur from pre-colonisation to the present day. The loss of otter habitat in urban areas in some ways mirrored the wider degradation of ecosystems in the region. The history of the koi is also important in this case. Koi is an important species in traditional Chinese culture and is seen by some to represent an acknowledgement of the importance of Chinese culture to the region. Support for Team Otter over Team Koi was thus experienced by some as a rejection of Chinese culture and newer immigrants. However, the incident also gives reason to hope for ongoing and future respectful intercultural relationships and exchange. As explained by Vincent Kwan:

> The koi, as with any other living things in the garden, can often trigger a bit of emotion if it's lost. And particularly, I think, for the koi, there's a lot

> of symbolic associations with Chinese culture. People tend to be naturally and understandably emotional if there's anything that happens to the koi. But I was also actually pleasantly surprised with the positive support and emotional response from the community as a result of that as well.

Even the otter showed some support for the koi, posting: "To be perfectly frank, I feel pretty bad. I didn't realise these koi were important … Can I go now?"

Imagining a way forward

Sitting under the Autumn Blaze maple in Town Centre Park, Lanny Englund expresses his hopes for the future of the region and the biocultural diversity it comprises: "So my hopes are that people will have a better understanding of the urban forest and the benefits that it provides and those values are more and more brought to the forefront and really impact how we build our communities".

The Vancouver metro area is a place of high biocultural diversity, as are many growing cities around the world. In addition, drawing on the examples of intercultural exchange and learning discussed above, and the evolving sustainability and green city discourses at the municipal and regional levels, it appears that new biocultural realities and discourses are emerging as the city grows and changes. Given the rich diversity present in the region, and the very real challenges of conflicting discourses, how can cities like Vancouver stay green and bioculturally diverse, and respect past and current populations while also welcoming more new residents, worldviews, needs and relationships? In particular, a central challenge will involve ensuring that the dominant institutional greening discourse does not obscure and diminish the rich biocultural diversity of the region. Sustainability-focused cities such as Vancouver run the risk of unintentionally promoting green gentrification, a process in which greening and sustainability policies lead to the displacement of (1) local populations due to unaffordability and (2) local biocultural values and discourses due to the dominance of institutional and private-sector discourses (Curran & Hamilton 2012, Wolch et al. 2014).

While these are complex challenges without clear solutions, we can learn from theories of environmental justice and urban forest governance, and the experiences of local neighbourhoods and case studies. A foundational theory underlying successful strategies for bioculturally diverse futures is the concept of recognitional equity and balancing collective and individual rights in urban space and decision-making. Theories of recognitional equity, and its application in practice, arose within social justice movements and analyses (Young 1990) and are subsequently being developed within the context of urban forestry and biocultural diversity (Nesbitt et al. 2018). Recognitional equity is a concept that can be used to evaluate whether urban forest decision-making and control of urban space are likely to be fair and representative of multiple biocultural discourses and perspectives. Key dimensions of recognitional equity in an urban forestry context include (1) representation (who is present in decision-making?), (2) procedure (are there systems in place to ensure that all voices are heard and valued?), (3) desire to participate (is a process or decision space of interest and welcoming to diverse perspectives?), and (4) ability to participate (can those who wish to participate do so and in a meaningful way?) (Nesbitt et al. 2018). A key consideration of recognitional equity is the relative power of different voices and discourses in a decision space. In the context of urban forestry and biocultural diversity, recognitional equity would look like, for example, green space design processes that solicit and value multiple perspectives, that are welcoming to those perspectives and facilitate their inclusion, and that result in spaces that reflect diverse biocultural discourses and allow them to grow.

Another useful concept in this discussion is mosaic governance. Mosaic governance is a context-sensitive style of governance that supports the diverse biocultural practices and discourses present in a landscape (Buijs et al. 2016, Gulsrud et al. 2018). It allows for and supports a mosaic of diverse practices, cultures, people and institutions within the urban landscape, and acknowledges that it is not necessary to have a "one size fits all" approach to sustainability or urban greening. While a mosaic of biocultural discourses and urban forest governance practices clearly exists already in Vancouver, a mosaic governance approach in the region would also require institutional commitments to govern in such a way from municipal and regional governments. In many ways, the mosaic governance approach expresses and operationalises the concept of balancing individual and collective rights articulated by theories of recognitional equity. It is an approach that acknowledges the important leadership role that formal institutions, such as municipal and regional governments, can play in urban greening, while providing support for diverse place-based biocultural relationships and discourses within the urban fabric (Gulsrud et al. 2018). Examples of mosaic governance or recognitional equity in urban greening are gradually emerging in cities around the world, such as the greening of the Greenpoint neighbourhood in Brooklyn, New York, USA, where greening is directed by a citizen's advisory council, and Melbourne, Australia's urban forest strategy, that includes unique neighbourhood precinct plans.

The municipal and regional governments in the Vancouver metro area appear to be committed to some aspects of recognitional equity and mosaic governance, but have yet to successfully operationalise these principles in such a way as to intentionally support biocultural diversity via urban greening and sustainability approaches. Vancouver's Urban Forest Strategy, for example, contains a clear goal to engage citizens in the urban forest (City of Vancouver and Vancouver Board of Parks and Recreation 2018) while one of the primary goals of the City of Surrey's Shade Tree Management Plan is to maintain strong community engagement, stewardship and education (City of Surrey

Parks Division 2016). Vancouver also has broad goals around equity, diversity and inclusion in urban decision-making that may inform future approaches to urban greening and sustainability (City of Vancouver 2019a). However, while these goals are laudable, they do not provide clear and specific guidance to operationalise these principles to promote recognitional equity and biocultural diversity, for example, through mosaic governance of urban greening. While biocultural diversity may not yet be a focus of Vancouver's urban greening efforts, there is some evidence that place-based planning and mosaic governance operate in other realms of municipal urban planning. For example, Vancouver has long had a practice of creating neighbourhood-specific community plans that guide development and decision-making in those neighbourhoods and reflect local values and priorities (City of Vancouver 2019b). These are living documents that are updated periodically to reflect the changing needs of the neighbourhoods. This neighbourhood-specific approach to planning, that works within broader city- and region-wide goals, could be a useful model for equitable urban greening and support for biocultural diversity.

Biocultural diversity is a key aspect of human wellbeing in urban environments, and equitable approaches to urban greening and sustainability can ensure such diversity continues to grow and thrive. As Jim explains,

> There's an old story that if you're out of centre and you're feeling off, you would sit yourself under a big cedar tree and just lay there and just kind of think or be, and ultimately the energy of the tree would centre you, you know, and bring you back to a healthier head space .... If there were more trees, I think, people would be more comfortable in their own skins, let alone their environment.

References

Beasley, L. 2019. *Vancouverism*. Vancouver, Canada: UBC Press.

Birnie, A.L. 1964. Time, tide wait for no man – seawall battle continues. *Vancouver Sun*, July 24.

Boyd, R. 1999. *The Coming of the Spirit of Pestilence: Introduced Infectious Diseases and Population Decline Among Northwest Coast Indians*. Vancouver, Canada: UBC Press.

Buijs, A.E., Mattijssen, T.J., Van der Jagt, A.P., Ambrose-Oji, B., Andersson, E., Elands, B.H. & Møller, M.S. 2016. Active citizenship for urban green infrastructure: Fostering the diversity and dynamics of citizen contributions through mosaic governance. *Current Opinion in Environmental Sustainability*, 22: 1–6. doi: 10.1016/j.cosust.2017.01.002.

Chan, C. 2018. Koi-munching otter still on the loose in downtown Vancouver garden. *Vancouver Sun*. Available at: https://vancouversun.com/news/local-news/koi-munching-otter-still-on-the-loose-in-vancouvergarden.

City of Surrey Parks Division. 2016. 2016 shade tree management plan. *Surrey*. Available at: https://dspace.library.colostate.edu/bitstream/handle/10217/183542/AEXT_055302013.pdf?sequence=10.

City of Vancouver. 2012. Greenest city 2020 action plan. Vancouver. Available at: http://vancouver.ca/files/cov/Greenest-city-action-plan.pdf.

City of Vancouver. 2018. First nations art and totem poles. Available at: https://vancouver.ca/parks-recreation-culture/totems-and-first-nations-art.aspx.

City of Vancouver. 2019a. Equity, diversity and inclusion. Available at: https://vancouver.ca/people-programs/equity-diversity-inclusion.aspx.

City of Vancouver. 2019b. Neighbourhood planning. Available at: https://vancouver.ca/people-programs/neighbourhood-planning.aspx.

City of Vancouver and Vancouver Board of Parks and Recreation. 2018. *Vancouver's Urban Forest Strategy*. Vancouver, Canada.

Crosby, A. 1986. *Ecological Imperialism: The Biological Expansion of Europe, 900-1900*. Cambridge, UK: Cambridge University Press.

Curran, W. & Hamilton, T. 2012. Just Green Enough: Contesting environmental gentrification in Greenpoint, Brooklyn. *Local Environment*, 17(9): 1027–1042. doi: 10.1080/13549839.2012.729569.

Dr. Sun Yat-Sen Classical Chinese Garden. 2019. *History*. Available at: https://vancouverchinesegarden.com/history/.

Green, A.G. & Green, D.A. 1999. The economic goals of Canada's immigration policy: Past and present. *Canadian Public Policy*, 25(4): 448–451.

Gulsrud, N.M., Hertzog, K. & Shears, I. 2018. Innovative urban forestry governance in Melbourne?: Investigating 'green placemaking' as a nature-based solution. *Environmental Research*, 161: 158–167. doi: 10.1016/j.envres.2017.11.005.

Heritage Vancouver. 2019. *10 Historic Street Trees*. Heritage Vancouver. Available at: http://heritagevancouver.org/top10-watch-list/2010/10-historic-street-trees/.

Hotte, N., Nesbitt, L., Barron, S., Cowan, J. & Cheng, Z.C. 2015. *The Social and Economic Values of Canada's Urban Forests: A National Synthesis*. Vancouver, Canada: Canadian Forest Service.

Jubas, K. 2018. Alberta's shameful pipeline politics ignores first nations. *The* Conversation, March 28.

Kataoka, S. 2009. Vancouverism: Actualizing the livable city paradox. *Berkeley Planning Journal*, 22: 42–57. doi: 10.5070/BP322112717.

Kheraj, S. 2013. *Inventing Stanley Park: An Environmental History*. Vancouver, Canada: UBC Press.

Kluckner, M. 2011. *Vancouver Remembered*. North Vancouver, Canada: Whitecap Books.

Konijnendijk, C.C., Ricard, R.M., Kenney, A. & Randrup, T.B. 2006. Defining urban forestry – a comparative perspective of North America and Europe. *Urban Forestry & Urban Greening*, 4(3–4): 93–103. doi: 10.1016/j.ufug.2005.11.003.

McCann, E. 2013. Policy boosterism, policy mobilities, and the extrospective city. *Urban Geography*, 34(1): 5–29. doi: 10.1080/02723638.2013.778627.

McPherson, E.G., Nowak, D., Heisler, G., Grimmond, S., Souch, C., Grant, R. & Rowntree, R. 1997. Quantifying urban forest structure, function, and value: The Chicago urban forest climate project. *Urban Ecosystems*, 1: 49–61.

Metro Vancouver. 2015. Connecting the dots: Regional green infrastructure network resource guide. *Burnaby*. doi: 10.1002/ejoc.201200111.

Musqueam First Nation. 2019a. Historic timeline: Musqueam teaching resource. Available at: http://www2.moa.ubc.ca/musqueamteachingkit/history.php.

Musqueam First Nation. 2019b. Stories: Musqueam teaching resource. Available at: http://www2.moa.ubc.ca/musqueamteachingkit/stories.php.

Nesbitt, L., Meitner, M.J., Sheppard, S.R & Girling, C. 2018. The Dimensions of urban green equity: A framework for analysis. *Urban Forestry & Urban Greening*, 34: 240–48. doi: 10.1016/j.ufug.2018.07.009.

Nowak, D.J., Crane, D.E. & Stevens, J.C. 2006. Air pollution removal by urban trees and shrubs in the United States. *Urban Forestry & Urban Greening*, 4: 115–23. doi: 10.1016/j.ufug.2006.01.007.

Park, H., Turner, N. & Higgs, E. 2018. Exploring the potential of food forestry to assist in ecological restoration in North America and beyond. *Restoration Ecology*, 26(2): 284–93. doi: 10.1111/rec.12576.

Pemberton, K. 2013. Vancouver's Stanley Park named among best in the world as it marks its 125th anniversary. *Vancouver Sun*, May 10, 2013. Available at: http://www.vancouversun.com/technology/Vancouver+Stanley+Park+named+among+best+world+marks+125th+anniversary/8363940/story.html.

Protect the Inlet. 2019. Kwekwecnewtxw: Protect the inlet. Available at: https://protecttheinlet.ca/structure/.

Squamish Nation. 2019. Xay Temixw (sacred land) land use plan: Squamish Nation. Available at: http://www.squamish.net/about-us/our-land/xay-temixw-sacred-land-land-use-plan/.

Stanley Park Ecology Society. 2019a. Programming with First Nations perspectives. Available at: http://stanleyparkecology.ca/education/first-nations-programs/.

Stanley Park Ecology Society. 2019b. Stanley Park Ecology Society: Our mission. Available at: https://stanleyparkecology.ca/.

Takeuchi, G. 2018. Chinatown garden attempting to save remaining koi from hungry otter in East Vancouver. Georgia Straight, 2018. Available at: https://www.straight.com/life/1169556/chinatown-garden-attempting-save-remaining-koi-hungry-otter.

Tseil-Waututh Nation. 2019. Our story – Tsleil-Waututh Nation. Available at: https://twnation.ca/our-story/.

UBC Communications & Marketing. 2016a. Maya in exile garden. Available at: https://www.ubc.ca/stories/2016-summer/good-soil-good-humans/micro-maya-garden.html.

UBC Communications & Marketing. 2016b. Tu'Wusht garden. Available at: https://www.ubc.ca/stories/2016-summer/good-soil-good-humans/micro-tuwusht-garden.html.

UBC First Nations Studies Program. 2009. Aboriginal title. Available at: https://indigenousfoundations.arts.ubc.ca/aboriginal_title/.

University of British Columbia. 2019. Home: CSFS at UBC farm. Available at: https://ubcfarm.ubc.ca/.

Whysall, S. 2013. How Vancouver got to be cherry blossom heaven. *Vancouver Sun*, March 28. Available at: https://vancouversun.com/news/staff-blogs/how-vancouver-got-to-be-cherry-blossom-heaven.

Wolch, J.R., Byrne, J. & Newell, J.P. 2014. Urban green space, public health, and environmental justice: The challenge of making cities 'just green enough'. *Landscape & Urban Planning*, 125: 234–244. doi: 10.1016/j.landurbplan.2014.01.017.

Young, I.M. 1990. *Justice and the Politics of Difference*. Princeton, NJ: Princeton University Press.

11 Urban realities of engaging with nature in Europe

Increasing diversity and consequences for wellbeing and social cohesion

Birgit Elands, Bianca Ambrose-Oji, Annegret Haase and Karin Peters

Engagements with nature in super-diverse cities

Increasing social diversity in European cities

Within the last decades, cities have become more heterogeneous in terms of people with different cultural backgrounds, lifestyles, values and attitudes all living together. Increased mobility and migration have made cities places where people from many different international and ethnic origins settle. For example, in Amsterdam (Netherlands) and Stockholm (Sweden), respectively 50 per cent and 33 per cent of the inhabitants are currently characterised as having a migration background (CBS 2019, Stockholms Stad 2019); in Berlin (Germany) 67 per cent of the inhabitants have a migration background or come from another country (Amt für Statistik Berlin-Brandenburg 2019); and in Manchester and Birmingham (England), between 30 and 40 per cent of the population come from Black, Asian and Minority Ethnic (BAME) backgrounds (Office of National Statistics 2011). Many Western European countries share this trend of increasing social and ethnic diversity, most of which can be attributed to intra-European and international migration. Historically, Europe has seen several "waves" of migration that have had different origins and affected certain countries in different ways. The most recent periods of immigration include the post-war period from the 1950s to the 1970s, which saw movement of people invited as workers to help develop Europe's infrastructure and economy from former colonised countries in Africa, Asia and South America; east to west migration of Europeans after the collapse of the socialist systems in Eastern Europe and the subsequent accession of these states into the European Union (EU); as well as the significant increase in refugee and economic migration, mainly from countries experiencing war, conflict and political instability, and from poorer regions including South-eastern Europe (e.g., former Yugoslavia), the Middle East and some countries of the Global South.

Migration status, national origin and ethnicity are not the only dimensions that are increasing the heterogeneity and diversity of European cities. Differences in socio-economic status (education, income), lifestyles and cultures are apparent as well. Consequently, some authors have pointed to the fact that urban societies are no longer dominated by a national majority but are made up of people from many minorities (Schneider et al. 2015). Others have conceptualised these changes as urban societies becoming "super-diverse" (Vertovec 2007), or "hyper-diverse" (Tasan-Kok et al. 2014). "Diversity", as a term, has also been criticised because of the different meanings given to it in different European countries, as well as how it has been used in some neoliberal policy environments to conceal existing inequalities and inequities between different inhabitants' groups (Großmann et al. 2018, Schuster 2018).

The social benefits of urban green

With the processes of continuing urbanisation and increasing urban diversity, there is a need to focus on developing attractive and healthy urban environments. It is generally acknowledged that urban green spaces (UGS) that range from woodlands and parks to allotment gardens and street greenery, contribute to the liveability of cities and the health and wellbeing of residents. UGS provide places for meeting and communication, recreation and leisure, learning and experimentation and experiencing nature (Pauleit et al. 2019). UGS have also been demonstrated as providing spaces for greater equality, by allowing people to move freely and behave free from discrimination (Low 2013). Moreover, UGS can contribute to social cohesion. This can be defined as "the extent to which a geographical place achieves 'community' in the sense of shared values, cooperation and interaction, positive and friendly relationships and feelings of being accepted and belonging" (Elands et al. 2018, p. 116). UGS provide spaces to meet friends and relatives or to engage in small talk with other visitors and users, but they can also create public familiarity which results in people feeling at home in a neighbourhood (Veen 2015, Elands et al. 2018). This is especially important in densely settled neighbourhoods, where housing and living conditions are below average and people have a greater need for public space to relax, meet and spend time (Berding 2013).

Although social cohesion may be an outcome of UGS, this is not guaranteed. In super-diverse societies, there are many different uses of and demands on UGS, which might be competitive or even raise conflicts. At the same time, existing power relations and practices of social exclusion are reproduced in UGS, leading to, at times, segregated use of some UGS or exclusion of some group(s) through the appropriation of an UGS by other group(s). If diverse voices and interests are not included in the decision-making, development, management and use of UGS, various forms of exclusion and marginalisation may result.

Shortcomings of knowledge about nature engagements in super-diverse cities

Although it is claimed that immigration increasingly characterises the composition and identity of today's societies (Foroutan 2018), scholarly knowledge about people with a migrant biography and their engagements with UGS is still scarce and shows some shortcomings. We know that people with a migration background do not visit urban green spaces as frequently as the non-migrant population. Managers of recreation areas have raised this as a concern, as they fear it may result in decreased commitment to environmental conservation among future, more diverse generations. It might also negatively affect all kinds of social benefits that individuals and communities might derive from nature engagements, such as improved health, wellbeing, social cohesion and environmental justice. Despite these concerns, Kloek (2015) concludes that the perceptions, wants and needs of people with a migration biography are under-represented in most UGS studies. She also concludes that there is a lack of knowledge about the heterogeneity between and within people with different national backgrounds and ethnicities, as they are often treated as one single and homogeneous group (ibid. 132). She argues that national or ethnic background is often just one of many criteria that shape perceptions or views a person has towards UGS and related wants and needs.

Focus of this chapter

Given the fragmented knowledge on the role and importance of UGS for an increasingly diverse urban population, this chapter therefore aims to illustrate and discuss how a variety of engagements with urban nature arises in heterogeneous, socially and culturally diverse urban environments. Engagement with nature refers to the variety of ways in which human beings live with nature, and how their wellbeing or livelihoods are dependent upon it. Citizens use UGS for enjoying the scenery, having walks or picnics, cycling through parks to go to their work or school, to worship or for psychological wellbeing and to grow vegetables and fruits in allotment gardens. Engagements with nature, however, are not only consumptive, in terms of enjoying nature for human wellbeing, but are also directed towards the protection of it. They may contribute to the management of UGS and raise their voices in cases where nature is under threat, such as from housing developments. In this chapter, we will focus on three main forms of active citizen engagement, in which consumptive engagements are gradually transformed into more productive engagements, namely, *recreation*, *community gardening* and *community-led nature management*.

In Europe, recreation is a major ecosystem service of urban green spaces. It mostly takes place in parks and forests that are owned and managed by the urban municipality, by NGOs or private landowners. Management of UGS is largely directed towards meeting recreational needs. Community gardening is another important function of urban green spaces. Gardening is, as recreation, directed towards fulfilling human needs as food production is its main purpose. However, due to growing interest in ecological gardening, it is no longer only about the utilitarian use of nature, but also about nature protection. Finally, we will discuss forms of community-led nature engagement in which citizen groups actively take the lead in developing and managing the urban environment for nature protection purposes. This type of engagement is also referred to as stewardship or active citizenship. For each of these engagements, we highlight the interactions between culturally diverse groups, as well as the social benefits that are derived from these engagements. In the final section, we draw conclusions and discuss how urban nature policies could become more inclusive in addressing cultural diversity to safeguard and enlarge citizens' engagement with urban nature in super-diverse cities.

Culturally diverse engagements with nature and its social benefits

Recreation in urban green spaces

Research in Europe has shown that the main activities carried out in UGS such as urban parks are relaxing, eating, walking, playing with children and taking part in sport. A large study conducted among 3,487 park users across the cities of Bari, Berlin, Edinburgh, Lisbon and Ljubljana revealed three major types of activities: *physical activities*, such as taking a walk, practising sports, passing through the park, walking a dog and playing with children; *social activities*, which included relaxing and meeting up with friends and relatives; and *nature-related activities*, which included experiencing nature as well as looking at specific plants or animals (Fischer et al. 2018). Although there were variations between the cities, physical activities were most frequently mentioned, followed by social and nature-related activities.

Numerous studies have indicated that citizens with a migration background visit green spaces less frequently than those without a migration background (e.g., Natural England 2019). This is, however, not or at least less the case for UGS. Therefore, Kloek (2015) argues that "under-participation" is a misleading term to typify the recreational behaviour of citizens with a migration background, because the extent to which immigrants participate differs according to ethnicity. She surveyed 1,057 young adults of Chinese, Turkish and non-immigrant descent in the Netherlands. The participation rate in urban recreation of respondents of Turkish descent was much higher (72 per cent) than the participation rate among respondents of Chinese descent and non-immigrant respondents (29 per cent and 56 per cent, respectively). Moreover, she found much variation within each group as personal identities, such as age, and to a lesser extent, gender and religion also influenced outdoor recreational behaviour. This is confirmed by other research which highlights that, although migration is an important explanatory variable, factors such as gender, age and religion and socio-economic position (e.g., level of education and income) also need to be taken into account when determining migrants' use of parks for recreational use (Figure 11.1).

Figure 11.1 Urban recreation in Europe. Source: Haase, Annegret.

Stodolska et al. (2017a, 2017b) investigated the different types of social benefits of UGS recreation by interviewing 70 people: 15 Ukrainian and 11 Vietnamese immigrants in Poland, nine Moroccan immigrants in the Netherlands and nine Turkish immigrants in Germany, complemented by 13 Latino and 13 Chinese immigrants in the USA. The results of the European interviews added insights to those of other researchers about the key benefits, as discussed below.

The restorative function of nature was identified as a main benefit. Although we know that this is important for everyone, it can play a particularly important role for recent immigrants because they often have to navigate the hardships of adjusting to everyday life in unfamiliar settings, deal with separation from their families and struggle with the constraints of a foreign language. Immigration-related stress might result in negative outcomes, such as depression and anxiety. The stress-relieving functions of natural environments can therefore be of importance for the quality of life among immigrants, regardless of their ethnicity, and country of settlement. The same study also showed that Turkish-Germans planted vegetable gardens, or Ukrainians in Poland visited wooded areas outside of the city. As such, they use leisure in nature to re-establish a connection to ways of life they lived in their home countries, and thus to achieve the desired balance between novelty and familiarity. It seemed that while immigrants have to adapt to the new social, cultural and natural environment of their host country, the benefits of being in natural environments are of critical importance for relaxation and stress relief.

Natural environments as settings where bonds with friends and family members can be strengthened is a second major social benefit that was identified. Citizens with a migration background use UGS for social gathering more often than non-immigrants do. They visit parks with their families or relatives (e.g., parents, siblings, aunts, uncles), and in larger groups which stresses the importance of family outings to them. Consequently, important activities are socialising and having picnics. Citizens with a non-migration background tend to visit parks more often on their own, with a dog, a friend or their partner (and children) (Peters 2011, Kloek 2015, Natural England 2019 see also Vierikko & Jaatsi (Chapter 5). Although also important for non-immigrant users, for immigrants it can be of greater significance since being in UGS can help to overcome the alien environment of the host country (Stodolska et al. 2017a). Here again differences among different migrant groups have been recorded. While immigrants from the Mediterranean region and Latin America tend to visit UGS in large family-oriented groups, Asian, Western European and North American migrants preferred more solitary visits with close friends and family members (Peters 2011, Kloek 2015).

Thirdly, UGS recreation led to an attachment to natural environments found in the host country. Such attachment was perceived to contribute to an overall sense of belonging that evolved over time, and coincided with achievements in other spheres of life such as moving to, or owning, a new house, getting a new job and starting a family. The interviewed newcomers were in the process of building their own and their families' histories in the host countries through engagement with natural places they found meaningful. For the first-generation immigrants and their families, UGS provided contexts for important milestones in their lives, such as the first date in a park, or places where their children took their first steps. These visits also helped them realise who they were (e.g., how their identities and behaviours have changed after immigration), and who they were not (e.g., feelings of alienation and lack of comfort they experienced while visiting their homelands). Urban parks served as settings for interacting with families and for children's birthdays and graduation parties, thus helping immigrants develop new memories, while preserving others along with their traditions and culture. Both frequent visits to urban parks, and more sporadic interactions with nature in more distant locations, helped to nurture their social relationships and to build a connection to the new environment at the destination. Interactions with natural environments in the host countries led to the development of new memories and at the same time, sparked comparison and often nostalgic feelings towards the environment of the home they had left behind. As such, it is a critical component in the process of establishing belonging at the destination.

Stodolska et al. (2017a) showed that natural environments in the host countries have a different meaning for those participants who had previously lived in the countryside and those who were urban dwellers before emigration. For instance, for the Moroccan-Dutch, Turkish-German and some of the Ukrainian interviewees in Poland, natural environments were closely tied to their everyday life in their home countries, but after immigration these spaces were considered more as leisure settings. The process of developing attachments to nature at the destination among these participants seemed to be slower; being in natural environments was no longer part of their everyday routine. Because the main focus for this group was on working to establish financial security, leisure activities were not undertaken on a regular basis. Conversely, immigrants from Vietnam in Poland considered natural environments in their home countries as either wild and uninviting, or as a scarce resource since so many UGS were small and situated in the midst of densely populated urban districts. They regarded urban natural environments in Poland as safer and more welcoming, and as a result visited them more often and evaluated them more positively.

A final social benefit of UGS relates to the increase of social interactions between people from different ethnic and cultural backgrounds. Interactions can be planned, such as meeting familiar or known people, or unplanned. These unplanned interactions can be more passive and one-off interactions, e.g., passing by other people while walking, but also more focussed, e.g., greeting or having a conversation with somebody. Since planned interactions fall within the realm of the discussion above about socialisation in UGS, here we only discuss the unplanned interactions.

Empirical studies have been executed with respect to unplanned interactions in urban parks (Peters 2011) and urban forests (Jay & Schraml 2009). With respect to fostering inter-ethnic contacts, the findings suggested that, in general, contacts with strangers and people from outside of one's ethnic group in natural settings were infrequent and rarely led to establishing long-term friendships. Peters (2011) concluded that most visitors talk with strangers only once in a while. Most conversations are brief and relate to specific issues such as one's children or dogs, or to everyday issues like the weather. Such settings helped to facilitate casual conversations with unknown others. The lack of inter-ethnic interactions are not perceived as a result of any prejudices held. Not being focussed on establishing interpersonal interactions in nature might lead to strengthening existing social ties within migrant communities. On the other hand, perceptions among non-migrants that immigrants are unwilling to interact with those outside their social or ethnic group might also increase (Stodolska et al. 2017b). Interestingly, Peters (2011) showed the opposite outcome; while some non-native Dutch people expressed their wish to have more interactions with other park visitors, native Dutch people did not. Some Turkish-Dutch women suggested that perhaps native Dutch people are less interested in conversations than Turkish people: "They're nice people, they say hello then walk away. Elderly people more often say things than youngsters. But in general, Dutch people do not easily strike up conversations". Having casual conversations with strangers in public settings is acceptable behaviour in most Western societies, but interactions with unfamiliar people among those of East Asian and Middle Eastern origin is often frowned upon, particularly among women (Stodolska et al. 2017b, p. 488). Moreover, Kloek (2015) concluded that ethnic groups engage in different levels of sociability in natural environments. These different results are interesting and ask for more research to take account of the variation within these groups.

Despite their superficial character, positive interactions do provide opportunities for people of different ethnic backgrounds to mingle, connect and observe each other's recreation patterns which, in turn, helps to build mutual understanding. Research showed that the spontaneous interactions did occur and were appreciated by immigrants. The interactions with non-immigrants that happened helped to create awareness about habits and social rules (Peters 2011, Stodolska et al. 2017b).

It is clear that public UGS offers limited opportunities for interaction between people of different ethnic and culturally diverse backgrounds. However, even without interactions between different groups of (non-)migrants, people could still value being together in parks, which may stimulate social cohesion. People come across others and have to relate to each other. As such, using urban parks and other green areas provides residents with a sense of what is going on. To conclude, UGS in which visitors have no or limited social contacts with each other can still have an impact, through positive identification with the place and a sense of belonging (Peters 2011, Veen 2015 also Vierikko & Mia Jaatsi (Chapter 5), Garcia et al. and de Mendonça (Chapter 9).

Community gardening

Despite a varied history, urban agriculture continues to be an integral part of European urban landscapes. Though it has always been present in and around towns and cities, the number of people growing their own food has recently and rapidly increased, as reflected by the following quote "during the last decade, urban agriculture has become ever more 'hot, hip and happening' in the Netherlands" (Veen 2015, p. 17). This resurgence of urban agriculture in Europe has many drivers, including revaluing growing food in cities, particularly where low intensity, organic farming methods have the potential to contribute to the biodiversity of urban nature. However, it may be about more than food, and driven by objectives connected with developing a green economy, providing cultural ecosystem services, reimagining urban landscapes and nature and meeting environmental justice needs. The outcome has led to a wide variety of European urban agricultural practices which include: city farms, community gardens and orchards, care farms, school farms, home gardens, allotments or small-scale market gardening. The community focus and communal forms of urban agriculture in particular, have the potential to bring together socially and culturally diverse groups of people and help build socially inclusive relationships (Van der Jagt et al. 2017; Veen 2015, Elands et al. 2018). Some forms of European urban agricultural practices such as care farms are aimed solely at providing therapeutic and socially inclusive outcomes. The focus in this section of the chapter is on urban community gardens, which are defined as "gardens with a non-profit structure, situated in an urban context, that are managed collectively or at least are managing some common areas and/or that a group of people is somehow involved in running the garden" (Van der Jagt et al. 2017, p. 265, Veen 2015, p. 17). This broad definition involves a wide range of gardening initiatives, including allotment gardens.

To illustrate different kinds of urban community gardening and the associated impacts, we use examples from the *Green Infrastructure and Urban*

Biodiversity for Sustainable Urban Development and the Green Economy (GREEN SURGE) research project (www.greensurge.eu) (Figure 11.2). The project was funded under the EU's 7th Framework Programme for research, and involved 24 partners, including small- and medium-sized enterprises (SMEs) and city municipalities, from 11 countries. In-depth studies were carried out of a range of community gardening initiatives, though desk-based study of documents relevant to each of the initiatives, and a series of interviews with the people involved (Buijs et al. 2016; Van der Jagt et al. 2017).

Community gardens found in urban areas in Europe consist of plots of land that are divided into varying numbers of smaller parcels (from a few to several hundred) which are assigned to individuals or households for use: the Igelbäcken allotment garden (Stockholm, Sweden) is an example of such an arrangement. The garden occupies 2.3 ha of land divided into 160 plots. It was founded in 1978 during a "Green Wave" period, mainly to provide a place for workers to grow food and flowers, to support health and wellbeing and to promote social interactions. The surrounding residential area is comprised of dense, high-rise apartment buildings (Buijs et al. 2016, pp. 26, 50, 56).

Figure 11.2 Overview of studied community gardening cases: (showing clockwise from top) The Stopping Place in Szeged (Luca Száraz), allotment garden network in Lisbon (Paula Gonçalves) Igelbäcken allotment garden, Granton Community Gardeners in Edinburgh (Granton Community Gardeners). Source: Haase, Annegret.

Sometimes the land is also cultivated collectively: the Granton Community Gardeners in Edinburgh (Scotland, UK, https://www.grantoncommunitygardeners.org/home) is a grassroots community gardening initiative in an area of multiple deprivation in north Edinburgh. It was started in 2010 by local people living in flats without a garden, and came from the desire of community members to grow vegetables near their homes. Other aims were to enhance green spaces, promote community cohesion, healthy habits and environmental awareness. Over the years, the group has gradually expanded activities from managing a single garden on a site at the end of a residential road, to nearly 10 gardens. Each of these gardens has been created on relatively small plots that had previously been maintained as mown grass. The gardeners do not have individual plots but work together, sharing the produce as well as the cultivation (Buijs et al. 2016, pp. 26, 49, 55).

Although food production is important, other plants might also be grown, such as for ornamental, ecological or sentimental reasons. Plots may also include a shed and in some instances even a cabin which is used for accommodation over weekends or seasonal holidays. As illustrated by the GREEN SURGE research, the aims of many community gardens include not only food production but also human health and social cohesion. To facilitate these social aims, urban community gardens increasingly create collective facilities, such as playgrounds for children, benches and picnic tables or even cafés for social gathering. Moreover, many of these gardens promote biodiversity protection through guidelines, or sometimes even rules, for ecological gardening, beekeeping, insect hotels and bird houses.

Community gardening is more established in some European countries than in others. In Western Europe (e.g., UK, Germany, Netherlands and Belgium) and Scandinavia (e.g., Denmark and Sweden), allotment gardens were created in the 19th century and served a key role in family food provision. Although by the end of the 20th century one could see a decline in allotment gardening and the number of sites, new allotments are once again being established to meet the growing demand (Bell et al. 2016). Contemporary drivers for this are demand for healthy food, wanting to be close to nature and leisure opportunities. In some East European cities, urban community gardens only became popular in the latter part of the 20th century. For example, in Szeged (Hungary), the first example of a community gardening project was established in 2014: The Stopping Place ("Megálló") (Szeged, Hungary) is a community garden (< 1 ha) of 12 plots of lands. The garden accommodates 11 users, mostly families with children, with one plot functioning as a shared herb and spice garden. Due to its popularity, there is a waiting list for joining the community garden. The garden belongs to a Community Centre which aims to create a living community for residents from within the area, reduce alienation and enlarge accessible green areas. It organizes many different activities (e.g., yoga classes, education about gardening, edible plants and medicinal herbs etc.) (Buijs et al. 2016, pp. 26, 51, 56).

Increasingly, community gardens are also being considered as contributing to the ecological network of cities. The biological diversity found in community gardens is often high because the gardening practices are often ecologically friendly. Cities are key promotors of this integrated urban green infrastructure principle. The most well-known example of the strategic implementation of allotment gardens to increase the connectivity of urban green space within a city is Lisbon, Portugal: the municipal allotment garden network (Lisbon, Portugal) was started by the municipality and provides allotments integrated in urban parks and gardens that are considered an important feature of the urban green infrastructure. The network of gardens occupies 38 ha. This network is promoted not only because it contributes to the city's biodiversity, but also to self-sufficiency in food production, the socio-economic stability of the city and the welfare of its residents. The project started in 2007 with the first allotment opening in 2011. Eleven allotments have been created thus far with seven more under construction and three in the planning phase (Buijs et al. 2016, pp. 26, 50, 56).

There is a significant body of research demonstrating the social benefits of community gardens. Buijs et al. (2016) and Van der Jagt et al. (2017) describe a number of benefits derived from an extensive literature review. Firstly, it is evident that gardening provides both physical and mental wellbeing benefits such as reduced stress levels, increased physical strength and healthy food and eating habits. Community gardens have also been shown to facilitate learning about sustainable farming practices and human-nature interdependencies in a local urban context. This can train citizens to be self-sufficient in times of food scarcity. Lastly, urban gardening has the potential to bring together culturally different groups of people; there is evidence from a number of European cities that urban gardening projects operate explicitly as "intercultural gardens". Consequently, community gardens can contribute to social cohesion within a neighbourhood (Elands et al. 2018, Veen 2015). The GREEN SURGE research, confirms this as illustrated by the examples of Edinburgh and Stockholm: among the Granton Community Gardeners in Edinburgh there are a very a high number of people from different nationalities, for example, Polish, French, Kenyan, Nepalese, Dutch, Bengali and Kurdish. The tradition of sharing produce with the wider community through the weekly free Community Meal means that other vulnerable people, older folk, the disabled and more deprived families also derive social and wellbeing benefits from engaging with the Granton Gardeners (Buijs et al. 2016, p. 57).

In Stockholm, the number of immigrant residents involved in the initiative has increased over the years, reflecting the changing demographics of the neighbouring estate. Countries of origin of the current gardeners include Brazil, Cape Verde, Chile, Finland, Gambia, Greece, Guinea Bissau, Kenya, Iran, Iraq, Pakistan, Senegal and Turkey. The garden also successfully engages younger, older and retired people as well as people who are ill and cannot work a formal job, or are unemployed (Buijs et al. 2016, p. 57).

Community gardening, however, does not guarantee cultural and social diversity. For example, the community garden in Szeged is characterised by families being ethnically Hungarian, young adults, moderately to highly educated, with most of them having young children (Buijs et al. 2016, p. 57).

Other examples from Leipzig and Berlin suggest that community gardening, indeed, is not always very inclusive as under certain circumstances only individuals sharing similar values and lifestyles take part.

The GREEN SURGE cases all show increased social interaction as a consequence of being involved in community gardening (Buijs et al. 2016, Van der Jagt et al. 2017). Such interactions took a variety of forms such as meeting up and engaging with family and friends as well as other gardeners. The latter involved not only greeting other gardeners but also sharing seeds, tools and knowledge about growing and harvesting different crops. In addition to more immediate interactions, community garden associations may also facilitate interactions with outside specialist individuals/groups and local schools. A number of associations were recorded such as arranging speciality courses for gardeners to participate in such as ecologically friendly farming practices, as well as for local school groups to visit the community garden. Community garden associations were also recorded organising social events for gardeners to come together and mix socially, i.e., to barbeque together. Such interactions were acknowledged as leading to a greater sense of understanding of each other and across different cultural groups. For example, in Stockholm and Edinburgh, the case studies provided evidence of community gardens creating opportunities for Swedish and British gardeners to interact with recently arrived immigrant resident gardeners, many of whom came from rural areas in their home countries (Buijs et al. 2016, Van der Jagt et al. 2017). Nesbitt et al. (Chapter 10) similarly describes how members of various migrant communities have been given opportunities to practice their gardening and agricultural skills in the city of Vancouver. In turn, this has provided opportunities to interact with community members in the communities they now reside in.

In addition, the cases made clear that community gardening often leads to an increased place attachment and people taking responsibility for their neighbourhood. The gardening strengthened a shared value and norm system to look after each other and to improve the look and feel of the area and instilled a sense of community pride. This also led to less vandalism and littering, better maintenance and more safety. This has benefitted especially the more deprived areas in which a garden was situated (Buijs et al. 2016, Van der Jagt et al. 2017). This effect is also stressed by Veen (2015) who argues that social cohesion is not only facilitated by social contacts between groups of people but also via place attachment.

It could therefore be argued, that urban community gardens have the potential to serve as a route to developing inclusive and safe urban bioculturally diverse places with the changing demographic of people in European cities.

Community-led nature management

Traditionally, citizens have frequently been involved in the protection and management of public UGS. This engagement includes not only physical activities, such as undertaking nature management activities, or conducting monitoring surveys, but also political activities, such as participation in policy processes or engagement in political protest. Striking though is that throughout Europe this engagement is increasingly carried out with some degree of citizen autonomy. There are an increasing number of examples where local communities have taken on the ownership or management of whole sites ranging from 100 m^2 to more than 1,000 ha (Mattijssen et al. 2018). Such "green self-governance" practices have increased across Europe for a number of reasons. The financial crisis of 2007–2008 saw a reduction in the budgets that local government, public agencies and charitable organisations had for the management of urban green space and nature. As a result, local residents, community groups and social enterprises were encouraged as alternative managers of these spaces. In addition, the parallel trend of decentralisation of territorial, institutional and public sector governance, has seen a corresponding rise in citizen involvement in a number of areas including the planning and management of UGS. Citizens have taken advantage of the opportunities provided, initiating their own "bottom-up" projects, or getting involved in a variety of citizen-led or multi-stakeholder, co-operative green space management practices (Mattijssen et al. 2018). It is likely that citizens and communities may have been prompted to take action for themselves because they believe government is failing to respond adequately to a variety of environmental issues, at both local level, e.g., public green space management, and global level, e.g., climate change.

The benefits of community-led nature management (CNM) relate not only to nature conservation, but also to the conservation of cultural landscapes, food production, environmental education and social cohesion (Mattijssen et al. 2018). Social cohesion indicates that urban green space under community-led management and ownership could be an inclusive space for community benefit. However, information about the engagement of people with a migrant or ethnic minority background in community-led initiatives, either as participants in management or as users of the resource, is scarce. We look at three CNM cases in England and the Netherlands, with which the authors are familiar through their work across several research projects (Figure 11.3) (Ambrose-Oji 2014, 2015, Lawrence and Ambrose-Oji 2014, Buijs et al. 2016, Mattijssen et al. 2017). We use these to demonstrate who takes part in nature management, and who benefits from accessing these natural spaces.

There are many reasons for citizens to start CNM. Quite often it starts from citizens perceiving problems and failures with existing management. One of the reasons for CNM is the concern of citizens about the neglect of an UGS. This was the case with Tower Hamlets Cemetery, in London England:

> The Friends of Tower Hamlets Cemetery (FTHC, https://www.fothcp.org) was established, in 1990, to protect and manage wildlife found on the 13 ha site, and to serve the local community who valued the area. After ten years of management by FTHC resulting in biodiversity improvements, in 2001 the cemetery was declared a Local Nature Reserve and Site of Metropolitan Importance for Nature Conservation.

Figure 11.3 Overview of studied community-led nature management cases: (showing from top to bottom) Tower Hamlets (Kenneth Greenway), Foundry Wood (Ambrose-Oji), De Ruige Hof (Martijn de Jonge). Source: Haase, Annegret.

Sometimes, the initiation of CNM can start from opportunities or dreams that citizens want to realise. An example of this is where citizens want to restore nature on former industrial sites and make the UGS publicly accessible and open to individual and community experiences. Foundry Wood in Leamington Spa, England (UK) is one such initiative: The Achieving Results in Communities group Foundry Wood (ARCFW, www.foundrywood.co.uk) is an urban community group, which was formed in 2012 to restore woodland on a post-industrial site of 2 ha, and build local community connection to urban nature for wellbeing benefits. To make sure that local representation was strong in the management of the woodland, the Friends of Foundry Wood group (FFW) was established.

An opportunity to develop nature in the city of Amsterdam (the Netherlands) was perceived by the initiators of De Ruige Hof: Nature Association De Ruige Hof (DRH; www.deruigehof.nl), was established in 1986, when a group of citizens stood up to protect the spontaneously emerging nature on a local site. They now manage 13 ha in a multicultural neighbourhood in the South east district of Amsterdam. Their aim is to bring citizens closer to nature as well as to bring nature closer to citizens.

The extent to which a diversity of social groups and individuals are involved in these initiatives differs. It is related to what social activities are organised, and the scope of those activities. At Foundry Wood, the "friends of" and the community group (FFW and ARCFW) are very active. For example, a restored building within the cemetery is being used to run education programmes and provide free outdoor and environmental learning programmes to local schools. They run over 170 fun and educational events each year, including nature volunteering, guided walks, festivals, forest school, history events and ecotherapy. Finally, DRH organises volunteering activities, nature excursions and a Kidsclub.

The CNM organisations have many volunteers donating time, skills and enthusiasm to make these activities happen. The social interaction, sense of community and social cohesion brought about is abundant. The FTHC are managed by a staff of three permanent people who are supported by over 3,000 volunteers, some come from the local community and others contribute through corporate social responsibility programmes. Around 30 people make up the FFW and there are around ten active regular volunteers. In addition, around 40 people a year take part in therapeutic ecotherapy activities which include conservation management work of the woodland. DHF has around 400 members (individuals or families), who pay membership fees and 50–60 regular volunteers. The former mostly include people who are retired, unemployed or who work part-time. Many of the volunteers at DHF experience mental health problems; their volunteering opportunities are designed to contribute to their integration in societal life.

CNM has the potential to strengthen social cohesion in surrounding neighbourhoods. To reach this aim, the engagement of both volunteers and visitors needs to represent the cultural diversity of the neighbourhoods. Our case studies show how this can be fulfilled in different ways. The *Tower Hamlets Cemetery* is located in a very diverse London Borough, with one of the largest Bangladeshi communities in the country (15 per cent of the Borough's population is Bangladeshi). The most recent project of the FTHC is called "Nature and Us" has been designed specifically as a social cohesion project to engage diverse communities through nature conservation. These and similar activities mean those who have benefitted from access to nature have included those with a minority ethnic or migrant background. *Foundry Wood* is located within Leamington Spa, a community that is home to people of Asian ethnicity (8.6 per cent), and European migrants (5.2 per cent). The establishment of the FFW was a main method to ensure social integration and representation. ARCFW has encouraged the use of the site as a place for members of the surrounding community to engage with more vulnerable, isolated and lonely members of the community. This often includes members from diverse backgrounds as well as people with mental health problems and substance abuse problems. The ecotherapy programme was started at the site as an idea from a member of the public who found coming to the woodland helped them to overcome their own anxiety problems. *De Ruige Hof* is located in a very multicultural neighbourhood in the South east district of Amsterdam. Inhabitants in this district hold more than 130 different nationalities, those with a migration background come mostly from Suriname, Dutch Antilles and Ghana. The volunteers, however, are according to the DRH coordinator predominantly "White" and "grey" people (Buijs et al. 2016). An exception are the volunteers of the Kids Club, who are much younger. The kids who participate in the Kids Club represent to some extent the multicultural diversity of the neighbourhood. All in all, both the cultural diversity and the number of people participating in the activities is relatively low.

The success of CNM in the long-term engagement of citizens depends on at least three factors: (i) a certain degree of formalisation, by means of established rules and procedures, which provides stability to citizens, (ii) adaptive capacity where citizens are able to adapt to contextual changes in order to deal with changing political, cultural and socio-economic developments over time and (iii) a constructive role of authorities, who legitimise and support place-keeping activities by citizens (Mattijssen et al. 2017). Whether our cases are successful in terms of inclusiveness can be interpreted by these factors. CNM *Tower Hamlets Cemetery* has been formalised as it has a nature reserve status and also permanent staff, it has grown and developed to move from a focus on nature conservation in the site itself to community engagement, and the Borough Council now have a formal contract with the group to continue with the site management and outreach activities. In the case of CNM *Foundry Wood* is a constitutional structure that involves a formal community-based organisation and a Friends of Group links the management of the site with the local community very strongly, while providing continuity of oversight for issues like fundraising and site ownership by local community organisation Achieving Results in Communities (ARC). The municipal authority has also provided legitimisation to the group through, for example, advertising the site as one of the town's UGS and including it on their official list of local volunteering opportunities. Finally, CNM *De Ruige Hof* scores perhaps the lowest on these three factors, especially as due to the low number of participating people the adaptive capacity may be limited. The more CNM is aimed at nature protection, the less social cohesion benefits (Mattijssen et al. 2018). This might also explain the difficulties DRH is experiencing with attracting people from the immediate residential areas.

In all three case studies, CNM has contributed to the delivery of more accessible green space, provided biodiversity benefits and maintained part of the larger ecological network of the city or town. Building social inclusion through nature connection with vulnerable, excluded or "disengaged" communities, including those from minority ethnic and migrant backgrounds, was a central objective in all three initiatives. While people from ethnic minority or migrant backgrounds may not have been members of the community-led organising committees, they were beneficiaries of the activities that took place on the sites and they were engaged in the practical management of the sites.

Conclusion and discussion

This chapter deals with the social benefits of diverse engagements with nature in European cities. We can conclude that the wellbeing effects in terms of relaxation, health and stress relief are abundant. Besides, UGS also provide many opportunities for residents to meet and interact with other people. These social interactions contribute to residents feeling that they belong in the community or neighbourhood they are living in. This chapter reveals that, although people generally do not visit UGS to meet strangers, they do like to engage in small talk with other visitors and users. These casual encounters appeared to be very relevant as they create public familiarity that can create feelings of being at home in a neighbourhood. From the case studies presented, we support optimistic views on green spaces as "zones of encounter" that may set the scene for meaningful contact across social differences as supportive and beneficial (Fincher & Iveson 2008). We do recognise that other research supports contrary views. Nast and Blokland (2014) argue that contacts are rather rare and do not necessarily have positive impacts. Next, UGS might be perceived as spaces of exclusion as the use of space by one group might exclude others, or where potential users don't access an area for fear of discrimination (Haase et al. 2019). Other scholars deliberately distinguish between different types of tolerance in situations of encounter. These range from intolerance, passive tolerance, a superficial acceptance of the presence of others and dynamic tolerance, i.e., the acknowledgement of the other as a legitimate subject on an equal footing (Bannister & Kearns 2013).

The likelihood of natural environments fostering interpersonal interactions is likely to depend on the type of the setting and its characteristics. We can speculate that large open spaces such as parks and forests provide fewer opportunities for interactions in contrast to community gardens, where people are encouraged to interact with one another by design (Elands et al. 2018). Our chapter highlights that, indeed, both urban gardening and community-led nature management, do lead to the strengthening of existing friendships and development of new ones.

Community gardening especially might lead to unexpected or unusual encounters between groups stemming from different cultural and social backgrounds as well as real world experiences. Community and allotment gardens may thus come to serve as places in which those people interact and (re-) negotiate a shared value system. This observation might be of particular interest in light of current levels of human migration on the European continent. Although this process of adaptation to a changing demographic is not always easy, the upside is that for every challenge there is an opportunity. While providing opportunities for social integration, community gardens may also be a way for people to affirm or express their culture (Van der Jagt et al. 2017, Elands et al. 2018).

Developing a sense of community and/or a sense of belonging is a bi-directional process. This is especially relevant for immigrants, though not exclusively. It involves both immigrants adjusting to the local culture of environmental use, but also of the local municipalities redesigning nature to cater to the newcomers' preferences and expectations. It also involves people who have lived in a certain neighbourhood for a long period of time as their needs need to be accommodated as well.

Although the role of UGS in providing places for encounter and communication, that may strengthen social cohesion across social and cultural differences and increase the wellbeing of inhabitants, has largely been recognised by urban policymakers and planners, it is also important to recognise that heterogeneous UGS are needed to accommodate not only multiple uses, but also multiple needs and wishes of an increasingly diverse society and people representing different real world experiences. Moreover, it is crucial to consider not just diversity, but also its interaction with equity issues to ensure that benefits are equitably shared and distributed. Equity does not come automatically with urban green engagement but needs to be deliberately included as a target into planning. UGS have a great potential for supporting equity but their planning and development need a framework that prioritises cohesion and equity at once and, not least, acknowledges diversity. In order to do so, it is more important than before to engage citizens from various living environments into the planning and stewardship of UGS.

References

Ambrose-Oji, B. 2014. Achieving results in communities (ARC) Warwickshire, England. Available at: http://smallwoods.org.uk/information/case-studies/.

Ambrose-Oji, B. 2015. *The Impacts of Woodlands on Individual and Community Wellbeing: Qualitative Evidence from Wales, Scotland and England*. Evidence Briefing Note 2 – Physical Activity and the Natural Environment. Cardiff: Welsh Government.

Amt für Statistik Berlin-Brandenburg. 2019. *Einwohnerinnen und Einwohner im Land Berlin am 31. Dezember 2018*. Available at: https://www.statistik-berlin-brandenburg.de/publikationen/stat_berichte/2019/SB_A01-05-00_2018h02_BE.pdf (Accessed: 25/10/2019).

Bannister, J. & Kearns, A. 2013. The function and foundations of urban tolerance: Encountering and engaging with difference in the city. *Urban Studies*, 50: 2700–2717.

Berding, U. 2013. Öffentliche Räume – Orte der gesellschaftlichen Integration? *Vhw Forum Wohnen und Stadtentwicklung*, 5: 247–250.

Bell, S., Fox-Kämper, R., Keshavarz, N., Benson, M., Caputo, S., Noori, S. & Voigt, A. 2016. *Urban Allotment Gardens in Europe*. London, UK: Routledge.

Buijs, A.E., Elands, B.H., Havik, G., Ambrose-Oji, B., Gerőházi, E., Van der Jagt, A.P., Mattijssen, T.J., Steen Møller, M. & Vierikko, K. 2016. Innovative governance of urban green spaces: learning from 18 innovative examples across Europe. Available at: http://greensurge.eu/working-packages/wp6.

CBS. 2019. *Population*. Available at: https://www.cbs.nl/en-gb/background/2018/47/population (Accessed: 22/10/2019).

Elands, B.H., Peters, K.B. & De Vries, S. 2018. Promoting social cohesion and social capital increasing wellbeing. In Van den Bosch, M. & Bird, W. (Eds.), *The Role of Nature in Improving the Health of a Population*. Oxford, UK: Oxford University Press. pp. 116–121.

Fincher, R. & Iveson, K. 2008. *Planning and Diversity in the City: Redistribution, Recognition and Encounter*. New York, NY: Red Globe Press.

Fischer, L.K., Honold, J., Botzat, A., Brinkmeyer, D., Cvejic, R., Delshammar, T., Elands, B.H., Haase, D., Kabisch, N., Karle, S.J., Lafortezza, R., Nastran, M., Nielsen, A.B., Van der Jagt, A.P., Vierikko, K. & Kowarik, I. 2018. Recreational ecosystem services in European cities: sociocultural and geographical contexts matter for park use. *Ecosystem Services*, 31: 455–467.

Foroutan, N., 2018. Die postmigrantische Perspektive. Aushandlungsprozesse in pluralen Gesellschaften. In Hill, M., Yildiz, E. (Eds.), *Postmigrantische visionen: erfahrungen – ideen – reflexionen*. Bielefeld, Germany: transcript. pp. 15–27.

Jay, M. & Schraml. U. 2009. Understanding the role of urban forests for migrants – Uses, perception and integrative potential. *Urban Forestry & Urban Greening*, 8: 283–294.

Großmann, K., Alexandri, G., Budnik, M., Haase, A., Haid, C., Hedtke, C., Kullmann, K. & Shokry, G., 2018. Living with diversity or living with difference? International perspectives on everyday perceptions of the social composition of diverse neighbourhoods. In Verschraegen, G., Oosterlynck, S. & van Kempen, R. (Eds.), *Divercities: Understanding Super Diversity in Deprived and Mixed Neighbourhoods*. Bristol, UK: Bristol University Press. pp. 211–234.

Haase, A., Budnik, M., Großmann, K., Haid, C., Hedtke, C., & K. 2019. Soziale differenzierung und zunehmende soziale Heterogenität in Leipzig im Spiegel von alltagswahrnehmung und politik. *Raumforschung & Raumordnung*, 77(5): 1–16.

Kloek, M.E. 2015. *Colourful green: immigrants' and non-immigrants' recreational use of greenspace and their perceptions of nature*. Wageningen, The Netherlands: Wageningen University.

Lawrence, A. & Ambrose-Oji, B. 2014. Beauty, friends, power, money: Navigating the impacts of community woodlands. *The Geographical Journal*, 181: 268–279.

Low, S. 2013. Public space and diversity: distributive, procedural and interactional justice for parks, In Young, G. & Stevenson, D. (Eds.), *The Ashgate Research Companion to Planning and Culture*. Surrey, UK: Ashgate. pp. 295–310.

Mattijssen, T.J., Van der Jagt, A.P., Buijs, A.E., Elands, B.H., Erlwein, S. & Lafortezza, R. 2017. The long-term prospects of citizens managing urban green space: From place making to place-keeping? *Urban Forestry & Urban Greening*, 26: 78–84.

Mattijssen, T., Buijs, A. & Elands, B.M. 2018. The benefits of self-governance for nature conservation: a study on active citizenship in the Netherlands. *Journal for Nature Conservation*, 43: 19–26.

Nast, J. & Blokland, T. 2014. Social mix revisited: neighbourhood institutions as a setting for boundary work and social capital. *Sociology*, 48(3): 482–499.

Natural England 2019. *Monitor of Engagement with the Natural Environment: The National Survey on People and the Natural Environment*. Peterbourgh: Headline Report 2019.

Office of National Statistics 2011. *Census Analysis, Ethnicity and National Identity in England and Wales: 2011. Data Tables*. Available at: https://www.ons.gov.uk/peoplepopulationandcommunity/culturalidentity/ethnicity/articles/ethnicityandnationalidentityinenglandandwales/2012-12-11#more-census-analysis (Accessed: 05/06/2019).

Pauleit, S., Ambrose-Oji, B., Andersson, E., Anton, B., Buijs, A., Haase, D., Elands, B., Hansen, R., Kowarik, I., Kronenberg, J., Mattijssen, T., Stahl Olafsson, A., Rall, E. &Van der Jagt, A.P. 2019. Advancing urban green infrastructure in Europe: Outcomes and reflections from the GREEN SURGE project. *Urban Forestry & Urban Greening*, 40: 4–16.

Peters, K.B. 2011. *Living together in multi-ethnic neighbourhoods: the meaning of public spaces for issues of social integration*. Wageningen, The Netherlands: Wageningen Academic Publishers.

Schneider, J., Crul, M. & Lelie, F. 2015. *Generation mix: die superdiverse zukunft unserer städte – und was wir daraus machen*. Münster: Waxmann.

Schuster, N. 2018. Diverse city. In Rink, D. & Haase, A. (eds.). *Handbuch stadtkonzepte*. Stuttgart: 63–86.

Stockholms Stad 2019. *Statistik*. Available at: http://statistik.stockholm.se/detaljerad-statistik. Accessed online 21/05/2019.

Stodolska, M., Peters, K.B. & Horolets, A. 2017a. Changes in recreation participation in natural environments after immigration among immigrants in the U.S., Netherlands, Germany and Poland. *Leisure Studies*, 36: 811–824.

Stodolska, M., Peters, K.B. & Horolets, A. 2017b. Immigrants' adaptation and interracial/interethnic Interactions in natural environments. *Leisure Sciences*, 39: 475–491.

Tasan-Kok, T., van Kempen, R., Raco, M. & Bolt, G. 2014. *Towards Hyper-Diversified European Cities: A Critical Literature Review*. Utrecht, The Netherlands: Utrecht University.

Van der Jagt, A.P., Szaraz, L.R, Delshammar, T., Cvejić, R., Santos, A., Goodness, J. & Buijs, A.E. 2017. Cultivating nature-based solutions: the governance of communal urban gardens in the European Union. *Environmental Research*, 159: 264–275.

Veen, E.J., 2015. *Community Gardens in Urban Areas: A Critical Reflection on the Extent to Which They Strengthen Social Cohesion and Provide Alternative Food*. Wageningen, The Netherlands: Wageningen University.

Vertovec, S. 2007. Super-diversity and its implications. *Ethnic & Racial Studies*, 30(6): 1024–1054.

12 Living in kinship within urban landscapes through equitable, multicultural, collaborative stewardship in New York City

Heather McMillen, Lindsay Campbell, Erika Svendsen, Christian Giardina, Kekuhi Kealiikanakaoleohaililani and Kainana Francisco

Introduction

> [Stewardship] means recognising the gifts that you see before you, recognising that others have protected it in the past, and holding hands with them from the past into the future to preserve and enhance something precious.
>
> (pers. comm. Community Gardener, New York City 2019)

Urban residents who cultivate relationships with nature tend to experience: enhanced wellbeing, cognitive function, social cohesion and sense of place; reduced stress; strengthened socio-ecological resilience; and, in some cases, improved sustenance via the gathering of foods and medicines as well as materials for cultural practices. The goal of strategic planning in urban areas is to design networks of green and blue spaces that deliver wide-ranging benefits across spatial scales (Hansen et al. 2017). By urban nature, we refer to the green (land) and blue (water) spaces in cities, suburbs and towns where plants grow and animals dwell, including parks, gardens, street trees, waterbodies and waterways. Urban nature, then, is vital not only for providing ecological services (e.g., water and air filtration, heat reduction, carbon sequestration) but also for supporting the "psychological, social, and, in some cases, spiritual needs of its residents" (Svendsen et al. 2016). While widely appreciated that precious psychological, emotional, spiritual, sustenance and health benefits can result from investments into urban green and blue infrastructure, a growing body of research documents inequitable access to these benefits (Gerrish and Watkins 2018, Rigolon et al. 2018). Surprisingly little is known about the conditions, programmes, policies and practices that support more equitable access to, influence over and governance of urban nature by diverse cosmopolitan communities (cf. Ernstson 2013, Nesbitt et al. 2019).

A foundational premise of this chapter is that sacred relationships with nature enhance the stewardship capacity of natural resource professionals and community members (Kealiikanakaoleohaililani et al. 2019), with profound impacts on how people interact with urban landscapes and seascapes (McMillen et al. 2020). We posit that having a shared sense of stewardship supports more equitable natural resource planning and access to the benefits of nature. Here, our use of "sacred" is not limited to any particular religious expression. It is inclusive and also embraces the "secular sacred" (cf. Anttonen 1996, Knott 2016), which encompasses elements of everyday life that are set apart based on their supreme value (Durkheim 1976) and spiritual nature, with which we share profound "love, respect, care, intimate familiarity, and reciprocal exchange" (Kealiikanakaoleohaililani & Giardina 2016). If the quality, depth and longevity of the relationships between urban residents and nature affect the extent to which residents benefit from nature, then understanding how these benefits are distributed is of paramount importance to successful urban planning. Specifically, sacred relationships with nature are formed when there is physical access to blue and green spaces and the policy environment fosters rather than hinders opportunities for residents to engage nature in socio-ecologically meaningful ways. These include co-creating with nature; shaping and being shaped by nature; taking refuge in nature; enhancing social relationships while spending time together in nature; growing, harvesting and sharing family foods; sharing knowledge, practices and stories that feature nature; and connecting to a power greater than oneself. In dense, complex and highly constructed urban landscapes, enlightened planning that prioritises equitable access to nature allows residents to cultivate sacred connections to place and with each other (Kealiikanakaoleohaililani et al. 2019).

Nature is valued in a range of ways, from an economic commodity to a beloved family member.

> While it is human nature to appreciate nature, there is no consensus on what we value most. What is beloved to some can be disregarded by others, and what is desirable can change over time, space, and context, even for the same person.
>
> (McMillen et al. 2019, p. 1)

And this extends to urban nature. Competing interests (e.g., economic, spiritual, aesthetic) often hinder urban planning efforts to provide equitable access to nature (McMillen et al. 2019), even while knowing that benefits (Ribot & Peluso 2009) are enhanced when residents can interact with, shape (e.g., through designing, planting, caring for and collecting from green spaces), and be shaped by nature (e.g., through being inspired, feeling connected, experiencing increased wellbeing).

Gentrification can impact equitable access to urban nature, and in most cities, investments into urban nature can drive gentrification (Pearsall 2010, Curran & Hamilton 2012) – ultimately influencing who can form and sustain sacred relationships with nature, and how, where and for whom these relationships will be degraded or even eliminated. We posit that effective planning needs to understand these dynamics as well as: how nature within the urban ecosphere is valued by diverse residents; how access to nature is negotiated; and how equitable access to, influence over and governance of urban nature by diverse urban communities can be facilitated. In this chapter, we rely on a political ecology of place – a useful framework for examining crucial questions about who decides how urban green and blue infrastructure is planned, built and re-created (Swyngedouw & Heynen 2003, Campbell & Gabriel 2016). We then ask: how are sacred relationships with urban nature formed in New York City (NYC) and what do these look like? To these ends, we: (1) establish the nature of sacred relationships in the mega-urban ecosphere that is NYC, the features and underlying conditions for these relationships, as well as the benefits of nature as described by residents; (2) describe research that engages the sacred as related to urban green and blue spaces; (3) offer preliminary insights into how relationships with nature in the city have been managed, promoted or degraded by powerful actors, economic shifts, programmes and policies over time and (4) explain actions that can enhance equitable access to urban nature.

The profound Biocultural diversity of New York City

Spanning three islands and the adjacent mainland on the Atlantic Coast of the United States (US), NYC covers nearly 800 km^2 and has a population of almost 8.5 million (US Census Bureau 2018), making it both the largest and densest city in the US. It is highly diverse in terms of ethnicity, national origin and languages spoken. The US Census Bureau (2018) reported that NYC was 43 per cent White, 29 per cent Hispanic or Latino, 24 per cent Black, 14 per cent Asian and 3 per cent multiracial. More than 37 per cent of NYC residents are foreign-born; 49 per cent speak a language other than English at home; and 23 per cent speak English less than "very well" (US Census Bureau 2010). NYC's urban park system covers 15,785 hectares (ha) representing 21 per cent of the city, the third highest percentage of open space among major US cities (The Trust for Public Land 2015). Of these parklands, approximately 8,220 ha consist of "natural areas" (The Trust for Public Land 2015), including forests, freshwater wetlands, meadows, rocky shorelines, beaches and salt marshes representing over 30 distinct vegetation associations (O'Neil-Dunne et al. 2014). According to Gómez-Baggethun et al. (2013, p. 215),

> Ensconced within these ecosystems are more than 40% of New York State's rare and endangered plant species. As a result, scientists are beginning to view New York City as an ecological hotspot – more diverse and richer in nature than the suburbs and rural counties that surround it.

In addition to these protected natural areas, half a million street trees in the public right of way and 2,100 ha of landscaped parkland provide additional nature benefits. While nature is abundant in NYC, it is not evenly distributed. Socioeconomic variables (i.e., education level and income) correlate positively with proximity to urban nature for many neighbourhoods (Nesbitt et al. 2019).

Sacred Connections to Place via Urban Nature in NYC

Human–nature connections have traditionally been enshrined within a concept of kinship, recognised as sacred, and codified in cultural practices founded on the intergenerational accumulation of knowledge about the social-ecological system that sustains them (Berkes 2012, Pascua et al. 2017, McMillen et al. 2017a). Some indigenous scholars have described human-nature relationships today as community-based and intergenerational guardianship that is sustained through reciprocal exchange (Johnson et al. 2016, Whyte et al. 2016). Reciprocal, kinship-based relationships with resources are not unique to rural, indigenous communities of the past; they continue to define contemporary relationships for indigenous and local communities and they can also be strong factors in other peoples' (e.g., settlers and migrants) relationships to nature (Nash 2014), including in urban areas (Elands et al. 2019).

For many residents, urban nature compels them into its care. Biophilia, the innate love and affinity humans have for nature (Wilson 1984), emphasises a reciprocal, kinship-based relationship between people and nature. Within academic and professional circles, the sacred elements of human relationships to nature are increasingly being recognised for driving decision-making about natural resources (Kohler et al. 2019). In our work in NYC, the sacred often emerges when residents share about their relationships with urban nature – features that inspire recovery and respite, facilitate connections with others and also lead to healing introspection (Svendsen et al. 2016, Sonti & Svendsen 2018). Residents use natural places for ceremonies, telling origin stories, connecting with natural cycles and experiencing wondrous phenomena. Nesbitt et al. (Chapter 10) provides similar examples of how elders from the Squamish First Nation use opportunities in natural places to reclaim and rediscover urban indigenous biocultural relationships, including among Squamish youth groups.

We draw upon our longitudinal research that includes large quantitative datasets and in-depth qualitative case studies to document how people are intimately connected to urban nature. While beyond the scope of this chapter, we are aware that strongly contrasting perceptions of urban nature exist. Some NYC residents view natural spaces as unsafe or inaccessible, especially for women and children (Sonti et al. 2020); others see the more "wild" (i.e., less landscaped) parklands as places where they can engage in private or transgressive acts, such as sleeping, sex, drug use or vandalism (Campbell et al. 2019).

Living memorials

For almost two decades, we have been documenting how living memorials created to commemorate the lives lost to the plane crashes in NYC on September 11th, 2001 (here after 9/11) also function (Svendsen & Campbell 2005, 2010, McMillen et al. 2017b) as sacred and therapeutic landscapes (Gesler 1993). A wide range of living memorials have been created – from the official September 11th Memorial and Museum built at "ground zero" where the twin towers collapsed to grassroots, guerrilla and informal spaces, such as public right of ways and vacant lots. We have documented hundreds of living memorials within public parks but also schools, hospitals, government agencies, fire stations, churches, and private gardens and woodlands. The scale and diversity of living memorials is enormous, ranging from single flowers to entire forests and dunescapes; from sunflowers to roses; from weeping cherry trees to columnar oaks (McMillen et al. 2017c) (Figure 12.1). These living memorials serve as public tributes to what was lost and because stewards incorporate natural, built, social and symbolic elements to communicate what they find important in their current lives, these memorials also provide ongoing therapeutic benefits. Further, care of living memorials assists with grieving, including through engaging nature, sharing of powerful stories and building social cohesion. Stewards indicate that planting and caring for plants helps them to process trauma and grief within a setting of strong social support, and so allowing a person to engage in something greater than oneself (McMillen et al. 2017b). As with community-based stewardship of green space along a street, in gardens and even in vacant lots (Campbell & Wiesen 2009, Sonti & Svendsen 2018), the healing power of nature in living memorials is bound up with the materiality of digging in the soil and weeding, but also the planting of particular species and the placing of meaningful objects (McMillen et al. 2017c).

Figure 12.1 Living memorial to 9/11: Healing garden of the federated garden clubs of New York State. Staten Island (NYC). Source: New York City Urban Field Station.

One steward at a living memorial in NYC (2017) explained: "Nature is one of the things greater than all of us. It's always here – as long as the world is here, nature is here".

A special kind of living memorial that evokes spiritual connections with nature is the survivor tree. Survivor trees are those that "have witnessed and withstood extreme disturbances and become compelling symbols for communities seeking to respond, recover, and reconnect following a tragedy" (McMillen et al. 2019, p. 3). The 9/11 survivor tree, a callery pear (*Pyrus calleryana*) that was growing near the World Trade Centre in NYC, survived the collapse of the Twin Towers. It has since become an internationally recognised symbol of resilience, strength and unity (McMillen et al. 2017c). Those who cared for the tree immediately following 9/11 describe this care as a healing process and spiritual connection, as might be used to describe relationships with family members. Those connected to this and other tragedies draw inspiration from the symbolism of this tree, finding hope, strength and resilience in the planting of saplings grown from its seeds, which are sought for commemorating recovery from the events of 9/11 and other traumatic events (McMillen et al. 2019).

Community gardens

Community gardens are a unique form of community-managed open space that provide multiple benefits to stewards, users and the public at large (Figure 12.2). Our work in NYC has shown how associated horticultural and agricultural activities can provide the practitioner with spiritual, psychological and emotional benefits (Campbell & Wiesen 2009, Sonti & Svendsen 2018). In surveying community gardeners from across NYC, Stone (2009) identified several key benefits of urban gardening: opportunities for unstructured recreation in a natural setting, dialogue about widely ranging topics (social, personal, political), and a place to develop community actions; intergenerational interaction, as well as programming specific for youth and seniors; and a strengthened sense of identity and pride. Similarly, Elands et al. (Chapter 11) describes how community gardening practices often lead to an increased place attachment and people taking responsibility for their neighbourhood.

In their research on motivations for community gardening in NYC, Sonti and Svendsen (2018, p. 1195) found that urban gardeners are also motivated by the "spiritual healing" they experience from planting, harvesting and sharing of produce. One person explained how gardening: "lifts the spirit and enables us to rise above life's tragedies" while another gardener explained: "I don't go to church to get in touch with God, I sit under a tree" (2018, p. 1198). Another shared:

> Personally, it's very relaxing to me. It is therapeutic, absolutely therapeutic and I get a lot out of it … it's very rewarding because my children are all grown, and it's a chance to nurture something from the seed to the point of harvest, and then prepare it.
>
> (Community gardener, NYC 2011)

Figure 12.2 Beach 41st Street community garden in public housing grounds. Rockaways, Queens (NYC). Source: New York City Urban Field Station.

Cocks et al. (Chapter 6) found similar sentiments being shared by gardeners gardening in their yard spaces adjacent to their newly acquired houses. The resulting reciprocal relationships between people and place (Sonti & Svendsen 2018) confer resilience to the community.

The profound meaning of nature is evidenced by so many new immigrants creating gardens reminiscent of their former agrarian lives – such as the strong tradition of Puerto Rican gardening in NYC, which leads to the building of *casitas* (little houses) and the naming of gardens after places or plants from home (Stone 2009). We have also seen how deeply rooted cultural traditions from distant homelands can be transported and adapted to a new home and then adopted by neighbours. At a community garden in Queens, NYC, a Korean family stored kimchi (traditional fermented, spiced cabbage) in crock pots buried in their garden, which became a vital food source for the family immediately following the devastating Hurricane Sandy. In response to multiple requests, this Korean family shared their practice with other community gardeners (McMillen et al. 2016). Nesbitt et al. (Chapter 10) shares a similar example of how Maya refugee families started up Maya in Exile Garden in the City of Vancouver in Canada. The garden provides the families with an opportunity to share Mayan knowledge and traditions with local people. Members of the garden also offer cooking and farming workshops to the public and volunteers are welcome to help out in the field.

Such interactions are building blocks for improved health and wellbeing, which foster social resilience at the community level (Svendsen 2009). Indeed, McMillen et al. (2016) identified that place attachment, social cohesion, social networks, and knowledge exchange and diversification are outcomes of stewardship at both living memorials and community gardens. Moreover, community gardeners are deeply attached to their plots, which was revealed in the late 1990s when across NYC gardens were threatened by a period of rapid development. In response to the resulting changes, gardeners organised and mounted a successful campaign to prevent the sale of these gardens without a proper, public review process (Von Hassel 2002, Stone 2009). The legacy of this grassroots social movement persists today in neighbourhood to citywide coalitions directed to supporting food justice and the equitable distribution of urban open space (Campbell 2017).

Parks

Parks, including natural areas such as forests and wetlands, are green and blue set-asides that serve a variety of functions, including: recreation, refuge, relaxation and socialisation (Auyeung et al. 2016, Campbell et al. 2016); psycho-social-spiritual wellbeing (Svendsen et al. 2016) and ecosystem services (Hansen et al. 2017, p. Iv). One visitor to Marine Park in Brooklyn commented: "I come here to get away, to relax and be with my thoughts. When I'm stressed out or something, I like to walk in these woods and explore". Many urban and national parks have a common origin in a 19th-century conservation movement focused on preserving landscape aesthetics and promoting specific uses for urban lands (Svendsen 2013 Campbell et al. 2019). A leader of this movement, Fredrick Law Olmsted, the designer of Central Park, hailed the development of urban parks as similar to that of preserving great areas of the American wilderness (Olmsted 1870).

As with intra and international migration across diverse geographies (Cocks & Dold 2006, Chan et al. 2016), residents of NYC have relationships with urban nature that are derived from their home culture before they, their parents or their grandparents came to NYC (McMillen et al. 2020). For example, in NYC Yoruba practitioners make offerings to Orisha at designated trees (Figure 12.3); Hindu practitioners make offerings to the Mother Ganga at urban waterfronts (Garcia et al. Chapter 8) and practitioners of Orthodox Judaism collect local grasses to construct temporary huts or sukkah for the celebration of Sukkot. What results is a complex assembly of diverse nature-based practices operating in close or even overlapping fashion. The emergence and dynamics of these relationships can bring about tensions, especially when emerging practices conflict with dominant models of land use and economic development. Importantly, these tensions can lead to novel and innovative integration and adaptation of sometimes ancient practices (McMillen et al. 2020).

Figure 12.3 *Orisha* offering at tree in Alley Pond Park. Queens (NYC). Source: New York City Urban Field Station.

Planning and paradigms governing urban nature

Here we explore how the dynamics of engaging nature shape and are shaped by governance and underlying power structures that influence the composition, abundance and distribution of nature across NYC. We offer a brief historical overview of the evolution of planning for and governance of nature in NYC. We explore the planning paradigms that have governed urban nature and examine how the balance of power between the state and the public shift in response to market forces and investment cycles.

Planning for nature in the capitalist city: role of elites in the 19th– early 20th century

Robert Foglesong (1986) examined planning in the late 19th century and used the establishment of public parks as an example of a "democratic-capitalist contradiction" whereby the desire for urban space competes with the flow of capital and private interests (see also Mellegard (Chapter 4)). The history of urban parks is fraught with examples of social and cultural divides and the pitting of privileged access against core principles of democracy (Cranz 1982). In the 19th century, the urban park concept shifted from private courtyards and green spaces for the wealthy to public spaces accessible to all. While important advances were made in realising this shift, many American cities were sites of inequitable, even unjust (e.g., removal of entire neighbourhoods), approaches to creating urban parks (Taylor 2009). Central Park was one of the first publicly financed and managed, large-scale capital projects undertaken by a city government in the US, and so presents a useful case study.

The establishment of Central Park was championed by an elite group of families who stood to benefit financially and socially from its construction (Taylor 2009). Comprised of local politicians and wealthy New Yorkers, the park's Board of Directors established and enforced public rules of social order and conduct in the park. The construction of public green space was a salve for ailments of a rapidly growing city and a new industrial working class. The notion was that the pastoral and artistic design of Central Park combined with proper rules of behaviour would "civilize and improve the lives of an untamed labor force" (Cranz 1982, Fein 1981). As planned, Central Park also became a real estate windfall for adjacent and nearby landowners. At times however, this park as with so many failed to serve the needs of a diverse public, which in the 20th century prompted a shift to the championing of smaller parks that could better serve the needs of multicultural and lower-income neighbourhoods (Taylor 2009). Today, Central Park remains one of the most iconic urban parks in the US if not the world, with a history that includes protest and power struggles, segregation and privilege, but also peace, beauty, joy and redemption. It continues to provide a reflection of many cultures and identities but retains a dominant design and narrative of ordered nature. Stanley Park also recognised as iconic park, i.e., "green jewel" of Vancouver shares a similar chequered history (Nesbitt et al. Chapter 10).

Caring for nature during crisis and austerity: transformation of stewardship in 1970s NYC

The 1970s financial crisis was a major turning point in the governance of NYC's nature. Much of the environmental work was linked to community development efforts designed to help NYC recover from the effects of divestments from public and private sector institutions that resulted from the political, economic and demographic shifts that began in the 1960s and culminated in the 1970s crisis. Recreation budgets were slashed, public restrooms were closed and urban green spaces became degraded. Budgets required that public authorities strongly limit park maintenance. Consequently, NYC parks became trash-strewn, overgrown and sometimes dangerous. Many civic stewardship groups responded to this decline by proactively taking on green space management. The resulting ethic of a highly engaged civic sector would birth a generation of powerful environmental stewardship groups (Connolly et al. 2014).

Central Park also experienced a downward trajectory but urban elite living near Central Park responded to the ensuing crisis by organising large amounts of private capital to rehabilitate and manage this iconic park. Since its founding as a non-profit in 1980, the Central Park Conservancy has invested US$500 million in restoration and maintenance. With US$390 million of this investment being raised from the private sector, the Conservancy has become the "gold standard" of urban park partnerships, inspiring similar initiatives in NYC including the Prospect Park Alliance and Friends of Hudson River and High Line Parks. There are now more than 40 park conservancies in NYC that have raised millions of dollars through private contributions, organised income generating schemes and mobilised a volunteer workforce for park management (Gentile 2009). Now mainstream across the US, this strategy has transformed the relationship between parks departments and civic groups.

Neighbourhood parks and open spaces throughout NYC were improved by these efforts, including the largely unsung efforts of small groups of neighbours and friends, block associations and social service organisations. Organised to "take back their parks", formal and informal organisations became involved in local environmental improvements (Svendsen 2013). Few local neighbourhood stewardship groups are comparable to the professionalised staffing and budget of the Central Park Conservancy, and while many of these groups continue to serve a critical role in public park and environmental stewardship (Taylor 2009), wealth discrepancies across NYC's uneven public-private partnerships have resulted in a "two-tiered" park system made up of the haves and the have nots (e.g., Kaminsky & Simonet 2017).

Supporting and being supported by nature in a 21st-century sustainable city

Heynen et al. (2006, p. 9) argue that from an urban political ecology perspective,

> there is no such thing as an unsustainable city in general, but rather there are a series of urban and environmental processes that negatively affect some social groups while benefitting others (See Swyngedouw & Kaika 2000). A just urban socio-environmental perspective, therefore, must consider the question of who gains and who pays.

Power imbalances and systemic and systematic exclusions – particularly based on class and race – are ongoing problems in NYC, but over the past two decades, there has been a rise of sustainability-oriented approaches to planning for and governing urban nature, including increasingly diverse stakeholders pursuing a more holistic and integrated set of strategies for managing water, waste, housing, energy and transportation (Pickett et al. 2013, Munoz-Erickson et al. 2016). First, the formulation and implementation of sustainability policies have necessitated the participation of numerous personnel across state agencies who interface with stakeholders (Brecher et al. 1993, Keil & Boudreau 2006). Second, hybrid governance approaches (Connolly et al. 2014), including public-private partnerships, continue to support implementation of sustainability efforts spanning multiple land jurisdictions, operating across decades (Campbell

2017), and engaging the public, especially volunteers and advocates for urban green infrastructure. Finally, increased attention is being given to the influence of civic stewardship groups in caring for and even shaping urban nature (Svendsen et al. 2016, Landau et al. 2019).

NYC's campaign to plant one million trees, MillionTreesNYC, provides a compelling case study of how public-private partnerships, in this case the NYC Parks Department and New York Restoration Project, can be directed to designing, investing in and implementing plans for the equitable creation or management of urban green space. Between 2007 and 2015, hundreds of staff and thousands of volunteers planted trees across private and public property in NYC, leveraging the complementary skillsets and resources of public and non-profit/private sectors. The effort was launched as part of the city's first long-term sustainability plan, PlaNYC, with a goal of improving quality of life in NYC. Whereas prior tree planting efforts had been triggered by requests, leading to an uneven distribution of urban forest, NYC Parks Department officials redefined trees as basic infrastructure, equivalent to stop signs and sidewalks, with the neighbourhoods with the least trees being targeted first for greening (Campbell 2017). This high-level change in mindset resulted in an equitable distribution of the 1 million trees across NYC's ethnically and economically diverse neighbourhoods. The campaign used multiple volunteer engagements and novel stewardship training programmes to teach residents how to care for these million new trees, which has resulted in a community-based urban forestry movement (Campbell 2017). Similarly, in 2015 other volunteers counted the city's trees in the census, TreesCountNYC! In the case of MillionTreesNYC, community tree plantings allowed residents to work together towards a common goal (Fisher et al. 2015). In the case of the tree census, participants enhanced their community awareness, heightened their observation skills and expanded their knowledge of nature (Johnson et al. 2018).

To better understand the occurrence and impact of civic stewardship, the Stewardship Mapping and Assessment Project (STEW-MAP) was established to map the scope and impact of environmental activities over time (Fisher et al. 2012; Campbell et al. in press). The Assessment was initiated in 2007 and repeated in 2017, during which STEW-MAP inventoried and geospatially defined the stewardship activities of organisations and their networks to better understand the extent and dynamics of urban stewardship (Svendsen et al. 2016, Landau et al. 2019). The 2017 assessment received responses from 800 groups representing budgets totalling US$800 million and an estimated 540,000 members. Analyses showed these groups made crucial contributions to urban stewardship, but because of the high degree of informality in the sector (half of these groups do not have legal non-profit status and about one-third have budgets under US $1,000), their impact and even their presence can be overlooked. Collectively, however, these stewardship groups wield enormous adaptive capacity (Connolly et al. 2013) and represent systems of change delivering on-the-ground stewardship, advocacy and data/knowledge production.

On this last point, these groups, often operating within complex collaborative learning and doing networks, develop and implement metrics to evaluate their work. Perhaps most importantly, these groups share resources and information across civic, public and private sectors (Landau et al. 2019), and bridge civic and public by playing "broker" within governance structures.

The STEW-MAP project has revealed the origin stories of many groups caring for NYC's nature, from highly professionalised, citywide broker NGOs to highly informal, local neighbourhood groups. Many of these stories are strikingly similar, beginning with friends, neighbours or colleagues responding to a hyper-local environmental threat, for example, development of a beloved park or declining condition of an empty lot, or a social threat, for example, neighbourhood violence and a lack of productive activities for youth. In many cases, these stewards do not own the property they steward but assert their rights and claims to place through acts of stewardship, including hands-on work and community organising (Peçanha Enqvist et al. 2019), or acts of place-making, knowledge exchange and collective action (McMillen et al. 2016). Elands et al. describes similar acts of knowledge exchange and collective action taking place in community steward programmes in European cities (Chapter 11). These acts often are founded on a shared understanding of the value, purpose, persistence and sacredness of urban nature. While contestation, conflict, protest and occupation have been key tactics used by social movement actors and civic groups to argue for the "right to the city" and "right to nature", these expressions highlight a need for new discursive arenas that foster trust and mutual understanding among people with radically different worldviews and life experiences but who share in the caring for place.

Strengthening kinship connections to place: the creation of new discursive arenas

In 2016, we began discussing how indigenous and local knowledge systems and relationships with nature might catalyse new approaches to the stewardship of urban nature. For this we relied on a partnership with Hālau ʻŌhiʻa, an intensive Native Hawaiian stewardship and lifeways training programme created to enhance the capacity of natural resources and conservation professionals to engage self, others and place (Kealiikanakaoleohaililani et al. 2018). It was created in 2015 with the goal to "transform the way we view and steward our lands and seas" and represents the formalised teaching on enhancing sacred relationships with the natural world (Kealiikanakaoleohaililani et al. 2018). The pedagogical approach is based on learning, practising, adapting and eventually creating rituals that open windows of opportunity for sacred, kinship-based relationships that connect us to place (Kealiikanakaoleohaililani & Giardina 2016).

While the conservation of biocultural diversity and biocultural stewardship have been written about widely, especially in indigenous and rural communities

(Gavin et al. 2015), the practice of applying them in multicultural, urban environments has rarely been explored. We have yet to realise the full potential of kinship-linked, place-based stewardship models in highly diverse and densely populated urban settings. We viewed the lessons of Hālau ʻŌhiʻa as being relevant and accessible beyond Hawaiʻi Island, where it was born, and so in 2017, participants in Hālau ʻŌhiʻa, USDA Forest Service researchers based in Hawaiʻi, worked with Kekuhi Kealiikanakaoleohaililani and USDA Forest Service researchers based in NYC to coordinate an exchange with stewards of nature in NYC. Our aims were to catalyse co-learning about connecting to place and cultivating stewardship across regional, professional, cultural and personal levels; identify and understand culturally based concepts and practices that drive environmental stewardship in NYC; and develop a framework for a pilot stewardship training programme for NYC. In the case study from Japan, similar approaches are being advocated in an attempt to restore notions of kinship with nature which are based on reinterpretation of traditional wisdom (Pastor-Ivars Chapter 2).

To achieve these goals, we organised *Learning from Place*, a two-day workshop (October 2017). It was led by Kekuhi Kealiikanakaoleohaililani, framed with the pedagogy of Hālau ʻŌhiʻa and based upon her approaches, exercises and concepts. These were adapted for NYC and included engaging in sacred ritual through song, traditional chant and dance (Figure 12.4); sharing sacred stories, creating poetic texts and art forms; and allowing time for personal

Figure 12.4 Hula illustrates kinship relationships to place, led by Kekuhi Kealiikanakaoleohaililani (right). Source: Ashford, Giles with permission.

reflection (Figure 12.5) and group discussions (for detailed description). By engaging in these practices and in relationships with surrounding green and blue spaces, participants (45 "stewards of NYC stewardship", – from land managers to educators and artists, as well as the authors of this chapter) were able to cultivate multiple ways of knowing their biology, minds, spirits and communities, while fostering deeper kincentric relationships with place.

In NYC, we identified a gap in the professional development and training currently available to natural resource managers, whose careers have been largely focused on the technical management of biophysical resources. This gap results in an under-developed capacity to engage spiritual dimensions of stewardship. Hālau ʻŌhiʻa directly addresses this gap by exposing participants to diverse knowledge-practice-belief exercises that help participants strengthen their connection to place and to their community of practice. Inspired by the success of this workshop, LC and ES created a series of Stewardship Salons (hereafter Salons) to build a community of practice among the "stewards of NYC stewardship". The Salons encourage participants to share their "origin stories" as a way to engage the secular sacred and reveal processes of personal transformation and then apply learning to stewardship practice (Figure 12.6). As our capacity is limited, this process cannot be all inclusive; however, the small group of Salon learners have brought their experiences back to their respective agencies, organisations and communities, raising our hope that the Salons are serving as a catalyst for change within broader personal and

Figure 12.5 Personal reflection during Learning from Place workshop. East River, Brooklyn (NYC). Source: Ashford, Giles with permission.

Figure 12.6 Small group discussion of origin stories. Bushwick Inlet Park, Brooklyn (NYC). Source: Ashford, Giles with permission.

professional networks. We recognised a need to transform training for stewardship professionals so that it would include listening with empathy; reflexively situating one's own perspectives, worldview and biases; and developing cultural competencies in interacting with diverse communities – including those with fundamentally different understandings of relationships with "nature." Bearing in mind histories of exclusion resulting in highly inequitable access to public space and trauma, resource managers of the future will become better stewards of the communities they serve through strengthening these cultural competencies.

The Salons have been led by diverse stewards filling both teacher and learner roles on a rotating basis. Each session is guided by "ground rules" that have been collaboratively developed by participants, and includes discussion with ample time for participant reflections. We have also instituted elements from Hālau ʻŌhiʻa – including an opening protocol, collectively creating a kuahu (altar) of natural and personal materials that bring certain intentions and energies to the space, informal sitting arrangements and the sharing of food. Content, topics, sites and speakers are sourced from our community of practice. For example, we have discussed Hindu ritual practices at Jamaica Bay (NYC) (see also Garcia et al. (Chapter 8)); examined the histories of Central Park and the displacement of Seneca Village; shared a Tu B'Shevat seder while exploring Jewish environmentalism; and reflected on the Great Migration (the 19th-century movement of African Americans from the Southern United States to Northern cities in often thwarted pursuit of greater economic opportunity and cultural freedom) and connections to Southern landscapes, including heirs' property in the Carolinas and urban farming and gardening in NYC (see also Schelhas et al. 2017). While reach and scope are limited, the Salon participants are highly diverse and we are demonstrating an effective approach to cultivating sacred stewardship in urban environments.

Through the first year of the Salons, we observed moments of transformation, which we consider to be the building blocks of change. With our guards down, hierarchies forgotten and roles as "agency representative" or "scientist" or "land manager" shed, participants are more comfortable recounting personal stories, reflecting on childhood memories, sharing deep emotions about stewardship and social change, and even partaking in laughter and play – modalities that are sorely lacking in many conventional meeting structures. Some have even reflected that they feel the Salons are a "safe space" where they can share insights they are not able to share in a workplace setting. And there has been deep emotional and intellectual wrangling with difficult issues that have surfaced during site visits and sharing of, for example, historical trauma, the meaning of land ownership, Indigenous rights and the need for significant work to diversify our communities of practice. Throughout, we have begun to treat each other, as well as the nonhuman others in our surrounding landscape, more as kin than commodity or fellow professional.

Conclusion

> Stewardship makes me think of being the captain of a ferryboat. It's about ferrying the ship in a way that helps me to arrive safely but also benefits all the people who are being stewarded along. And in a larger sense we're all being stewarded so the responsibility of stewarding is on us. I believe in a vision of a society that is interdependent.... I would much rather live in a world where everyone is taking care of one another than a world where everyone is just out for themselves. I think we're stronger that way, more resilient and it's a more beautiful world. My hope for the future of stewardship work in New York City is that all citizens of New York start to take it seriously as their job, it's not someone else's job, it's our job.
>
> (Steward at an educational programme about sustainable sanitation in NYC, 2019)

The dominant narratives around the stewardship of urban nature have been historically framed by economic development, social order and privilege. To overcome these historic legacies, we need a new framework from which coalitions of diverse bioculturally oriented stewards of place can operate and grow.

Based on our shared experiences with applying a model from Hawai‘i in NYC, we propose that more equitable access to and governance of urban nature can be realised by building communities that share a common desire for sacred relationships within multicultural, collaborative efforts to care for NYC's green and blue spaces. When these communities include natural resource management professionals and all types of residents, we are best able to equitably connect these biocultural coalitions to urban nature. Through Learning From Place, the Stewardship Salons, STEW-MAP and numerous other experiences, we have learned that people find intense meaning in urban nature, meaning often founded on sacred relationships with nature.

By engaging in traditional rituals and adapting rituals for new settings, communities work to cultivate and express personal connections to place. Today, urban stewards desire to more deeply engage these topics, to more effectively foster the sacred within an economic–social–political–programmatic–legal matrix that often determines who accesses urban nature and how. Future work in NYC and beyond might address (1) Can sacred relationships with nature be cultivated and nurtured, to reach a larger number and greater diversity of people, including resource managers, planners and policy makers? (2) Can doing so improve the economic conditions, funding programmes, public policies and law enforcement practices that support equitable access to urban nature? And (3) How can a heightened awareness of the sacred help us better care for urban nature and foster the type of social innovation that will improve land use planning and urban economic models? In the process of doing that work, we may grow as biocultural stewards of our familial relationships with each other and with our places.

References

Anttonen, V. 1996. *English Summary: Ihmisen ja maan rajat. 'Pyhä' kulttuurisena ategoriana* (The Making of Corporeal and Territorial Boundaries. 'The Sacred' as a Cultural Category). Helsinki: Finnish Literature Society.

Auyeung, D.S.N., Campbell, L.K., Johnson, M., Sonti, N.F. & Svendsen, E. 2016. *Reading the Landscape: Citywide Social Assessment of New York City Parks and Natural Areas in 2013–2014*. Social Assessment White Paper No. 2. New York, NY: New York Department of Parks and Recreation. 69 p. Available at: https://www.nrs.fs.fed.us/pubs/506170.

Berkes, F. 2012. *Sacred Ecology: Traditional Ecological Knowledge and Resource Management* (3rd ed.). New York, NY: Routledge.

Brecher, C., Horton, R.D., Cropf, R.A. & Mead, D.M. 1993. *Power Failure: New York City Politics and Policy Since 1960*. New York: Oxford University Press.

Campbell, L.K. 2017. *City of Forests, City of Farms: Sustainability Planning for New York City's Nature*. Ithaca, NY: Cornell University Press.

Campbell, L.K. & Gabriel, N. 2016. Power in urban social-ecological systems: Processes and practices of governance and marginalization. *Urban Forestry & Urban Greening*, 19(1): 253–254.

Campbell, L.K. & Wiesen, A. (Eds.). 2009. *Restorative Commons: Creating Health and Well-being through Urban Landscapes, General Technical Report*. Newtown Square, PA: U.S. Department of Agriculture, Forest Service, Northern Research Station.

Campbell, L.K., McMillen, H. & Svendsen, E.S. 2019. The written park: Reading multiple urban park subjectivities through signage, writing, and graffiti. *Space and Culture*, 1–19. doi: 10.1177/1206331218820789

Campbell, L.K., Svendsen, E.S., Sonti, N.F. & Johnson, M.L. 2016. A social assessment of urban parkland: Analyzing park use and meaning to inform adaptive management and resilience planning. *Environmental Science and Policy*, 62: 34–44.

Chan, J., Pennisi, L. & Francis, C.A. 2016. Social-ecological refuges: Reconnecting in community gardens in Lincoln, Nebraska. *Journal of Ethnobiology 36*(4): 842–860.

Cocks, M. & Dold, T. 2006. Cultural significance of biodiversity: The role of medicinal plants in urban African cultural practices in the Eastern Cape, South Africa. *Journal of Ethnobiology*, 26(1): 60–81.

Connolly, J.J., Svendsen, E.S., Fisher, D.R. & Campbell, L.K. 2013. Organizing urban ecosystem services through environmental stewardship governance in New York City. *Landscape and Urban Planning*, 109: 76–84.

Connolly, J.J.T., Svendsen, E.S., Fisher, D.R., & Campbell, L.K. 2014. Networked governance and the management of ecosystem services: The case of urban environmental stewardship in New York City. *Ecosystem Services*, 10: 187–194.

Cranz, G. 1982. *The Politics of Park Design: A History of Urban Parks in America*. Cambridge, MA: MIT Press.

Curran, W., & Hamilton, T. 2012. Just green enough: Contesting environmental gentrification in Greenpoint, Brooklyn. *Local Environment*, 17: 1027–1042.

Durkheim, É. 1976 [1912]. *The Elementary Forms of the Religious Life*. London, UK: George Allen and Unwin Ltd.

Elands, B.H.M., Vierikko, K., Andersson, E., Fischer, L.K., Gonçalves, P., Haase, D., Kowarik, I., Luz, A.C., Niemelä, J., Santos-reis, M. & Wiersum, K.F. 2019. Biocultural diversity: A novel concept to assess human-nature interrelations, nature conservation and stewardship in cities. *Urban Forestry & Urban Greening*, 40: 29–34.

Ernstson, H. 2013. The social production of ecosystem services: A framework for studying environmental justice and ecological complexity in urbanized landscapes. *Landscape and Urban Planning*, 109: 7–17.

Fein, A. 1981. *Landscape into Cityspace: Fredrick Law Olmsted's Plans for a Greater New York City*. New York, NY: Van Nostrand Reinhold Company.

Fisher, D.R., Campbell, L., Svendsen, E.S. 2012. The organisational structure of urban environmental stewardship. *Environmental Politics*, 21(1): 26–48.

Fisher, D.R., Svendsen, E.S. & Connolly, J. 2015. *Urban Environmental Stewardship and Civic Engagement: How Planting Trees Strengthens the Roots of Democracy*. London, UK: Routledge Press, Explorations in Environmental Studies Series, 24 February.

Foglesong, R.E. 1986. *Planning the Capitalist City: The Colonial Era to the 1920s*. Princeton, NJ: Princeton University Press.

Gavin, M.C., McCarter, J., Mead, A., Berkes, F., Stepp, J.R., Peterson, D. & Tang, R. 2015. Defining biocultural approaches to conservation. *Trends in Ecology and Evolution*, 30(3): 140–145.

Gentile, S. 2009. Public officials grapple over city's approach to public-private partnerships. Available at: www.cityhallnews.com, January 28, 2009.

Gerrish, E. & Watkins, S.L. 2018. The relationship between urban forests and income: A meta-analysis. *Landscape and Urban Planning* 170: 293–308. doi: 10.1016/J.LANDURBPLAN.2017.09.005.

Gesler, W.M. 1993. Therapeutic landscapes: Theory and a case study of Epidauros, Greece. *Environment and Planning D: Society and Space*, 11(2): 171–189.

Gómez-Baggethun, E., Gren, A., Barton, D.N., Langemeyer, J., McPhearson, T., O'Farrell, P., Andersson, E., Hamstead, Z. & Kremer, P. 2013. Urban ecosystem services. 2013. In Elmqvist, T., Fragkias, M., Goodness, J., Guneralp, B., Marcotullio, P.J., McDonald, R.I., Parnell, S., Schewenius, M., Sendstad, M., Seto, K.C. & Wilkinson, C. (Eds.), *Urbanization, Biodiversity and Ecosystem Services: Challenges and Opportunities*. New York: Springer. pp. 175–252.

Hansen, R., Rall, E., Chapman, E., Rolf, W. & Pauleit, S. 2017. Urban green infrastructure planning: A guide for practitioners. Retrieved December 13, 2017 from GREEN SURGE, http://greensurge.eu/working-packages/wp5/.

Heynen, N., Kaika, M. & Swyngedouw, E. 2006. *In the Nature of Cities: Urban Political Ecology and the Politics of Urban Metabolism*. New York, NY: Routledge.

Johnson, J.T., Howitt, R., Cajete, G., Berkes, F., Louis, R.P. & Kliskey, A. 2016. Weaving indigenous and sustainability sciences to diversify our methods. *Sustainability Science* 1(11): 1–11.

Johnson, M., Campbell, L.K., Svendsen, E.S. & Silva, P. 2018. Why count trees: Volunteer motivations and experiences in tree monitoring efforts. *Arboriculture and Urban Forestry*, 44(2): 59–72.

Kaminsky, J. & Simonet, M. 2017. *Who Cleans the Parks: City Work and Urban Governance in NYC*. Chicago, IL: University of Chicago Press.

Keil, R. & Boudreau, J. 2006. Metropolics and metabolics: Rolling out environmentalism in Toronto. In: Heynen, N., Kaika, M. & Swyngedouw, E. (Eds.), *In the Nature of Cities: Urban Political Ecology and the Politics of Urban Metabolism*, (pp. 41–62). London, UK: Routledge.

Kealiikanakaoleohaililani, K., McMillen, H., C., & Francisco, K. 2019. Cultivating sacred kinship to strengthen resilience. In Campbell, L.K., Svendsen, E., Sonti, N.F., Hines, S.J., Maddox, D. (Eds.), *Green Readiness, Response, and Recovery: A Collaborative Synthesis. General Technical Report, NRS-P-85*, (pp. 188–204). Newtown Square, PA: Department of Agriculture, Forest Service.

Kealiikanakaoleohaililani, K., Kurashima, N., Francisco, K., Giardina, C., Louis, R., McMillen, H., Asing, C., Asing, K., Block, T., Browning, M., Camara, K., Camara, L., Dudley, M., Frazier, M., Gomes, N., Gordon, A., Gordon, M., Heu, L., Irvine, A., Kaawa, N., Kirkpatrick, S., Leucht, E., Perry, C., Replogle, J., Salbosa, L.-L., Sato, A., Schubert, L., Sterling, A., Uowolo, A., Uowolo, J., Walker, B., Whitehead, A., & Yogi, D. 2018. Ritual + Sustainability Science? A Portal into the Science of Aloha. *Sustainability*, 10(10): 3478.

Kealiikanakaoleohaililani, K. & Giardina, C.P. 2016. Embracing the sacred: An indigenous framework for tomorrow's sustainability science. *Sustainability Science*, 11(1): 57–67. doi: 10.1007/s11625-015-0343-3.

Kohler, F., Holland, T.G., Kotiaho, J.S., Desrousseaux, M. & Potts, M.D. 2019. Embracing diverse worldviews to share planet Earth. *Conservation Biology*, 33(5): 1014–1022.

Knott, K. 2016. The secular sacred: In between or both/and? In Day, A., Vincett, G., & Cotter, C.R. (Eds.), *Social Identities Between the Sacred and the Secular*. London, UK: Routledge. pp. 145–160.

Landau, L., Campbell, L.K., Johnson, M., Svendsen, E., & Berman, H. 2019. *STEW-MAP in the New York City Region: Survey Results of the Stewardship Mapping and Assessment Project. General Technical Report, NRS-189*. Newtown Square, (pp. 145–160). PA: Department of Agriculture, Forest Service, Northern Research Station. 69 p. doi: 10.2737/NRS-GTR-189.

McMillen, H., Campbell, L.K., Svendsen, E.S., Kealiikanakaolehaililani, K., Francisco, K., & Giardina, C.P. in press. Biocultural stewardship, indigenous and local ecological knowledge, and the urban crucible. *Ecology and Society*.

McMillen, H., Campbell, L.K. & Svendsen, E.S. 2019. Weighing values and risks of beloved invasive species: The case of the survivor tree and conflict management in urban green infrastructure. *Urban Forestry and Urban Greening*. 40 (December): 44–52. doi: 10.1016/j.ufug.2018.06.023.

McMillen, H., Ticktin, T. & Springer, H.K. 2017a. The future is behind us: Traditional ecological knowledge and resilience over time on Hawai'i Island. *Regional Environmental Change*, 17(2): 579–592. doi: 10.1007/s10113-016-1032-1.

McMillen, H.L., Campbell, L.K. & Svendsen, E.S. 2017b. Co-creators of memory, metaphors for resilience, and mechanisms for recovery: Flora in living memorials to 9/11. *Journal of Ethnobiology*, 37(1): 1–20. doi: 10.2993/0278-0771-37.1.1.

McMillen, H., Campbell, L.K., & Svendsen, E.S. 2017c. The power of living things: Living memorials as therapeutic landscapes. *Medicine Anthropology Theory*, 4(1): 185–192.

McMillen, H., Campbell, L., Svendsen, E. & Reynolds, R. 2016. Recognizing stewardship practices as indicators of social resilience: In living memorials and in a community garden. *Sustainability*, 8(8): 775. doi: 10.3390/su8080775.

Munoz-Erickson, T., Campbell, L.K., Childers, D.L., Grove, J.M., Iwaniec, D.M., Pickett, S.T.A., Romolini, M. & Svendsen, E.S. 2016. Demystifying governance and its role for transitions in urban social-ecological systems. *Ecosphere*, 7(11): 1–11.

Nash, R. 2014. *Wilderness and the American Mind*. London, UK: Yale University Press.

Nesbitt, L., Meitner, M.J., Girling, C., Sheppard, S.R.J. & Yuhao, L. 2019. Who has access to urban vegetation? A spatial analysis of distributional green equity in 10 US cities. *Landscape and Urban Planning*, 181: 51–79. doi: 10.1016/j.landurbplan.2018.08.007.

O'Neil-Dunne, J.P.M., MacFaden, S.W., Forgione, H.M. & Lu, J.W.T. 2014. *Urban Ecological Land-Cover Mapping for New York City. Unpublished Report to the Natural Areas Conservancy. Spatial Informatics Group, University of Vermont, Natural Areas Conservancy, and New York City Department of Parks & Recreation.*

Olmsted F.L. 1870. *Public Parks and the Enlargement of Towns*. Cambridge, MA: American Social Science Association.

Pascua, P., McMillen, H., Ticktin, T., Vaughan, M. & Winter, K.B. 2017. Beyond services: A process and framework to incorporate cultural, genealogical, place-based, and Indigenous relationships in ecosystem service assessments. *Ecosystem Services*, 26: 465–475.

Pearsall, H. 2010. From brown to green? Assessing social vulnerability to environmental gentrification in New York city. *Environment and Planning C*, 28(5): 872–886.

Peçanha Enqvist, J., Campbell, L.K., Stedman, R.C. & Svendsen, E.S. 2019. Towards a typology of stewardship pathways: Sense of place and civic engagement for urban waterfronts. *Sustainability Science*, 14(3): 589–605.

Pickett, S.T.A., Boone, C.G., McGrath, B.P., Cadenasso, M.L., Chilers, D.L., Ogden, L.A., Mc Hale M. & Grove, M.J. 2013. Ecological science and transformation to the sustainable city. *Cities* 32: 510–520.

Ribot, J.C. & Peluso, N.L. 2009. A theory of access. *Rural Sociology*, 68(2): 153–181. doi: 10.1111/j.1549-0831.2003.tb00133.x.

Rigolon, A., Browning, M., & Jennings, V. 2018. Inequities in the quality of urban park systems: An environmental justice investigation of cities in the United States. *Landscape and Urban Planning, 178*(June): 156–169.

Schelhas, J., Hitchner, S., Johnson G.C. & Jennings, V. 2017. *Sunshine, Sweat, and Tears: African-American Ties to Land and Forests in the South. e-Gen.* Technical Report, SRS-220. Asheville, NC: Department of Agriculture Forest Service, Southern Research Station. P. 154.

Sonti N.F., Campbell K.L., Svendsen E.S., Johnson M.L., Auyeung D.S.N. 2020. Fear and fascination: Use and perceptions of New York City's forests, wetlands, and landscaped park areas. *Urban Forestry & Urban Greening*, 49. doi.org/10.1016/j.ufug.2020.126601.

Sonti, N.F. & Svendsen, E.S. 2018. Why garden? Personal and abiding motivations for community gardening in New York City. *Society & Natural Resources*, *31*(10): 1–17. doi: 10.1080/08941920.2018.1484971.

Stone, E. 2009. The benefits of community-managed open space: Community gardening in New York City. In Campbell, L. & Wiesen, A. (Eds.), *Restorative commons: Creating health and well-being through urban landscapes, General Technical Report*, (pp. 122–137). Newtown Square, PA: Department of Agriculture, Forest Service, Northern Research Station.

Svendsen, E. 2009. Cultivating resilience: Urban stewardship as a means to improving health and well-being. In Campbell, Lindsay; Wiesen, Anne, eds. *Restorative Commons: Creating Health and Well-being Through Urban Landscapes. Gen. Tech Rep. NRS-P-39*, (pp. 122–137). Pennsylvania, PA: U.S. Department of Agriculture, Forest Service, Northern Research Station. pp. 122–137.

Svendsen, E.S. 2013. Storyline and design: How civic stewardship shapes urban design in New York City. In Pickett, S.T.A., Cadenasso, M.L., McGrath, B. (Eds.), *Resilience in Ecology and Urban Design: Linking Theory and Practice for Sustainable Cities* Vol. 3, (pp. 267–287). Dordrecht: Springer Netherlands.

Svendsen, E.S. & Campbell, L.K. 2005. *Living Memorials Project: Year 1 Social and Site Assessment*. General Technical Report. Newtown Square, PA: Department of Agriculture, Forest Service, Northeastern Research Station. p. 123.

Svendsen, E.S. & Campbell, L.K. 2010. Living memorials: Understanding the social meanings of community-based memorials to September 11, 2001. *Environment and Behavior*, 42: 318–334. doi: 10.1177/0013916510361871.

Svendsen, E.S.; Campbell, L.K. & McMillen, H.L. 2016. Stories, shrines, and symbols: Recognizing psycho-social-spiritual benefits of urban parks and natural areas. *Journal of Ethnobiology*, 36(4): 881–907.

Swyngedouw, E. & Heynen, N.C. 2003. Urban political ecology, justice and the politics of scale. *Antipode* 35(5): 898–918.

Swyngedouw, E. & Kaika, M. 2000. The environment of the city or … the urbanisation of nature. In Bridge, G. & Watson, S. (eds) *Reader in Urban Studies* (pp. 567–580). Oxford: Blackwell Publishers.

Taylor, D.E. 2009. *The environment and the people in American cities, 1600s-1990s: Disorder, inequality, and social change*. Durham, NC: Duke University Press.

The Trust for Public Land. 2015. 2015 city park facts. Available at: https://www.tpl.org/sites/default/files/files_upload/2015-CityPark-Facts-Report.pdf. Accessed on April 7, 2015.

U.S. Census Bureau. 2018. *Quick facts*. New York City, New York. Available at: https://www.census.gov/quickfacts/newyorkcitynewyork (Accessed: 25 April 2019).

U.S. Census Bureau. 2010. Census PL and SF1 Files. Population division – NYC Department of City Planning (July 2001). [online] Available at: http://www.nyc.gov/html/dcp/html/neigh_info/nhmap.shtml (Accessed: August 4, 2013).

Von Hassel, M. 2002. *The Struggle for Eden: Community Gardens in New York City*. Westport, CT: Bergin and Garvey.

Wilson, E.O. 1984. *Biophilia*. Boston, MA: Harvard University Press.

Whyte, K.P., Brewer J.P. & Johnson, J.T. 2016. Weaving Indigenous science, protocols and sustainability science. *Sustainability Science*, 1(11): 25–32.

13 Situating biocultural relations in city and townscapes

Conclusion and recommendations

Michelle Cocks and Charlie Shackleton

Introduction

The different geographic regions represented in the book have brought to the fore the diversity of ways in which nature is conceptualised, which have in turn influenced the types of nature found in urban areas. Through processes of urbanisation, colonialism, immigration and migration a diversity of cultural groups now live in urban areas and consequently, biocultural relationships have been suppressed, reshaped or enriched. Accordingly, a diversity of uses, experiences, cosmologies, interactions and engagement with the nature are now found which, for many, offer opportunities to strengthen a sense of wellbeing and belonging. Within these diversities of ontological framings of nature and ways of being, conflicting tensions emerge which are further impacted upon by micro and macro social, economic and political processes.

Despite the diversity of contexts and subjects of the different chapters, several themes emerge from the preceding chapters which have meaning for the study and theory of bioculture as well as urban planning and the management and provision of urban nature. These themes include (1) embeddedness of emotion in biocultural relationships; (2) colonial legacies and rupture of biocultural relationships; (3) migration in shaping and enriching biocultural realities; (4) harvesting nature and (5) a wide diversity of spaces are required and used. Each of these is discussed in relation to the contributions made from individual chapters and existing knowledge. These are discussed in the first half of this chapter.

The second half of the chapter also considers conceptual points for advancing and using a biocultural diversity lens within urban studies. We conclude by providing a set of more practical considerations to enhance the opportunities for more and deeper biocultural relationships in urban settings. These include (1) engaging local voices; (2) providing sufficient urban nature; (3) ensuring an equitable distribution of and access to urban nature; (4) fostering social access to urban nature; (5) decolonialisation of nature offerings and experiences; (5) celebrating and embracing local natural heritage; (6) rebuilding severed biocultural links and (7) embedding a plural valuation approach to decision-making and the need to embed biocultural understandings within legislative frameworks.

Emergent themes

Embeddedness of emotion in biocultural relationships

In reading the narratives in the chapters of this compilation, one is immediately struck by the plethora of words used to describe aspects of urban human-nature relationships. To mention only a few, some examples are happiness, enjoyment, satisfaction, pain, sadness, responsibility, inspiration, sense of inner peace, wellbeing, restoration, love, belonging and connectedness. Many heartfelt emotions are portrayed, see Mellegård (Chapter 4), Cocks et al. (Chapter 6), Lagunes Gasca & Merçon (Chapter 7), Nesbitt et al. (Chapter 10) and McMillen et al. (Chapter 12).

Within certain discourses, particularly psychological writings, many authors have provided ample evidence to show how social interaction with and in natural environments fosters positive emotional outcomes (Arnould & Price 1993, Farber & Hall 2007). Particular attention has been drawn to the restorative properties of human-nature relationships that in turn have been identified as contributing to a sense of physical and mental wellbeing. The seminal works of Ulrich (1984) and Fuller et al. (2007) have resulted in attention being given to the positive physical and psychological benefits people derive from being in nature across varied urban settings and contexts (e.g., Nutsford et al. 2013, Alcock et al. 2014, Mukherjee et al. 2017, Yigitcanlar et al. 2020). More recent studies are attempting to explain how the positive effects of being in nature affect regulatory processes and indeed, the way that the human brain functions (Richardson et al. 2016, Olszewska-Guizzo et al. 2020). These associations have developed over evolutionary time as it has been argued that emotions evolved and helped to promote human survival in nature. These evolutionary processes, together with ancestral experiences of living for approximately 2.2 million years in natural environments, continue to influence our mental health (Mantler & Logan 2015).

Within the field of emotional geographies, attempts are being made to make explicit how emotional experiences contribute to a sense of self and being. It is increasingly understood that the body and experiences of embodiment are central to one's sense of being, who we think we are and what others attribute to us, which is constantly reshaped by how we feel (Davidson & Milligan 2004). Indeed, Davidson & Milligan (2004) point out that emotions often only become understandable – or "sensible" – in the context of particular places and cultures. Likewise, place must be felt to make sense. Recording these details allows one, as Davidson & Milligan (2004) explain, to appreciate the emotionally dynamic spatiality of social life. Calls have been made to make more explicit the engagement with the spatiality of emotion (e.g., Simonsen 2007). Within such a framing, the collection of personal narratives is seen as key to appreciate the physical and cultural contexts within which emotions are experienced. Here Mellegård (Chapter 4) illustrates beautifully how the sensations she experienced in her parent's garden evoked many emotions which led to the creation of happy childhood memories. These included hearing the sounds of certain bird calls and insects and the rustling of leaves. Feeling the wind and sun's rays on her skin and soft soil under her feet. Smelling the rich perfume of roses her grandmother had planted. Seeing the detailed intricacies of certain plants, their colours and shape.

It is also important to highlight that each of ones' s memories are also deeply embedded within cultural contexts. Studies within cognitive anthropology are beginning to show that culture affects both what people remember and how they use their memories in everyday life, as the content and functions of memories are shown to be shaped by differences in physical environments, self-views, concerns for behavioural and emotional regulation, socialisation and language. The bidirectional relations between culture and memory are beginning to be examined (Ross & Wang 2010). Emotions are a part of human evolutionary history and impact every aspect of being, including the biological, psychological, social and cultural. Until the seminal works of Damasio (1999) and Thrift (2004) emotion was largely kept out of modern academic research and discourse (Thien 2005). An engagement with the everyday emotional subjectivities was discouraged as emotions were related to the realm of personal feeling and distinct from wider (public) agendas. It is argued that such framings have had a devastating impact of regulating emotion as negatively positioned and in opposition to reason; and as objectively soft and implicitly feminised and thus furthering the distinction between personal and political (Thien 2005, p. 452). Scholars within the field of emotional geographies are increasingly revealing the falseness of such distinction and in response argue that emotions consciously foreground embodied experiences and show how power permeates social life and inevitably science (Thien 2005). Building on the seminar works of Deleuzes (1988), whose writing on Spinoza has been one of the cornerstones of current geographical forging of theories of affect, scholars are increasingly highlighting the importance of engaging with notions of power as power provides individuals with different capacities "to affect and be affective" (Tolia-Kelly 2006, p. 215).

In recognition of these processes, a historicist account becomes imperative to be attentive to these differences of affect and be affective as further argued by Tolia-Kelly (2006) not only are undifferentiated bodies caught up in networks of affect but specific bodies carrying (representational) burdens of meaning, expectation and ability. Within such a framing, the emotions, memories and ways of being held by marginalised communities take on a very profound significance for the social and cultural differences they may embody. For example, in chapter 6 the township resident Kwanele expresses how cultural practices, such as growing traditional medicinal plants in his garden, are a part of his identity. Yet his ability to carry out traditional practices and daily rituals is constrained by Eurocentric ideological framings still upheld in the post-colonial town where he lives. Such framings continue to have a devastating impact on urban resident's ability to embrace realities that do not conform to such ideals and to attain notions of wellbeing which are often linked to certain kinds of activities and certain places within nature which evoked a sense of pride, being linked to one's roots, being with one's ancestors, and a sense of inner peace and wellbeing; see Cocks et al. (Chapter 6).

Similarly, in Chapter 7, the migrant worker Carmen expressed a deep sadness at the inability to perform pre-Hispanic rituals in the city of Tapachula in Mexico. One such ritual includes decorating the water spring during the fiesta patronal (annual patron saint festival). Seeds, machetes and hoes are taken to the spring and presented as gifts to thank God for all that he gives to eat and be alive. Carmen remembers partaking in these rituals while she was growing up, which she described as being very beautiful. It fills her with great sadness that she can no longer partake in them as her family has done for so many generations. Such sadness stands in stark contrast to Robert's ability (Nesbitt et al. Chapter 10) to stand under the cedar trees that were brought and planted by the Squamish Nation in Oppenheimer Park, Vancouver. He feels a great sense of pride standing under these trees as he knows this place is where his grandfather was born. He enjoys visiting because he can take time out and sit under the trees and breathe the fresh air. He finds this clears his thoughts and feelings. He can laugh or cry under the trees as they create balance in his life.

Emotions attributed to biocultural practices are deeply embedded in notions of self and are well acknowledged and appreciated by people who engage with and carry out such practices. For example, Kwanele and Bonsile, two township male residents (see Cocks et al. (Chapter 6)), explained that the ability to carry out their cultural practices contributed to their identity as Xhosa men. The making of certain *amayeza* (medicinal) remedies was also considered as reinforcing such associations. Cocks et al. (2012, 2016) have presented similar findings among rural and urban women who described how the ability to collect firewood (*ukutheza*) reinforced their identity as hard-working women and as *umama* (a mother; hard-working adult woman). Similarly, the ownership of a woodpile (*igoqo*) is considered an important status symbol for married women as it signifies their status as hard-working women in their households and their commitment to their families and ancestral veneration (Cocks et al. 2006).

Examples in the chapters also reveal the levels of agency taken to (re)create emotions found in biocultural practices in urban settings. Individuals, particularly from more marginalised communities, often do so at great odds, which in turn bring them a great sense of pride and joy. Examples include the establishment of community gardens by migrant and refugee communities in Vancouver (Nesbitt et al. Chapter 10) and home gardens by migrant women in Tapachula, Mexico (Chapter 7) and the nurturing of medicinal and cultural plants in yards in South African townships (Cocks et al. Chapter 6). Through their own actions they have created a place to carry out farming traditions engaged in by their elders.

Colonial legacies and rupture of biocultural relationships

The very rationale for this book (Chapter 1) was to reveal the diverse and intricate relationships between urban citizens and urban nature, which extend well beyond those most commonly portrayed in the academic literature and which are most provided for by urban planners and management agencies. As previously stated, such literature and discourses are dominated by perspectives from the Global North (Roy et al. 2012, Shackleton 2012) and are frequently unquestioningly prescribed or voluntarily adopted by planners and management agencies in the Global South, with little regard to crucial differences. These contextual differences include aspects of history, culture, ecology, biodiversity richness, livelihood needs and political conditions.

This disjuncture between the nature needs of many urban citizens and what is planned for and provided is not a modern phenomenon. Several chapters in this book reveal that in many Global South settings the disjunction was catalysed during the period of colonial domination. This encompassed alienation of indigenous lands, resources, cultures and knowledge systems (Johnson & Murton 2007, Mashford-Pringle 2015). This might have been accelerated further with the increasing reach of mass education and more recently globalisation, but it was the original colonial regimes that set the prescripts that transformed what was adjudged as "civilised", "sophisticated" and "modern" that penetrated all facets of the colonial administrations and hence lives of the subjugated people. Indigenous worldviews, cosmologies, beliefs, practices and opinions were systematically undermined, sidelined, suppressed or denigrated (Johnson & Murton 2007) through the apparatus of colonial planning and administration, the creation of "accepted" norms through social inclusion or exclusion, mass education systems premised solely on Western knowledge systems and religious bodies' advocacy of monotheism and that humans were separate from nature. The latter laid the foundation for colonial administrations' exploitation of nature, through the rich economic resources that were plundered and exported (timber, medicines, foods, horticultural species and genetic resources). As Lagunes Gasca & Merçon (Chapter 7) comment, perhaps controversially for some, based on a focus group participant's statement, "the church also desacralises nature". This discord includes the types of urban spaces that were provided for nature experiences, the types of nature provided in such spaces and the types of activities or behaviours that were accepted (socially or legally) in such spaces. In some settings, it even prescribed who could access particular urban natures, such as the racially prescribed dictates in South Africa (Chapter 6) and Australia.

What this book reveals is that the colonial imprint and mindset can still be found in many cities and towns to the current day. This is evident in multiple dimensions, from (1) urban planning with respect to where and how urban parks are envisaged and provided, (2) the species planted or maintained, to (3) what types of nature experiences are planned for and are legally or socially permitted or excluded. With respect to planning, there is a growing body of literature exposing the inequities in the provision of urban green spaces or trees between different parts of a city, with the most marginalised, typically indigenous communities, having much less than areas frequented by the social, political and economic elites (Heynen et al. 2006, McConnachie & Shackleton

2010, Astell-Burt et al. 2014, Nero 2017, Sathyakumar et al. 2019, Nesbitt et al. 2019). The chapter by Cocks et al. (Chapter 6) in this book describes how little formal public green space is provided in the poorer areas of Qonce and Komani in South Africa even in the current deracialised, post-apartheid era, while de Mendonça (Chapter 9) reveals how some local communities were marginalised from the more spectacular beaches in Rio de Janerio, and the descendants of the original fishers are restricted to a peripheral area at the end of the beach. On a more personal level, Adedeji (Chapter 3) comments on how the traditional courtyard style of open spaces was lost under the dictates of colonial architecture, severing or supressing the close spatial relationship between people and nature. This also meant a loss of immediate access to the useful species that used to be nurtured within the courtyards.

Many of the parks originally designed and planted during the colonial periods were populated with species from around the world, being a mix of the familiar from the colonising country or "motherland", and those swapped between colonies, subjugated and administered by the same colonial power. This legacy remains evident for green spaces with long-lived tree species, with the older parts of former colonial cities still being dominated by non-native street trees. For example, Moro & Castro (2015) reported that 72 per cent of the tree species in public spaces in Fortaleza (Brazil) were non-native, and most of the indigenous species were represented by only a few individuals. In Karachi (Pakistan) 75 per cent of the street trees are of exotic species (Shams et al. 2020), with corresponding figures for Bangalore (India) of 67 per cent (Nagendra & Gopal 2010) and Makhanda (South Africa) 59 per cent (Kuruneri & Shackleton 2011). While non-native species provide a suite of benefits or ecosystem services, many fail to meet some local cultural needs, while also contributing to the insidious erosion of local knowledge and identity related to the natural heritage and biodiversity of a particular region or country. They may also support less indigenous biodiversity of other taxa such as birds and hemiparasites (Shackleton 2016). This colonial legacy is further translated into elements such as biodiversity representations in street names where, in South Africa, there are more streets named after non-native plant species than native plant species, and there are more non-native plant species represented than native ones (Shackleton 2018).

The colonial legacy also shaped the formal or informal "rules" of what types of behaviour are "acceptable" by dominant social norms in public green spaces. More passive and culturally or morally enriching pursuits (such as art displays, sculptures or musical concerts) were elevated above more rough or indigenous cultural or spiritual activities. This colonial disenfranchisement of certain uses of or activities in nature remains embedded in formal and informal policies, by-laws and socially prescribed behaviours to this very day. In many cities, the gathering of biodiversity resources such as flowers, mushrooms and edible plants is expressly disallowed, and yet collection and use of such resources is a very intimate interaction with nature that also retains local ecological knowledge. In other localities, certain groups revere particular green spaces or species or a specific tree, and wish to undertake religious or other ceremonies in these spaces, but are often barred from doing so, or constrained by other users (Ngulani & Shackleton 2019).

Migration in shaping and enriching biocultural realities

Since the beginning of human existence individuals and groups have moved across lands, escaping from dangers and searching for better life conditions. The reconstruction of a sense of place and belonging, as well as the mixing of cultures and languages are human experiences that have cut across time and space. Although migration has always characterised human life on the planet, it had never reached the magnitude witnessed today. It is estimated that in 2019 there were 272 million migrants around the globe, 51 million more than in 2010. International migrants already encompass 3.5 per cent of the global population, of which approximately 31 per cent reside in Asia, 30 per cent in Europe, 26 per cent in the Americas, 10 per cent in Africa and 3 per cent in Oceania (United Nations 2020).

Geographers have long studied the meanings that migrants attach to their mobility and settlement experiences, and how they manage to maintain links to their places of origin while simultaneously adapting to their new environments (Ahmed et al. 2003, Blunt & Varley 2004, Ralph & Staeheli 2011). Extensive work has been written on migrants' settlement experiences in their host cities or countries. Journeys taken away from home, whether chosen or imposed, are often associated with feelings of risk, social exclusion and often marginalisation. Such experiences are often felt by labour migrants, refugees and people living in exile (Mallett 2004) and often lead to feelings of alienation and disorientation, which in turn can evoke emotions of wanting to be located in specific places (Staeheli & Nagel 2006) so as to feel a sense of belonging (Massey 1992).

Intersectional theorists such as Yuval-Davis, draw attention to differentiating between belonging and the politics of belonging (2011). Within the former, belonging is described as an emotional attachment, whereby one feels the need to feel "at home" (2011). Previous writings on home, inspired by philosophical writings on place attachment, examined home as a place that grounded people to a particular place (Ralph & Staeheli 2011). In more recent years, scholars have been criticised for placing a too sedentary analytic bias on constructions of home. Such responses emerged from research among transnational migrants' sense of home. Authors within this field have highlighted that the very act of moving throws into question the ability to locate people to specific places and homes. They have argued that feelings of belonging are constructed in time and space rather than a specific place, emerging gradually and in response to an engagement with daily spatial activities which in turn contribute to feelings of safety, comfort and commitment. Such arguments, as noted by Ralph and Staeheli (2011), have contributed to the emergence of concepts such as "bifocality" that gives recognition to a binational state that certain migrants create and maintain.

Ralph and Staeheli (2011, p. 4) have drawn attention to the need to recognise that not all migrants experience the same level of integration into the host society. Integration is often aligned to either race, class or gender or all three; thus, it is not merely a personal matter but one that is embedded in social factors (Antonsich 2010). Belonging to a social group and to a place – political belonging, or "citizenship" – is not simply an individual's choice to make but depends on being recognised as part of the community where they live. Studies have shown that maintaining contact with one's place of origin has provided migrants with coping strategies, particularly in countries where they experience hostile or unreceptive host contexts. For example, Njwambe et al. (2019) has shown that migrants from rural, economically impoverished communities of Centane migrate to Cape Town in search of work opportunities. Even after many years of working in Cape Town, many actively retain their rural identity and return home frequently to experience the more positive emotional and spiritual dimension of home. Such findings echo calls made for a sense of place to be included in notions of belonging. As noted by Antonsich (2010) discourses and practices of belonging do not exist in a geographical vacuum.

McMillen et al. (Chapter 12) show how migrant communities' relationships with urban nature are derived from their home culture before they, their parents or their grandparents came to New York City (NYC). They provide examples of how Yoruba practitioners make offerings to Orisha at designated trees in NYC and how practitioners of Orthodox Judaism collect local grasses to construct temporary huts or sukkah for the celebration of Sukkot. In a similar fashion, descendants of Indian Hindu migrants now appropriate parts of the foreshore in Queens, New York for sacred rituals reminiscent of those practised on the banks of the Ganges River in India (Garcia et al. Chapter 8). More recent migrants may be deterred from public green spaces for fear of social-cultural or racial discrimination (Elands et al. Chapter 11).

Recognition given to emotions and place within migration studies provides the opportunity to discuss how biocultural relations contribute to feelings of home within the cityscape. Again, we turn to migration studies which have highlighted the importance of material objects in creating a sense of home and of sustaining relationships that help to constitute home (Ralph & Staeheli 2011). In such a context, material objects have been described as serving "both a buffer for individuals from the pressures of outside cultures, but also to help forge a feeling of identity and belonging somewhere" (see Tolia-Kelly 2004, Elands et al. (Chapter 11)). Such framings help in appreciating the significance attached to displaying one's cultural artefacts within the yard space of newly constructed state-provided housing in South Africa townships Cocks et al. (Chapter 6). Notions of belonging not only contribute to an individual's sense of attachment but are linked to an ability to lead "a life that is meaningful, a life worth living". The absence of place-belongingness is likened to a sense of loneliness, isolation, alienation and displacement which can lead to a loss of motivation and mental health problems (Antonsich 2010).

The relationship between people's food and their identity has long been the subject of social research (Gerodetti & Foster 2016). More recent years have seen the emergence of gastrodynamic studies within which food is acknowledged as being "imbued with meaning", and "identifies and symbolizes who we are" (Oyangen, 2009, p. 325). Such understandings have largely emerged from studies carried out among migrant communities. The relationship between the actual production of food and identity has also received attention, for example, research on gardens as places that express personal, social, cultural and environmental or ecological identities (Gerodetti & Foster 2016, p. 2). Gardening has also been shown to contribute to a sense of homemaking and provides opportunities to create or reconnect to past memories, as shown by migrant women living in T'ja Xuj, Mexico. The women who occupy T'ja Xuj, despite many personal and financial difficulties and physical constraints of limited living space, grow and prepare foods that they are accustomed to (Lagunes Gasca & Merçon Chapter 7). Where gardening is more of a community endeavour, it can prompt social interactions, and the beginnings of a shared identity place. As Elands et al. (Chapter 11) state it "might lead to unexpected or unusual encounters between groups stemming from different cultural and social backgrounds as well as real world experiences".

Similarly, McMillen et al. (Chapter 12) draw attention to how newly arrived Puerto Rican families in NYC create gardens reminiscent of their former agrarian lives and give them names after places or plants from their place of origin. Migrant communities also often observe ancient practices and experiences in nature in new places. For example, members of Indo-Caribbean Hindu communities who live in NYC have adopted the waterfront of Jamaica Bay in Queens, New York as a cultural keystone place. Particular sites have become places of workship, replicating those performed on sacred rivers in India such as the Ganga. Such places have become critical to their identity and wellbeing (see Garcia et al. (Chapter 8). Such insight illustrates how biocultural relations enrich residents' lives within cityscapes and contribute to a sense of belonging (Elands et al. (Chapter 11). These examples reflect those that the migrants find that they are "permitted" to do in their adoptive city or country. However, there may well be needs and expressions of bioculture that they find they cannot practice but hanker after. This might include harvesting of certain plants or worshipping in public green space, or practising of certain traditions or rituals (e.g., initiation into manhood).

Harvesting nature

An important use or desired use of urban nature in many cities across the world is for tangible products such as wild foods (berries, fruits, mushrooms and indigenous leafy vegetables), medicinal bulbs and bark, craft fibres and even firewood (Shackleton et al. 2017). The demand for these sorts of products spans from the purely consumptive, such as firewood (Davenport et al. 2012, Schlesinger et al. 2015), through to the more culturally mediated needs for particular wild foods, medicines and craft materials (Cocks & Dold 2004, Grabbattin et al. 2011, Hurley & Emery 2018), as well as for recreation and knowledge transfer (Poe et al. 2014, Synk et al. 2017, Landor-Yamagata et al. 2018). Despite the wide diversity of products harvested, and the varied reasons underlying their harvest, such needs and practices are largely overlooked in most cities, most noticeably in the Global North, and consequently are rarely planned for in terms of the spaces and species available for harvesting (Shackleton et al. 2017). Indeed, in some cities, foraging of wild foods and the like from urban spaces is at best frowned upon, or at worst actively regulated against (Poe et al. 2014, Charnley et al. 2017), even though Fischer & Kowarik (2020) conclude that urban foraging is an important pathway for connecting urban citizens and nature, echoing the arguments of Martin (2018) in Australia.

The chapters of this book provide rich insights into the more biocultural basis of much urban foraging. In the case of Helsinki in Finland, Vierkko & Jaatsi (Chapter 5) record how some residents value the more natural forests throughout the city for collecting mushrooms and berries. In contrast, Cocks et al. (Chapter 6) allude to the importance of peri-urban commonages in providing a wide range of products from firewood to fruits and medicinal plants. Indeed, previous work from three towns in the same part of South Africa estimated the value of products collected from urban commonages to range between 7 and 18 per cent of total household income (cash and non-cash) (Davenport et al. 2012). Commonages are also crucial for grazing livestock, which themselves are an important part of Xhosa identity (Cocks et al. Chapter 6). Other works emphasise that urban foraging is not just about the plants or products collected, but that the process itself can be culturally shaped and reinforced (Hall 2013, Poe et al. 2014, Garekae & Shackleton 2020), foster a sense of identity and belonging (Hall 2013, McLain et al. 2014, Poe et al. 2014), being an activity for socialisation and education (McLain et al. 2014, Poe et al. 2014, Garekae & Shackleton 2020) and potentially therapeutic.

The case studies also reveal that different sites may be used by urban citizens to procure resources of consumptive and cultural interest or necessity. This mirrors studies internationally (Hurley et al 2008, McLain et al. 2014, Charnley et al. 2017, Landor-Yamagata et al. 2018, Fischer & Kowarik 2020, Garekae & Shackleton 2020) revealing that a diversity of different urban sites and tenure systems are used to collect these resources. Such spaces range from domestic gardens and corporate grounds under private tenure, to formal parks under municipal governance and informal spaces such as road verges and disused lots seemingly under public tenure but where access is not constrained in any way.

Not all plants used by urban residents are sourced from the broader urban matrix, but some may be planted in or harvested from domestic spaces (Kaoma

& Shackleton 2015, Charnley et al. 2017). This may be to ensure a secure and continued supply free from competition from other harvesters, or it may be because cultural beliefs require that certain plants are located close to the home. For example, Cocks et al. (Chapter 6) reveal that 45 per cent of households in Qonce have plants in their gardens that protect them from malevolent forces. The Yorba in Nigeria also plant protective plants around the homestead (Adedeji, Chapter 3). The converse also applies, namely where plants are cultivated around the home for positive experiences associated with their symbolism, beauty, scent, food and the birds and insects they attract, as described by Pastor-Ivars (Chapter 2) in Kanazawa in Japan. Additionally, migrants transport and cultivate plants that are familiar to them as a means of place-making and providing culturally valued ingredients, as evidenced in Tapachula, Mexico (Lagunes Gasca & Merçon, Chapter 7).

But many urban citizens lack sufficient home space in which to develop a garden, such as in high-density suburbs or those who live in multi-story apartments. In such contexts, container pots and community or allotment gardens gain in significance as revealed in several of the case chapters. Once again, such gardens are not exclusively for food to eat, but also to grow particular species that are not readily available in local shops, or the actual process of growing in itself provides opportunities for social interaction and a sense of oneness with nature and a greater spirituality (Rogge et al. 2018, Kingsley et al. 2019). The chapter by McMillen et al. (Chapter 12), goes further to describe the process of gardening as aiding grieving from a personal or family loss, while Mellegård (Chapter 4) shows that Iranian culture and identity are intertwined with the "ideal of the Persian garden" and the "love-story" of Iranians and roses spans over three millennia. Pastor-Ivars (Chapter 2) reveals even deeper biocultural aspects of gardening in terms of how the symbolism and the spatial arrangement of different elements in the gardens hold deep and ingrained meanings, echoing the precise design of the Yorba courtyards in Nigeria (Adedeji, Chapter 3). Similarly, Molebatsi et al. (2010) describe the various elements of what constitute a Tswana tshimo garden in northwest South Africa and how they are arranged.

The widespread practices of urban foraging and gardening are one manifestation of the deep biocultural links between urban residents and local or managed nature. Several propositions emerge from this realisation, most notably the need for urban planners to ensure there is sufficient and accessible urban nature. However, it is not just the area or quantity, and its distribution and accessibility, but also the fostering of species held in significant demand or held in high regard because of marked biocultural affinities. For example, the almost irreplaceable need for branches of wild olive on which to serve ceremonial meat amongst the Xhosa in South Africa (Cocks et al., Chapter 6).

A wide diversity of spaces is required and used

The case studies assembled in this book reveal unequivocally that urban dwellers' interactions with nature comes in many forms, guises and spaces, and it is not restricted solely to recreation or sports in public parks, which dominate the literature (which is largely from the Global North (Roy et al. 2012, Shackleton 2012). However, although private spaces also offer opportunities for the establishment and maintenance of biocultural relationships (Mellegård, Chapter 4, Pastor-Ivars, Chapter 2), they fall outside the remit of urban planners, decision-makers and urban parks departments. The focus therefore for this section is on public spaces, because that is where public officials (elected or appointed), policies and funds are directed to respond to the needs of local citizens. That said, it is still necessary to observe and caution that in many settings plans do not provide sufficient space within private residential plots because of increasing housing densities. This is common in poorer neighbourhoods, which paradoxically also often suffer from low proportions and accessibility to public green space, thereby exacerbating the ill effects of limited private green space (McConnachie & Shackleton 2010).

In considering specifically public green space and nature, it is clear from the chapters presented in this book that they exist in diverse forms, each of which is used by varying sectors of urban society. Formal public parks are appreciated and used by many different urban residents, provided that the parks are accessible, safe, reasonably well maintained and not dominated by particular groups of people, cultures or users, i.e., they are welcoming of diverse groups and cultures (e.g., Vierkko & Jaatsi Chapter 5, Nesbitt et al. Chapter 10, Elands et al. Chapter 11). Several of the chapters describe how urban authorities in many countries have provided open, lawned, "English-style" parks, which are favoured localities by some for family outings and exercise such as walking, jogging and perhaps a casual game of football. As such, these sorts of formal parks potentially foster an enjoyment of being outdoors and a break from the busyness of life (Vierkko & Jaatsi Chapter 5), but more intimate relationships with nature and specific elements within it are more constrained. Moreover, there may also be prescriptions of what is allowed or not, in terms of "the right kind of behavior" (Vierkko & Jaatsi Chapter 5). For example, many such parks used to, and some still do, disallow visitors or users from walking or sitting on the lawns – an anachronism from colonial times. Similarly, many formal parks caution against what is deemed "excessive" noise or disturbance. Such designs or prescriptions may mean that formal parks may fail to meet the needs of certain cultures or sectors of urban society. The Eurocentric form of the memorial gardens in Komani and Qonce (South Africa), which are hardy visited by local residents, is a case in point (Cocks et al. Chapter 6). Similar restrictions are applied to green urban areas in Mexico, where indigenous groups are not allowed to perform certain rituals (Lagunes Gasca & Merçon Chapter 7).

In contrast to the manicured formal parks are the designated forests or "wild" natural areas retained or preserved from some assumed manifestation of the original vegetation of the area. Examples in the book include the natural forests in Osun, Nigeria (Adedeji Chapter 3), Helsinki, Finland (Vierkko & Jaati Chapter 4) and Vancouver, Canada (Nesbitt et al. Chapter 10). These are often important habitat spaces for indigenous plants and animals, much appreciated by those seeking deeper connections with nature. Indeed, the cases of Helsinki, Finland (Vierkko & Jaatsi Chapter 4), South Africa (Cocks et al. Chapter 6) and Vancouver, Canada (Nesbitt et al. Chapter 10) show that some urban residents obtain more benefits from and prefer more "natural" spaces, but they may be visited less frequently because there are fewer of them and hence are harder to access. This is also supported by studies elsewhere (e.g., Gómez -Baggethun & Barton 2013, Rupprecht et al. 2015). Commonly, some more natural, wild spaces will offer formal environmental education experiences too, such as nature trails and signboards to assist identification of the more common species of plants and animals, or school outings. Collection or picking of wildflowers, berries and mushrooms may also be permitted, but collection of wood (for fire or construction) is usually prohibited. The more natural nature of these spaces, and perhaps them being quieter than formal, open parks, may also encourage more spiritual connections, contemplations and worship (Ngulani & Shackleton 2019). Some may be well known and revered by locals as sacred spaces, such as in Osun, Nigeria (Adedeji Chapter 3). However, their more "wild" nature can lead to some urban citizens being afraid of entering them, or assuming that they are undeveloped lands that can be used for dumping rubbish, antisocial activities or even building upon.

Informal green spaces are typically well used, although because they are not being formally zoned or designated by urban planners, their value to urban citizens and their relationships with nature are likely to be relatively underappreciated by urban authorities and certainly under-researched (Rupprecht et al. 2015). The same applies in our case chapters where informal green spaces and nature are hardly mentioned, other than by Cocks et al. (Chapter 6). The sheer diversity of types of informal spaces undermines possible generalisations about their use, value and management. Informal green spaces span a considerable range of sizes, location and "qualities". They may be a few square metres in between some buildings, long thin corridors under electricity pylons, derelict or abandoned lots, or large areas of currently undeveloped land belonging to the city or private individuals that, through time, some neighbourhood residents have appropriated for particular uses. Because they are not formally designated as green spaces, they rarely receive much formal management from urban authorities – maybe some removal of litter or dumped materials if it accumulates beyond certain limits, and maybe some control of invasive alien species. But there is little management of species composition, abundance or mowing of the herbaceous layer. There are rarely public amenities or information boards. Nonetheless, this informality attracts some urban residents and affords them an opportunity to be with nature for pleasure, mental and spiritual wellbeing or for privacy. Because of their informality, these green spaces are often vital for urban foragers seeking wild foods, firewood, medicinal plants and craft materials, with reduced fears of sanction (Grabbattin et al. 2011, Garekae & Shackleton 2020). They may also be favoured for urban agriculture by urban residents wishing to grow food or medicinal plants but who lack space around their home (Drescher 2004), or for grazing livestock.

Alongside these sorts of green spaces, are a host of others, such as road and railway verges, churchyards and cemeteries, sports fields, school and public institutional grounds (such as libraries, police stations, hospitals), riparian zones, memorial gardens, commonages, botanical and zoological gardens, and allotments or community gardens. Each of these types offers different opportunities for connections with nature that need to be recognised and encouraged. Some researchers or authorities have developed typologies of different urban green spaces (e.g., Bell et al. 2007, Rupprecht & Byrne 2014) but these require further development in relation to the types of species and human-nature relationships each supports or engenders.

Appreciation of the different forms of public green spaces and their respective uses will prompt urban planners to move beyond just supplying green space in the legislated ratios (see Shackleton & Cocks (Chapter 1), to contemplating what type of green space is required, and where. Doing so will enhance the range of nature relationships available to all urban citizens irrespective of age, gender or culture and the particular uses they seek or favour in urban nature.

Advancing the framing of biocultural diversity

The concept of biocultural diversity was originally coined to give recognition to the inextricable link between biodiversity and cultural diversity among traditional and indigenous communities (Maffi & Woodley 2010, Buizer et al. 2016). Later it was modified and broadened to include other social and ecological contexts (see Shackleton & Cocks (Chapter 1)); including how communities live with and need nature and biodiversity in urban contexts as covered in this book. Such a framing has been useful in illustrating the diversity of ways that urban residents view, interact, experience and engage with nature in urban areas and has brought several considerations to the fore, notably the embeddedness of emotions, ways of being and wellbeing, power and rural urban linkages within biocultural diversity relations. The implications of these are discussed in relation to some of the key concepts placed into context in Shackleton & Cocks (Chapter 1).

Emotions, ways of being and wellbeing

The embeddedness of emotions within biocultural relations is clearly articulated. Culture has been described as being key in offering meaning to emotions generated from interactions and engagements with nature, which in turn, have become embodied through their effect on regulatory and neurophysiological processes. The effects of these have been further strengthened through human evolutionary and ancestral history. Loss of languages and the knowledges contained in different cultures would mean a loss of such experiences and an impoverishment of peoples' lived experiences.

These findings also highlight the complexities of notions of self and wellbeing, including aspects of identity and spirituality that have being identified as the least understood aspects of wellbeing (see Shackleton & Cocks (Chapter 1)). Notions of identity are described as being enhanced by an ability to carry out practices that are perceived as being aligned to a particular set of knowledges (see Cocks et al. (Chapter 6) and Nesbitt et al. (Chapter 10)). Notions of spirituality are facilitated by being able to visit places in which one can feel connected to ones' ancestors and gods (see Cocks et al. (Chapter 6), Garcia et al. (Chapter 8), de Mendonça (Chapter 9)). For many, a sense of wellbeing is heightened by memories of family and ancestral histories situated in particular nature settings (see Cocks et al. (Chapter 6), Garcia et al. (Chapter 8), de Mendonça (Chapter 9) and Nesbitt et al. (Chapter 10)). The inclusion of biocultural relations in wellbeing studies and practice would go a long way in enhancing understanding of how people find meaning and satisfaction in their lives as previously noted by Diener & Lucas (2000) and Møller & Theuns (2013).

Power

Although not a central theme of the chapters in this book, power was found to be deeply embedded within biocultural realities. Such influences were captured in the personal account of how broader political and economic factors in Iran impacted a family's ability to retain their family garden. The narrative puts into context how such processes have destroyed many of the century-old Persian Gardens in the city of Tehran (Mellegård Chapter 4). At a broader scale, Pastor-Ivars (Chapter 2) provides an analysis of how different ideals contained in different historical periods in Japan have influenced the structures and elements represented and incorporated into Japanese gardens, while Cocks et al. (Chapter 6) and Nesbitt et al. (Chapter 10) reveal how ideals held by colonial authorities have influenced the conceptualisation and design of urban natures found in current-day post-colonial towns. de Mendonça (Chapter 9) provides an account of how dominant political and economic power structures have dictated what types of activities are acceptable along Copacabana beach in Rio de Janeiro. An analysis of these cases reveals that only those activities that have direct financial benefits to the political and economic elite are sanctioned. The personal narratives in Gasca & Merçon (Chapter 8) provide insight into the impact that broader political and economic processes have on marginalised communities.

The case material in the chapters reiterates calls to include power relations in biocultural studies. For example, Boonstra (2016) notes that despite the central role of power, analyses of such are often lacking in human-nature relationships, thus offering limited and naive understandings of bioculturality. It has been further noted that construction of symmetrical power relations and more equitable urban biocultural realities can significantly benefit from decolonialised thinking and action. The living legacy of colonialism in contemporary societies is manifested by the systematic discrimination, subjugation and exclusion of certain social groups, worldviews and activities. This coloniality of power operates through hierarchical orders based on racial classification, Eurocentric forms of knowledge and the capitalist economic system (Quijano 2000, Mignolo 2007).

An analysis of power dynamics within biocultural diversity relations can also throw light on the knowledge and action needed to foster social and environmental justice in the cities (Anders Sandberg et al. 2015, Nesbitt et al. 2018).

Rural and urban linkages

Although the focus of this book has been on urban nature and wellbeing, rural-urban linkages play a significant role in providing opportunities to foster and to reconnect to nature through biocultural relations. Such notions are evident in migrant communities newly established in urban settings. Many see such a move as fulfilling economic needs only and frequently return to their former rural home. Within such a context, the notion of home, for many, is encapsulated through biocultural relations. These include a desire for many of the natural and biodiversity features found in rural areas such as rivers and forests. They also yearn for or speak fondly of the positive sensory experiences of such places which are often in stark contrast to mostly negative sensory experiences of urban settings – the crowded living conditions, unpleasant smells, unattractive and unhomely shacks of the urban areas they endure in search of economic security. Opportunities to return to home were described as restorative and reviving. On returning to cities, memories of visited places were described as assisting them during difficult times (Njwambe et al. 2019).

In Japan, certain landscapes are well recognised for representing examples of where humans and nature have coexisted for many centuries (Satoyama landscapes). In response to rapid industrialisation during the last century, there was a rapid decline in land use practices and depopulation of rural landscapes. Programmes such as the Ownership System were initiated to encourage cooperation between rural and urban communities to restore landscape use based on traditional practices. The first such initiative was started in Yusuhara on Shikoku Island, which today is a national movement. Urban owners have been described as being motivated to participate because of their love for such landscapes and because it provides them an opportunity to carry out manual work in rural landscapes, which they described as rewarding. Active participation in such activities was also described as an opportunity to educate their children and protect the heritage of their ancestors (Yokohari & Bolthouse 2011). Academic institutions in Japan also make use of traditional rural landscapes as living classrooms to expose learners to ways of integrating traditional ecological knowledge with modern science to find solutions for regional sustainability and to understand the nature-human-culture interactions (Mammadova 2017).

It has also been well documented that many natural resources are harvested from rural areas and through various networks are sold either in their natural state or as products in cities and towns and even across regional and international borders to meet the demands that exist for cultural resources and products. Such resources and products include medicinal plant- and animal-based products, wild foods including fruits, insects and bushmeat as well as cultural artefacts such as grass mats and brooms, to name a few (e.g., Cunningham 2011).

Such insights assist in recognising that movement into urban areas is not a linear progression. In contrast there is continuous movement between rural and urban areas. The movement between the two helps maintain biocultural diversity relations.

Moving forward

Drawing from the rich and varied empirical chapters of this book, and the emerging themes in the first half of this chapter, we now turn to what can be done to rediscover, reconnect, maintain and enhance biocultural relationships in urban settings around the world. In doing so, we acknowledge that not all the suggestions we make will be fitting or useful for every setting, but we hope that some will be, and even if only one is taken up by planners, urban authorities or civil society groups, then we feel that the lives and wellbeing of urban residents will be more fulfilled.

Engaging local voices

The main rationale for this book (Chapter 1) was the readily apparent mismatch between the forms and extent of urban nature that many city authorities provide for their citizens without, or rarely, meaningfully and respectfully engaging with their citizens about their views and needs of nature and heritage. Revealing and understanding local grounded views will not be achievable through short and quick surveys, but rather will require deep and genuine engagement over time and with the full diversity of urban constituencies and citizens. Listening to, engaging with and respecting local voices, concerns and suggestions are alien measures in many urban settings of the world, even though they are not complicated, nor do they demand significant (and often typically scarce) budgetary resources. But it does require mutual respect, commitment and time. It is frequently a matter of ethical and political positioning (inclusive relationships, social learning, reciprocal respect of multiple worldviews and plural valuations), which lays the foundations for continued dialogue and collaboration, rather than technical approaches that involve spatial planning, design or management for specific green areas.

Providing sufficient urban nature

At a very practical level, urban policies and authorities can aim to provide sufficient quantity and quality of urban nature for the citizens they serve. Precisely what is "sufficient" and who determines what is "sufficient" will vary from place to place based on locally relevant criteria and contexts. Global agencies and authorities in the Global North are increasingly setting guides prescribing target ratios or areas of urban green space per person. These are set with the best of intentions and certainly help prompt urban authorities to at least start to realise that they should, and can, be providing urban green space in pursuit of better living conditions, human wellbeing and urban sustainability. However, such guidelines rarely consider the quality, heritage and attached values to nature or citizen's desired uses of the spaces. Hence, the notion of "sufficient" spaces should embrace all that they embody and offer, and not just the amount or ratio. This relates equally to private as well as public spaces. In some societies, cultures and worldviews, having access to sufficient private space around one's home is vitally important to foster the planting of sacred, useful or revered species and to practice important rituals, ceremonies and activities that define the individual or the culture. This requires that urban planners are knowledgeable about the heritage, cultures and customs of the people they are planning for and provide "sufficient" space around the home, and within the city.

Ensuring an equitable distribution of and access to urban nature

As previously stated, with respect to urban green spaces and nature, it is not just about the amount or ratio. Physical access must be ensured through an equitable distribution and social access through appropriate measures (see below). Unfortunately, it is increasingly evident that the spatial distribution of public and private green spaces is far from equitable in many cities in both the Global North and South. This indicates the need for planners and local authorities to make a comprehensive inventory of the formal public green spaces in the city and to determine the physical accessibility to residents per smallest governance or census unit. They need to then engage with citizens and consider how to address any inequities exposed and put in place strategies to avoid entrenching or replicating such in future spatial and development plans. This must include listening to local voices with respect to designating and protecting particular sacred or valued spaces for spiritual, customary or traditional needs.

Fostering social access to urban nature

Equitable physical access does not guarantee social access. Social access is shaped by the formal and informal rules and norms of who can use what spaces, for what purposes, as well as the design of the spaces and amenities that attract or deter particular groups. For example, Lagunes Gasca & Merçon (Chapter 7) describe how migrants of different origins do not feel equally welcome in particular spaces. In some areas, the regular presence of noisy or boisterous groups may deter those requiring a quieter and contemplative experience. In many settings women may feel unsafe, which consequently constrains which spaces they can visit and enjoy. Fostering social access is complex, especially so in increasingly multicultural cities. It is likely to be best achieved through the involvement of local citizens in monitoring and raising community awareness of the various needs of different groups.

Decolonialisation of nature experiences

Deeply linked to social access is the necessity to embrace local worldviews and heritage in the design of public urban green spaces and the activities deemed acceptable within them. For example, urban parks should include places for local and indigenous communities to display examples of their cultural heritage, such as the erection of totem poles (see Nesbitt et al. (Chapter 10)). Cocks et al. (Chapter 6), also draw attention to how the inclusion of memorials depicting significant historical events of previously marginalised communities does not always result in urban green spaces becoming valued. Attempts to include additional heritages within green spaces need to be undertaken with care and sensitivity and in consultation with relevant communities. Such attempts should also be accompanied by providing indigenous local communities and cultural groups access to carry out practices they deem important, such as cultivating food crops and medicinal plants to produce medicinal remedies. Spiritual and cultural groups should be granted permission to use urban nature to carry out practices aligned to their faith and or cultural practices. Such opportunities also create possibilities to celebrate and share knowledge and heritages.

Celebrating and embracing local natural heritage

Part of the decolonisation process is to embrace and celebrate local natural heritage. An important strategy to achieve this is to promote native species over exotic ones in public urban green spaces and along streets, and local knowledge alongside global. Native species are more widely known by local citizens, especially those who recently migrated from rural surrounds, and they are also more embedded in local folklore, languages, traditions and rituals. Promoting native species will therefore contribute to a greater sense of identity and potentially place attachment and be better aligned with local knowledges and heritage. Coincidentally, they better provide habitat and food resources to support other native species, such as birds, insects and small vertebrates.

Diversity of urban nature knowledge, values, spaces and experiences

Another very practical strategy is for urban planners and authorities to acknowledge the need for a diversity of types of nature and the types of activities and experiences to be catered for, and that a "one size fits all" approach is insufficient to foster and build biocultural relationships. Urban citizens have a greater diversity of knowledges, values, needs of and relationships with urban nature beyond just exercise and passive recreation. Urban citizens have varied needs in and from nature; to be able to sit and contemplate, to show reverence for some greater entity, to collect resources of cultural or material value, to plant and cultivate, to interact with and experience nature in a very spiritual or visceral manner, to grieve and remember, to touch, feel, taste and learn, to explore and invigorate, to simply be one with nature. To address such a diversity of values, relationships and needs requires a diversity of spaces and species, based on plural values, from the wild and natural through to the manicured and lawned park, to the corner allotment where residents can plant herbs, medicinal plants, vegetables and the like, all drawing on a diversity of knowledges and values. Precisely what form the diversity of spaces and relationships may take must be informed by the process of engaging and working with local voices and values, as mentioned above.

Rebuilding severed biocultural links

The value of nature to human identity and wellbeing is increasingly being realised across a wide diversity of disciplines from anthropology to medicine to psychology to restoration ecology. This realisation brings to the fore that in many cultures and settings the intimate and emotional links humans have with nature have been entirely severed or horribly deformed through a variable suite of local and global processes operating at different scales, through various means and in different regions. The process of severing continues as particular knowledge forms and values dominate over others and with ever more urban citizens being required to live in extremely dense conditions with limited access to (and meaningful experiences with) any nature for months or even years at a time. There is an urgent need to rebuild or reconnect the severed links for the very real wellbeing benefits that they bring to urban citizens, urban sustainability and species conservation. Doing so will require a shift in mind-sets and values. Schools are a good place to start because school-based programmes can reach a large number of citizens in a short period and in their learning years. Institutions of higher learning are key thought spaces which can innovate about the rebuilding processes and values while simultaneously doing so amongst the next generation of future business, political, legal and cultural leaders. Indigenous cultural groups are important in bringing to the fore local knowledges and heritage. Nevertheless, the rebuilding processes should not and cannot be restricted to any particular group, but should draw in all actors across civil society, religious bodies and environmental groups, indeed, communities and society at large.

Embedding a plural valuation approach

Embedded throughout the preceding suggestions is the notion of engaging with local communities to learn from their views, knowledge, values and opinions with respect to nature and integrate them in decision-making processes. In doing so, it is important that the engagement processes are participatory, respectful, inclusive and responsive to all views and not dominated by those of the city authority or other groups with power. Indeed, a plural valuation approach as advocated in the 2020 Oaxaca statement (Jacobs et al. 2020) is a useful guide to provide inclusive processes, that actively seek to limit or remove power imbalances in the present and move towards addressing those from the past. A way to achieve full inclusivity is also to acknowledge identity and difference. Recognition of others has been presented as a fundamental part of human integrity and recognised as a vital human need (Nesbitt et al. 2018, see Young 1990, Schlosberg 2007).

Embedding biocultural understandings within legislative frameworks

Many indigenous peoples and local communities' ontological understandings also embrace a co-existence between themselves and nature. Such ontological understandings do not perceive humans as being separate from nature and stand in stark contrast to Western ontological understanding within which nature is not given the same status as a human (see Shackleton & Cocks (Chapter 1)). Attempts to remedy such are beginning to appear. For example, in Ecuador's constitution nature "has the right to exist, persist, maintain and regenerate its vital cycles, structure, functions and its processes in evolution'"(Cailon et al. 2017, p. 29). The enshrining of such principles has made it possible for Ecuador's courts to uphold the rights of nature. The Whanganui River in New Zealand has been granted legal personhood. Wellbeing is highlighted repeatedly in the legal agreement between Whanganui and the Crown. Placing an emphasis on how the health and wellbeing of the Whanganui River is intrinsically interconnected with the health and wellbeing of people (Cailon et al. 2017). Such advancements are opening up new ontological and normative possibilities which would previously have been thought of as "unthinkable" beyond the paradigms of indigenous philosophies (Smith 2014, Cailon et al. 2017).

Biocultural research for the future

This book has presented 11 case studies from around the world illustrating the deep and intimate links that urban citizens have with or desire from nature. In some of the case studies, urban planners and authorities have recognised some of these needs and are working in various ways to meet them. In other cases, there has been little or no acknowledgement of the disjuncture between citizens' needs and what planners or authorities are providing. Thus, it is very clear that future research on the biocultural links of urban society and citizens will take different trajectories in different settings. Some of what we mention below may be well researched and considered in some societies and settings but will be totally novel in others. Yet, through asking and answering similar questions in different cultural and social contexts a more global understanding (and opportunities for hypothesis testing) can be shaped and debated.

It is also not just about the research needs and questions, but also about the processes. Bioculture as a concept or a way of thinking transcends single disciplines and consequently requires multi-, inter-, and transdisciplinary approaches to truly reveal and appreciate the diversity, intimacies and expressions of the relationships. Moreover, in accepting the need to restore severed biocultural relations and shift mind-sets, more participatory action, or transformative, research approaches are required. We identified the following areas as enhancing understandings of biocultural diversity relations and facilitating the maintenance of biocultural relations within urban environments.

Enhancing understandings

- by capturing ontological framings of human and nature relationships
- by capturing the meanings attached to animals, birds and insects found in cityscapes, as well as other forms of nature such as rivers, mountain ranges, beaches and oceans
- by articulating the emotions contained in biocultural diversity relations and how biocultural relationships contribute to everyday lived realities
- by investigating the impact that biocultural diversity relations have on body and mind, i.e., regulatory and physiological processes
- by providing a deeper contextual and historical analysis to biocultural relations studies
- by including an analysis of the power dynamics held in biocultural relations
- by exploring how linkages between rural and urban areas contribute to the maintenance of biocultural relations

Facilitating the maintenance of biocultural diversity relations

- by moving beyond the common and static surveys of who goes to public green spaces to ask why they go and to what extent does it meet their needs; if it does not, what other sorts of nature and nature experiences would they like?
- by exploring planning possibilities that enhance opportunities for biocultural relations to take place
- by ensuring that the levels of diversity of green spaces and species best meet the wants and needs of urban citizens
- by encouraging legal philosophers to engage with ontological framings of biocultural relations and develop legislative frameworks to uphold the values contained in them
- by incorporating components of biocultural relations into wellbeing indictor studies

Acknowledgements

The authors wish to extend their thanks to Julian Merçon for her comments and suggestions on the drafting of the chapter.

References

Ahmed, S., Castaneda C., Fortier, A.M. & Sheller M. 2003. *Uprootings/Regroundings: Questions of Home and Migration*. Oxford, UK: Berg.

Alcock, I., White, M.P., Wheeler, B.W., Fleming, L.E. & Depledge, M.H. 2014. Longitudinal effects on mental health of moving to greener and less green urban areas. *Environmental Science & Technology*, 48(2): 1247–1255.

Anders Sandberg, L., Bardekjian, A. & Butt, S. 2015. *Urban Forests, Trees, and Greenspace: A Political Ecology Perspective*. New York, NY: Routledge.

Antonsich, M. 2010. Searching for belonging – An analytical framework. *Geography Compass*, 4(6): 644–659.

Arnould, E.J. & Price, L.L. 1993. River magic: Extraordinary experience and the extended service encounter. *Journal of Consumer Research*, 20: 24–45.

Astell-Burt, T., Feng, X., Mavoa, S., Badland, H.M. & Giles-Corti, B. 2014. Do low-income neighbourhoods have the least green space? A cross-sectional study of Australia's most populous cities. *BMC Public Health*, 14(1): 292.

Bell, S., Montarzino, A. & Travlou, P. 2007. Mapping research priorities for green and public urban space in the UK. *Urban Forestry & Urban Greening*, 6: 103–115.

Blunt, A. & Varley, A. 2004. Introduction: Geographies of home. *Cultural Geographies*, 11: 3–6.

Boonstra, W.J. 2016. Conceptualizing power to study social-ecological interactions. *Ecology & Society*, 21(1), 21. doi: 10.5751/ES-07966-210121.

Buizer, M., Elands, B. & Vierikko, K. 2016. Governing cities reflexively: The biocultural diversity concept as an alternative to ecosystem services. *Environmental Science & Policy*, 62: 7–13.

Caillon, S., Cullman, G., Verschuuren, B. & Sterling, E.J. 2017. Moving beyond the human–nature dichotomy through biocultural approaches: Including ecological well-being in resilience indicators. *Ecology and Society*, 22(4): 27. doi: 10.5751/ES-09746-220427.

Charnley, S., McLain, R.J. & Poe, M.R. 2017. Natural Resource access rights and wrongs: Non-timber forest products gathering in urban environments. *Society & Natural Resources*, 31: 734–750.

Cocks, M., Alexander, J., Mogano, L. & Vetter, S. 2016. Ways of belonging: Meanings of 'nature' among Xhosa-speaking township residents in South Africa. *Journal of Ethnobiology*, 36(4): 820–841.

Cocks, M.L., Bangay, L. & Wiersum, K.F. 2006. Seeing the wood for the trees: The role of woody resources for the construction of gender specific household cultural artefacts in non-traditional communities in the Eastern Cape, South Africa. *Environment Development Sustainability*, 8: 519–533.

Cocks, M.L. & Dold., A.P 2004. A new brush sweeps clean: The economic and cultural values of grass brushes in the Eastern Cape Province, South Africa. *Forests Trees & Livelihoods*, 13: 33–42.

Cocks, M.L., Dold, T. & Vetter, S. 2012. 'God is my forest' – Xhosa cultural values provide untapped opportunities for conservation. *South African Journal of Science*, 108,(5/6): 1–8.

Cunningham, A.B. 2011. Non-timber products and markets: Lessons for export-oriented enterprise development from Africa. In Shackleton, S.E., Shackleton, C.M. & Shanley, P. (eds). *Non-Timber Forest Products in the Global Context*, (pp. 83–106). Heidelberg, Germany: Springer.

Damasio, A.R. 1999. *The Feeling of What Happens: Body and Emotion in the Making of Consciousness*. Orlando, FL: Harcourt.

Davenport, N.A., Shackleton, C.M. & Gambiza, J. 2012. The direct use value of municipal commonage goods and services to urban households in the Eastern Cape, South Africa. *Land Use Policy*, 29: 548–557.
Davidson, J. & Milligan, C. 2004. Embodying emotion sensing space: Introducing emotional geographies. *Social & Cultural Geography*, 5(4): 523–532.
Deleuze, G. 1988. *Spinoza: Practical Philosophy*. San Francisco, CA: City Lights Books.
Diener, E., & Lucas, R.E. 2000. Explaining differences in societal levels of happiness: Relative standards, need fulfilment, culture, and evaluation theory. *Journal of Happiness Studies*, 1(1): 41–78.
Drescher, A.W. 2004. Food for the cities: Urban agriculture in developing countries. *Acta Horticulturae*, 643: 227–231.
Farber, M.E. & Hall, T.E. 2007. Emotion and environment: Visitors' extraordinary experiences along the Dalton Highway in Alaska. *Journal of Leisure Research*, 39(2): 248–270.
Fischer, L.K. & Kowarik, I. 2020. Connecting people to biodiversity in cities of tomorrow: Is urban foraging a powerful tool? *Ecological Indicators*, 112, 106087.
Fuller, R.A., Irvine, K.N., Devine-Wright, P., Warren, P.H. & Gaston, K.J. 2007. Psychological benefits of greenspace increase with biodiversity. *Biological Letters*, 3(4): 390–394.
Garekae, H. & Shackleton, C.M. 2020. Urban foraging of wild plants in two medium-sized South African towns: People, perceptions and practices. *Urban Forestry & Urban Greening*, 49: 126581.
Gerodetti, N. & Foster, S. 2016. Growing foods from home: Food production, migrants and the changing cultural landscapes of gardens and allotments. *Landscape Research*, 41: 808–819
Gómez-Baggethun, E. & Barton, D.N. 2013. Classifying and valuing ecosystem services for urban planning. *Ecological Economics*, 86: 235–245.
Grabbattin, B., Hurley, P.T. & Halfacre, A. 2011. "I still have the old tradition": The co-production of sweetgrass basketry and coastal development. *Geoforum*, 42: 638–649.
Hall, C.M. 2013. Why forage when you don't have to? Personal and cultural meaning in recreational foraging: A New Zealand study. *Journal of Heritage Tourism*, 8: 224–233.
Heynen, N., Perkins, H.A. & Roy, P. 2006. The political ecology of uneven urban green space: The impact of political economy on race and ethnicity in producing environmental inequality in Milwaukee. *Urban Affairs Review*, 42(1): 3–25.
Hurley, P.T. & Emery, M.R. 2018. Locating provisioning ecosystem services in urban forests: Forageable woody species in New York City, USA. *Landscape & Urban Planning*, 170: 266–275.
Hurley, P., Halfacre, A., Levine, N. & Burke, M. 2008. Finding a "disappearing" nontimber forest resource: Using grounded visualization to explore urbanization impacts on sweetgrass basket making in Greater Mt. Pleasant, SC. *Professional Geographer*, 60(4): 1–23.
Johnson, J.T. & Murton, B. 2007. Re/placing native science: Indigenous voices in contemporary constructions of nature. *Geographical Research*, 45(2): 121–129.
Jacobs, S., Zafra-Calvo, N., Gonzalez-Jimenez, D., Guibrunet, L., Benessaiah, K., Berghöfer, A., Chaves-Chaparro, J., Díaz, S., Gómez-Baggethun, E., Sharachchandra, L., Martín-López, B., Masterson, V., Merçon, J., Moersberger, H., Muraca, B., Norström, A., O'Farrell, P., Ordonez, J., Prieur-Richard, A-H., Rincón-Ruiz, A., Sitas, N., Subramanian, S., Tadesse, W., van Noordwijk, M., Pascual, U. & Balvanera, P. 2020. Use your power for good: Plural valuation of nature – The Oaxaca statement. *Global Sustainability*, 3: 1–7, e8.

Kaoma, H. & Shackleton, C.M. 2015. The direct use value of urban tree non-timber forest products to household income in poorer suburbs in South African towns. *Forest Policy & Economics*, 61: 104–112.

Kingsley, J., Foenander, E. & Bailey, A. 2019. You feel like you're part of something bigger: Exploring motivations for community garden participation in Melbourne, Australia. *BMC Public Health*, 19: 745.

Kuruneri-Chitepo, C. & Shackleton, C.M. 2011. The distribution, abundance and composition of street trees in selected towns of the Eastern Cape, South Africa. *Urban Forestry & Urban Greening*, 10: 247–254.

Landor-Yamagata, J.L., Kowarik, I. & Fischer, L.K. 2018. Urban foraging in Berlin: People, plants and practices within the metropolitan green infrastructure. *Sustainability (Switzerland)*, 10(6): 1873.

Maffi, L. & Woodley, E. (Eds.). 2010. *Biocultural Diversity Conservation: A Global Sourcebook*. London, UK: Earthscan.

Mallett, S. 2004. Understanding home: A critical review of the literature. *The Sociological Review*, 52(1): 62–89.

Mammadova, A. 2017. Development of fieldwork activities to educate the youth for the biological and cultural preservation in rural communities of Ishikawa Prefecture, Japan. *International Journal of Environmental & Science Education*, 12(3): 441–449.

Mantler, A. & Logan, A.C. 2015. Natural environments and mental health. *Advances in Integrative Medicine*, 2(1): 5–12.

Martin, M. 2018. Urban foraging: Rethinking the human-nature connection in cities. *eTropic*, 17(1): 149–163.

Mashford-Pringle, A. 2015. Indigenous peoples and bioculturedness. *The Canadian Journal of Native Studies*, 35: 137–152.

Massey, D. 1992. A place called home? *New Formations*, 21(17): 3–15.

McConnachie, M.M. & Shackleton, C.M. 2010. Public green space inequality in small towns in South Africa. *Habitat International*, 34: 244–248.

McLain, R.J., Hurley, P.T., Emery, M.R. & Poe, M.R. 2014. Gathering "wild" food in the city: Rethinking the role of foraging in urban ecosystem planning and management. *Local Environment*, 19: 220–240.

Mignolo, W.D. 2007. Introduction: Coloniality of power and de-colonial thinking. *Cultural Studies*, 21(2–3): 155–167.

Molebatsi, L.Y., Siebert, S.J., Cilliers, S.S., Lubbe, C.S. & Davoren, E. 2010. The Tswana tshimo: A homegarden system of useful plants with a particular layout and function. *African Journal of Agricultural Research*, 5(21): 2952–2963.

Møller, V. & Theuns, P. 2013. What are the best and worst times in the lives of South African township dwellers? A content analysis of the self-defined end-anchors for Bernheim's ACSA Scale of subjective well-being. *Social Indicators Research*, 112(3): 611–640.

Moro, M.F. & Castro, A.S. 2015. A check list of plant species in the urban forestry of Fortaleza, Brazil: Where are the native species in the country of megadiversity? *Urban Ecosystems*, 18(1): 47–71.

Mukherjee, D., Safraj, S., Tayyab, M., Shivashankar, R., Patel, S.A., Narayanan, G., Ajay, V.S., Ali, M.K., Narayan, K.V., Tandon, N. & Prabhakaran, D. 2017. Park availability and major depression in individuals with chronic conditions: Is there an association in urban India? *Health & Place*, 47: 54–62.

Nagendra, N. & Gopal, D. 2010. Street trees in Bangalore: Density, diversity, composition and distribution. *Urban Forestry & Urban Greening*, 9: 129–137.

Nero, B.F. 2017. Urban green space dynamics and socio-environmental inequity: Multi-resolution and spatiotemporal data analysis of Kumasi, Ghana. *International Journal of Remote Sensing*, 38: 6993–7020.

Nesbitt, L., Meitner, M.J., Girling, C., Sheppard, S.R.J. & Lu, Y. 2019. Who has access to urban vegetation? A spatial analysis of distributional green equity in 10 US cities. *Landscape & Urban Planning*, 181: 51–79.

Nesbitt, L., Meitner, M.J., Sheppard, S.R. & Girling, C. 2018. The dimensions of urban green equity: A framework for analysis. *Urban Forestry & Urban Greening*, 34, 240–248.

Ngulani, T. & Shackleton, C.M. 2019. Use of public urban green spaces for spiritual services in Bulawayo, Zimbabwe. *Urban Forestry & Urban Greening*, 38: 97–104.

Njwambe, A. Cocks, M. & Vetter, S. 2019. *Ekhayeni*: Rural–urban migration, belonging and landscapes of home in South Africa. *Journal of Southern African Studies*, 45(2), 413–431.

Nutsford, D., Pearson, A.L. & Kingham, S. 2013. An ecological study investigating the association between access to urban green space and mental health. *Public Health*, 127(11): 1005–1011.

Olszewska-Guizzo, A., Sia, A., Fogel, A. & Ho, R. 2020. Can exposure to certain urban green spaces trigger frontal alpha asymmetry in the brain? Preliminary findings from a passive task EEG study. *International Journal of Environmental Research & Public Health*, 17(2): 394.

Oyangen, K. 2009. The gastrodynamics of displacement: Place-making and gustatory identity in the immigrants' midwest. *Journal of Interdisciplinary History*, 39(3): 323–348.

Poe, M.R., LeCompte, J., McLain, R. & Hurley, P. 2014. Urban foraging and the relational ecologies of belonging. *Social & Cultural Geography*, 15: 901–919.

Quijano, A. 2000. Coloniality of power and Eurocentrism in Latin America. *International Sociology*, 15(2): 215–232.

Ralph, D. & Staeheli, L., 2011. Home and migration: Mobilities, belongings and identities. *Geography Compass*, 5(7): 517–530.

Richardson, M., McEwan, K., Maratos, F. & Sheffield, D. 2016. Joy and calm: How an evolutionary functional model of affect regulation informs positive emotions in nature. *Evolutionary Psychological Science*, 2(4): 308–320.

Rogge, N., Theesfeld, I. & Strassner, C. 2018. Social sustainability through social interaction – a national survey on community gardens in Germany. *Sustainability (Switzerland)*, 10(4): 1085.

Ross, M. & Wang, Q. 2010. Why we remember and what we remember: Culture and autobiographical memory. *Perspectives on Psychological Science*, 5: 391–400.

Roy, S., Byrne, J. & Pickering, C. 2012. A systematic quantitative review of urban tree benefits, costs, and assessment methods across cities in different climatic zones. *Urban Forestry & Urban Greening*, 11: 351–363.

Rupprecht, C. D. & Byrne, J. A. 2014. Informal urban greenspace: A typology and trilingual systematic review of its role for urban residents and trends in the literature. *Urban Forestry & Urban Greening*, 13: 597–611.

Rupprecht, C. D., Byrne, J. A., Ueda, H. & Lo, A. Y. 2015. 'It's real, not fake like a park': Residents' perception and use of informal urban green-space in Brisbane, Australia and Sapporo, Japan. *Landscape & Urban Planning*, 143: 205–218.

Sathyakumar, V., Ramsankaran, R. & Bardhan, R. 2019. Linking remotely sensed urban green space (UGS) distribution patterns and socio-economic status (SES) – A multi-scale probabilistic analysis based in Mumbai, India. *GIScience & Remote Sensing*, 56(5): 645–669.

Schlesinger, J., Drescher, A. & Shackleton, C.M. 2015. Socio-spatial dynamics in the use of wild natural resources: Evidence from six rapidly growing medium-sized cities in Africa. *Applied Geography*, 56: 107–115.

Schlosberg, D., 2007. *Defining Environmental Justice: Theories, Movements, and Nature*. New York, NY: Oxford University Press.

Shackleton, C.M. 2012. Is there no urban forestry or greening in the developing world? *Scientific Research & Essays*, 7(40): 3329–3335.

Shackleton, C.M. 2016. Do indigenous street trees promote more biodiversity than alien ones? Evidence using mistletoes and birds in South Africa. *Forests*, 7, 134.

Shackleton, C.M. 2018. Urban street names: An opportunity to examine biocultural relationships? *PLoS One*, 13(7): e0200891. doi: 10.1371/journal.pone.0200891.

Shackleton, C.M., Hurley, P.T., Dahlberg, A., Emery, M.R. & Nagendra, H. 2017. Urban foraging: A ubiquitous human practice but overlooked by urban planners, policy and research. *Sustainability*, 9: 1884.

Shams, Z.I., Shahid, M., Nadeem, Z., Naz, S., Raheel, D., Aftab, D., Fraz, T.R. & Roomi, M.S. 2020. Town socio-economic status and road width determine street tree density and diversity in Karachi, Pakistan. *Urban Forestry & Urban Greening*, 47: 126473.

Simonsen, K. 2007. Practice, spatiality and embodied emotions: An outline of a geography of practice. *Human Affairs*, 17(2): 168–181.

Smith, J.C.A. 2014. '"...diversity within unity:" sovereignty, recognition and ecology in indigenous interpretations of modernity.' *Asian Journal of Latin American Studies*, 27(1): 65–87.

Staeheli, L. A. & Nagel, C.R. 2006. Topographies of home and citizenship: Arab-American activists in the United States. *Environment and Planning, A*, 38: 1599–1614.

Synk, C.M., Kim, B.F., Davis, C.A., Harding, J., Rogers, V., Hurley, P.T., Emery, M.R. & Nachman, K.E. 2017. Gathering Baltimore's bounty: Characterizing behaviors, motivations, and barriers of foragers in an urban ecosystem. *Urban Forestry & Urban Greening*, 28: 97–102.

Tolia-Kelly, D. 2004. Locating processes of identification: Studying the precipitates of re-memory through artefacts in the British Asian home. *Transactions of the Institute of British Geographers*, 29(3): 314–329.

Tolia-Kelly, D. 2006: Affect – an ethnocentric encounter? Exploring the 'universalist' imperative of emotional/affectual geographies. *Area*, 38: 213–17.

Thien, D. 2005. After or beyond feeling? A consideration of affect and emotion in geography. *Area*, 37: 450–454.

Thrift, N. 2004. Intensities of feeling: Towards a spatial politics of affect. *Geografiska Annaler*, 86B(1): 57–78.

Ulrich, R.S. 1984. View through a window may influence recovery from surgery. *Science*, 224(4647): 420–421.

United Nations. 2020. Shaping our future together. Migration. Online, retrieved 28 March 2020: https://www.un.org/en/sections/issues-depth/migration/index.html.

Yigitcanlar, T., Kamruzzaman, M., Teimouri, R., Degirmenci, K. & Alanjagh, F. 2020. Association between park visits and mental health in a developing country context: The case of Tabriz, Iran. *Landscape & Urban Planning*, 99: 103805.

Yokohari, M., & Bolthouse, J. 2011. Keep it alive, don't freeze it: A conceptual perspective on the conservation of continuously evolving satoyama landscapes. *Landscape and Ecological Engineering* 7: 207–216.

Young, I.M., 1990. *Justice and the Politics of Difference*. Princeton, NJ: Princeton University Press.

Yuval-Davis, N. 2011. *The Politics of Belonging. Intersectional Contestations*. London, UK: Sage.

Index

Page numbers in **bold** denote tables, in *italic* denote figures

For Product Safety Concerns and Information please contact our EU representative GPSR@taylorandfrancis.com
Taylor & Francis Verlag GmbH, Kaufingerstraße 24, 80331 München, Germany

www.ingramcontent.com/pod-product-compliance
Ingram Content Group UK Ltd.
Pitfield, Milton Keynes, MK11 3LW, UK
UKHW022220190926
475702UK00003B/457
* 9 7 8 0 3 6 7 4 2 7 5 7 3 *